Maritime Welding

Level Three

Trainee Guide

Pearson

Boston Columbus Indianapolis New York San Francisco Amsterdam
Cape Town Dubai London Madrid Milan Munich Paris Montreal Toronto Delhi
Mexico City Sao Paulo Sydney Hong Kong Seoul Singapore Taipei Tokyo

NCCER

President and Chief Executive Officer: Don Whyte
President: Boyd Worsham
Vice President: Steve Greene
Chief Operations Officer: Katrina Kersch
Maritime Welding Project Manager: Chris Wilson
Senior Development Manager: Mark Thomas
Senior Production Manager: Tim Davis
Managing Editor: Natalie Richoux
Desktop Publishing Manager: James McKay
Art Manager: Kelly Saddler
Desktop Publishing Specialists: Gene Page, Eric Caraballoso
Editors: Graham Hack, Jordan Hutchinson

Writing and development services provided by Topaz Publications, Liverpool, NY

Lead Writer/Project Manager: Troy Staton
Desktop Publisher: Joanne Hart
Art Director: Alison Richmond
Permissions Editor: Alice Garner
Writer: Troy Staton

Pearson

Director of Alliance/Partnership Management: Kelly Trakalo
Content Producer: Alexandrina B. Wolf
Assistant Content Producer: Alma Dabral
Digital Content Producer: Jose Carchi
Senior Marketing Manager: Brian Hoehl
Composition: NCCER
Printer/Binder: LSC Communications
Cover Printer: LSC Communications
Text Fonts: Palatino and Univers

Credits and acknowledgments for content borrowed from other sources and reproduced, with permission, in this textbook appear at the end of each module.

ScoutAutomatedPrintCode

Pearson

Perfect Bound ISBN-13: 978-0-13-560382-6
ISBN-10: 0-13-560382-X

Preface

To the Trainee

Welding is a high-tech industry that can take you to places all over the world. Welds are everywhere — just look around. From ladders to aircraft carriers; from NASCAR to national defense; from the laboratory to sales and repair; the welding trade impacts virtually every industry.

Few career choices offer as many options for employment and opportunities for growth. According to the Department of Labor, "the basic skills of welding are similar across industries, so welders can easily shift from one industry to another, depending on where they are needed most." Welders will continue to be in high demand since there is a continual need to rebuild aging infrastructure such as highways and buildings. In addition, the creation of new power generation facilities will yield new job prospects for trained and skilled welders.

Of course, one of the places to find a prosperous career as a welder is in the maritime industry. Shipyards and repair facilities are plentiful in coastal areas and on major rivers and lake systems. Maritime welding can be done on land or sea and above or underneath the surface of the water. It includes the work on ships, boats, and off-shore oil rigs. It includes the installation and repair of metal structures and components.

Maritime Welding introduces the fundamentals of the welding trade. The three levels of this curriculum, plus the Maritime Aluminum Welding curriculum, present an apprentice approach and will help you be knowledgeable, safe, and effective on the job.

Maritime Welding

This *Maritime Welding* curriculum is a product of collaboration between NCCER and the National Maritime Education Council (NMEC). Several maritime welding subject matter experts determined which of NCCER's *Welding* modules are suitable to maritime industry training. *Maritime Welding* can be used for welding training and can be supplemented by instructors with information specific to the maritime industry.

We invite you to visit the NCCER website at **www.nccer.org** for the latest releases, training information, newsletter, and much more. You can also reference the Pearson product catalog online at **www.nccer.org**. Your feedback is welcome. You may e-mail your comments to **curriculum@nccer.org** or send general comments and inquiries to **info@nccer.org**.

NMEC

The National Maritime Education Council (NMEC) is a multi-regional, industry-centered organization working to address critical workforce issues facing its member yards and the maritime industry. Established in 2012 to provide a centralized voice for the industry as it relates to workforce development, NMEC's primary mission is to lead the maritime industry in the development, promotion, and implementation of a national workforce development system that includes standardized craft training, portable credentials, and journey-level assessments.

Through its partnership with NCCER, NMEC is working in collaboration with its member yards to:

- Establish a system of career awareness and recruiting that will allow the industry to speak with one voice and deliver a consistent message to students, student influencers, the underemployed, women, and returning military.
- Establish and implement an industry-wide, comprehensive, standardized training and credentialing system for maritime craft professionals.
- Grow a pipeline of skilled craft professionals trained and credentialed through an industry-driven, uniform process.
- Improve industry outcomes by developing a better skilled, safer, more productive workforce.

NCCER Standardized Curricula

NCCER is a nonprofit 501(c)(3) education foundation established in 1996 by the world's largest and most progressive construction companies and national construction associations. It was founded to ad-dress the severe workforce shortage facing the industry and to develop a standardized training process and curricula. Today, NCCER is supported by hundreds of leading construction and maintenance companies, manufacturers, and national associations. The NCCER Standardized Curricula was developed by NCCER in partnership with Pearson, the world's largest educational publisher.

Some features of the NCCER Standardized Curricula are as follows:

- An industry-proven record of success
- Curricula developed by the industry for the industry
- National standardization providing portability of learned job skills and educational credits
- Compliance with the Office of Apprenticeship requirements for related classroom training (*CFR 29:29*)
- Well-illustrated, up-to-date, and practical information

NCCER also maintains a Registry that provides transcripts, certificates, and wallet cards to individuals who have successfully completed a level of training within a craft in NCCER's Curricula. *Training programs must be delivered by an NCCER Accredited Training Sponsor in order to receive these credentials.*

Special Features

In an effort to provide a comprehensive and user-friendly training resource, this curriculum showcases several informative features. Whether you are a visual or hands-on learner, these features are intended to enhance your knowledge of the construction industry as you progress in your training. Some of the features you may find in the curriculum are explained below.

Introduction

This introductory page, found at the beginning of each module, lists the module Objectives, Performance Tasks and Trade Terms. The Objectives list the skills and knowledge you will need in order to complete the module successfully. The Performance Tasks give you an opportunity to apply your knowledge to real-world tasks. The list of Trade Terms are words and phrases that will become part of your vocabulary as you study this module.

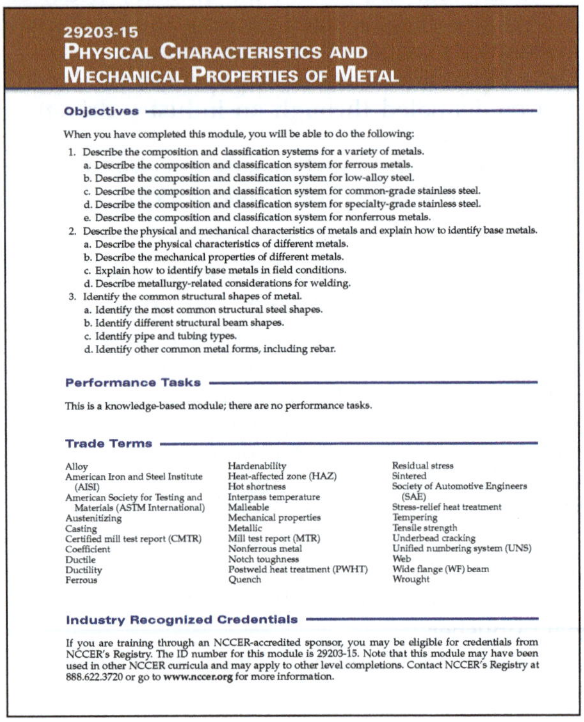

29203-15
PHYSICAL CHARACTERISTICS AND MECHANICAL PROPERTIES OF METAL

Objectives

When you have completed this module, you will be able to do the following:

1. Describe the composition and classification systems for a variety of metals.
   a. Describe the composition and classification system for ferrous metals.
   b. Describe the composition and classification system for low-alloy steel.
   c. Describe the composition and classification system for common-grade stainless steel.
   d. Describe the composition and classification system for specialty-grade stainless steel.
   e. Describe the composition and classification system for nonferrous metals.
2. Describe the physical and mechanical characteristics of metals and explain how to identify base metals.
   a. Describe the physical characteristics of different metals.
   b. Describe the mechanical properties of different metals.
   c. Explain how to identify base metals in field conditions.
   d. Describe metallurgy-related considerations for welding.
3. Identify the common structural shapes of metal.
   a. Identify the most common structural steel shapes.
   b. Identify different structural beam shapes.
   c. Identify pipe and tubing types.
   d. Identify other common metal forms, including rebar.

Performance Tasks

This is a knowledge-based module; there are no performance tasks.

Trade Terms

Alloy	Hardenability	Residual stress
American Iron and Steel Institute (AISI)	Heat-affected zone (HAZ)	Sintered
American Society for Testing and Materials (ASTM International)	Hot shortness	Society of Automotive Engineers (SAE)
Austenitizing	Interpass temperature	Stress-relief heat treatment
Casting	Malleable	Tempering
Certified mill test report (CMTR)	Mechanical properties	Tensile strength
Coefficient	Metallic	Underbead cracking
Ductile	Mill test report (MTR)	Unified numbering system (UNS)
Ductility	Nonferrous metal	Web
Ferrous	Notch toughness	Wide flange (WF) beam
	Postweld heat treatment (PWHT)	Wrought
	Quench	

Industry Recognized Credentials

If you are training through an NCCER-accredited sponsor, you may be eligible for credentials from NCCER's Registry. The ID number for this module is 29203-15. Note that this module may have been used in other NCCER curricula and may apply to other level completions. Contact NCCER's Registry at 888.622.3720 or go to **www.nccer.org** for more information.

Trade Features

Trade features present technical tips and professional practices from the construction industry based on real-life scenarios similar to those you might encounter on the job site.

Bowline Trivia

Some people use this saying to help them remember how to tie a bowline: "The rabbit comes out of his hole, around a tree, and back into the hole."

Figures and Tables

Photographs, drawings, diagrams, and tables are used throughout each module to illustrate important concepts and provide clarity for complex instructions. Text references to figures and tables are emphasized with *italic* type.

Figure 31 Marking a cutting line.

Notes, Cautions, and Warnings

Safety features are set off from the main text in highlighted boxes and categorized according to the potential danger involved. Notes simply provide additional information. Cautions flag a hazardous issue that could cause damage to materials or equipment. Warnings stress a potentially dangerous situation that could result in injury or death to workers.

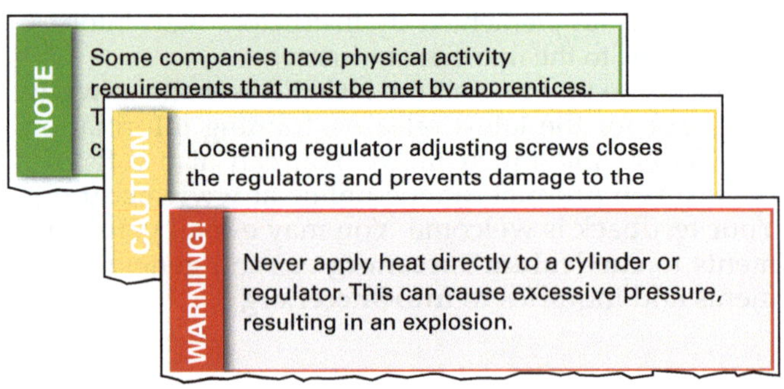

Case History

Case History features emphasize the importance of safety by citing examples of the costly (and often devastating) consequences of ignoring best practices or OSHA regulations.

Case History

Requesting an Outage

An electrical contractor requested an outage when asked to install two bolt-in, 240V breakers in panels in a data processing room. It was denied due to the 24/7 worldwide information processing hosted by the facility. The contractor agreed to proceed only if the client would sign a letter agreeing not to hold them responsible if an event occurred that damaged computers or resulted in loss of data. No member of upper management would accept liability for this possibility, and the outage was scheduled.

The Bottom Line: If you can communicate the liability associated with an electrical event, you can influence management's decision to work energized.

Going Green

Going Green features present steps being taken within the construction industry to protect the environment and save energy, emphasizing choices that can be made on the job to preserve the health of the planet.

GOING GREEN

Reducing Your Carbon Footprint

Many companies are taking part in the paperless movement. They reduce their environmental impact by reducing the amount of paper they use. Using email helps to reduce the amount of paper used,

Think About It

The Think About It features introduce historical tidbits or interesting and sometimes surprising facts about the construction trades.

Think About It

Shape Shifters

Trigonometry is the study of triangles, primarily right triangles. How can trigonometry also be applied to circles?

Step-by-Step Instructions

Step-by-step instructions are used throughout to guide you through technical procedures and tasks from start to finish. These steps show you not only how to perform a task but also how to do it safely and efficiently.

Perform the following steps to erect this system area scaffold:

Step 1 Gather and inspect all scaffold equipment for the scaffold arrangement.

Step 2 Place appropriate mudsills in their approximate locations.

Step 3 Attach the screw jacks to the mudsills.

Trade Terms

Each module presents a list of Trade Terms that are discussed within the text and defined in the Glossary at the end of the module. These terms are denoted in the text with **bold, blue type** upon their first occurrence. To make searches for key information easier, a comprehensive Glossary of Trade Terms from all modules is located at the back of this book.

During a rigging operation, the **load** being lifted or moved must be connected to the apparatus, such as a crane, that will provide the power for movement. The connector—the link between the load and the apparatus—is often a sling made of synthetic, chain, or **wire rope** materials. This section focuses on three types of slings:

Section Review

Each section of the module wraps up with a list of Additional Resources for further study and Section Review questions designed to test your knowledge of the Objectives for that section.

Additional Resources

Materials Handling Handbook, The American Society of Mechanical Engineers (ASME) and The International Material Management Society (IMMS), Raymond A. Kulwiec, Editor-in-Chief. 1985. New York, NY: Wiley-Interscience.

Manufacturing Facilities Design & Material Handling, Matthew P. Stevens, Fred E. Meyers. 2013. West Lafayette, IN: Purdue University Press.

1.0.0 Section Review

1. For material handling tasks, it is just as important to be mentally fit as it is to be _____.
   a. physically fit
   b. physically aggressive
   c. closely supervised
   d. over 200 pounds

2. Which of the following is a type of knot that is often used to join the ends of two ropes in non-critical, low-strain applications?
   a. Bowline
   b. Clove hitch
   c. Half hitch
   d. Square knot

Review Questions

The end-of-module Review Questions can be used to measure and reinforce your knowledge of the module's content.

Review Questions

1. Identification tags for slings must include the _____.
   a. type of protective pads to use
   b. type of damage sustained during use
   c. color of the tattle-tail
   d. manufacturer's name or trademark

2. The type of wire rope core that is susceptible to heat damage at relatively low temperatures is the _____.
   a. fiber core
   b. strand core
   c. independent wire rope core
   d. metallic link supporting core

3. Synthetic slings must be inspected _____.
   a. once every month
   b. visually at the start of each work week
   c. before every use
   d. once wear or damage becomes apparent

4. An alloy steel chain sling must be removed from service if there is evidence that _____.
   a. the sling has been used in different hitch configurations
   b. replacement links have been used to repair the chain
   c. the sling has been used for more than one year
   d. strands in the supporting core have weakened

5. A piece of rigging hardware used to couple the end of a wire rope to eye fittings, hooks, or other connections is a(n) _____.
   a. eyebolt
   b. hitch
   c. shackle
   d. U-bolt

6. A lifting clamp is most likely to be used to move loads such as _____.
   a. steel plates
   b. piping bundles
   c. concrete blocks
   d. plastic tubing

7. Chain hoists are able to lift heavy loads by utilizing a _____.
   a. rope and pulley system
   b. rigger's strength
   c. stationary counterweight
   d. gear system

8. Before attempting to lift a load with a chain hoist, make sure that the _____.
   a. hoist is secured to a come-along
   b. load is properly balanced
   c. tag lines are properly anchored
   d. tackle is connected to its power source

9. A hitch configuration that allows slings to be connected to the same load without using a spreader beam is a _____.
   a. double-wrap hitch
   b. choker hitch
   c. bridle hitch
   d. basket hitch

10. To make the emergency stop signal that is used by riggers, extend both arms _____.
   a. horizontally with palms down and quickly move both arms back and forth
   b. directly in front and then move both arms up and down repeatedly
   c. vertically above the head and wave both arms back and forth
   d. horizontally with clenched fists and move both arms up and down

NCCER Standardized Curricula

NCCER's training programs comprise more than 80 construction, maintenance, pipeline, and utility areas and include skills assessments, safety training, and management education.

Boilermaking
Cabinetmaking
Carpentry
Concrete Finishing
Construction Craft Laborer
Construction Technology
Core Curriculum: Introductory Craft Skills
Drywall
Electrical
Electronic Systems Technician
Heating, Ventilating, and Air Conditioning
Heavy Equipment Operations
Highway/Heavy Construction
Hydroblasting
Industrial Coating and Lining Application Specialist
Industrial Maintenance Electrical and Instrumentation Technician
Industrial Maintenance Mechanic
Instrumentation
Insulating
Ironworking
Masonry
Millwright
Mobile Crane Operations
Painting
Painting, Industrial
Pipefitting
Pipelayer
Plumbing
Reinforcing Ironwork
Rigging
Scaffolding
Sheet Metal
Signal Person
Site Layout
Sprinkler Fitting
Tower Crane Operator
Welding

Maritime

Maritime Industry Fundamentals
Maritime Pipefitting
Maritime Structural Fitter

Green/Sustainable Construction

Building Auditor
Fundamentals of Weatherization
Introduction to Weatherization
Sustainable Construction Supervisor
Weatherization Crew Chief
Weatherization Technician
Your Role in the Green Environment

Energy

Alternative Energy
Introduction to the Power Industry
Introduction to Solar Photovoltaics
Introduction to Wind Energy
Power Industry Fundamentals
Power Generation Maintenance Electrician
Power Generation I&C Maintenance Technician
Power Generation Maintenance Mechanic
Power Line Worker
Power Line Worker: Distribution
Power Line Worker: Substation
Power Line Worker: Transmission
Solar Photovoltaic Systems Installer
Wind Turbine Maintenance Technician

Pipeline

Control Center Operations, Liquid
Corrosion Control
Electrical and Instrumentation
Field Operations, Liquid
Field Operations, Gas
Maintenance
Mechanical

Safety

Field Safety
Safety Orientation
Safety Technology

Supplemental Titles

Applied Construction Math
Tools for Success

Management

Fundamentals of Crew Leadership
Project Management
Project Supervision

Spanish Titles

Acabado de concreto: nivel uno (*Concrete Finishing Level One*)
Aislamiento: nivel uno (*Insulating Level One*)
Albañilería: nivel uno (*Masonry Level One*)
Andamios (*Scaffolding*)
Carpintería: Formas para carpintería, nivel tres (*Carpentry: Carpentry Forms, Level Three*)
Currículo básico: habilidades introductorias del oficio (*Core Curriculum: Introductory Craft Skills*)
Electricidad: nivel uno (*Electrical Level One*)
Herrería: nivel uno (*Ironworking Level One*)
Herrería de refuerzo: nivel uno (*Reinforcing Ironwork Level One*)
Instalación de rociadores: nivel uno (*Sprinkler Fitting Level One*)
Instalación de tuberías: nivel uno (*Pipefitting Level One*)
Instrumentación: nivel uno, nivel dos, nivel tres, nivel cuatro (*Instrumentation Levels One through Four*)
Orientación de seguridad (*Safety Orientation*)
Paneles de yeso: nivel uno (*Drywall Level One*)
Seguridad de campo (*Field Safety*)

Acknowledgments

This curriculum was revised as a result of the farsightedness and leadership of the following sponsors:

Austal USA
Bollinger Shipyards
Claddagh Enterprises
Central Louisiana Technical College
Continental Maritime
Exelon Generation
Gulf States, Inc., Michigan Division
Ingalls Shipbuilding
KBR Industrial Services
Lee College
Lincoln Electric
Northeast Community College
Northland Pioneer College
Robins & Morton
Tecnico Corporation
TIC-The Industrial Company
Toledo Refining Co. LLC
Zachry Industrial, Inc.

This curriculum would not exist were it not for the dedication and unselfish energy of those volunteers who served on the Authoring Team. A sincere thanks is extended to the following subject matter experts:

Tom Ashley
Paul Bennett
Gerald Bickerstaff
Curtis Casey
Bill Cherry
Gerald Guthrie
Rod Hellyer
Frank Johnson
John Knapp
Jason Langford
Jerry McDaniel
Tim Riley
Kyle Savoie
Jason Scales
Dan Sterry
Holley Thomas
Jerry Trainor
Adam Webb

NCCER Partners

American Council for Construction Education
American Fire Sprinkler Association
Associated Builders and Contractors, Inc.
Associated General Contractors of America
Association for Career and Technical Education
Association for Skilled and Technical Sciences
Construction Industry Institute
Construction Users Roundtable
Design Build Institute of America
GSSC – Gulf States Shipbuilders Consortium
ISN
Manufacturing Institute
Mason Contractors Association of America
Merit Contractors Association of Canada
NACE International
National Association of Women in Construction
National Insulation Association
National Technical Honor Society
NAWIC Education Foundation
North American Crane Bureau
North American Technician Excellence
Pearson
Prov
SkillsUSA®
Steel Erectors Association of America
U.S. Army Corps of Engineers
University of Florida, M. E. Rinker Sr., School of Construction Management
Women Construction Owners & Executives, USA

NCCER Business Partners

ISNETWORLD DATA PROVIDER ISN
JUDGMENT INDEX
NACB NORTH AMERICAN CRANE BUREAU, INC.
PEARSON
Prov

Contents

Module Six

SMAW – Stainless Steel Plate and Pipe Groove Welds

This module explains stainless steel metallurgy, how to select SMAW electrodes for stainless steel welds, and how to weld different types of stainless steels. It covers safety issues associated with welding on stainless steels, how to prepare weld coupons, and how to set up SMAW equipment for welding stainless steel. It provides procedures for making open-root V-groove welds with SMAW equipment on stainless steel plate in the 1G, 2G, 3G, and 4G positions, and procedures for making open-root V-groove welds with SMAW equipment on stainless steel pipe in the 1G-ROTATED, 2G, 5G, and 6G positions. (Module ID 29306-16; 100 Elective Hours)

Glossary

Index

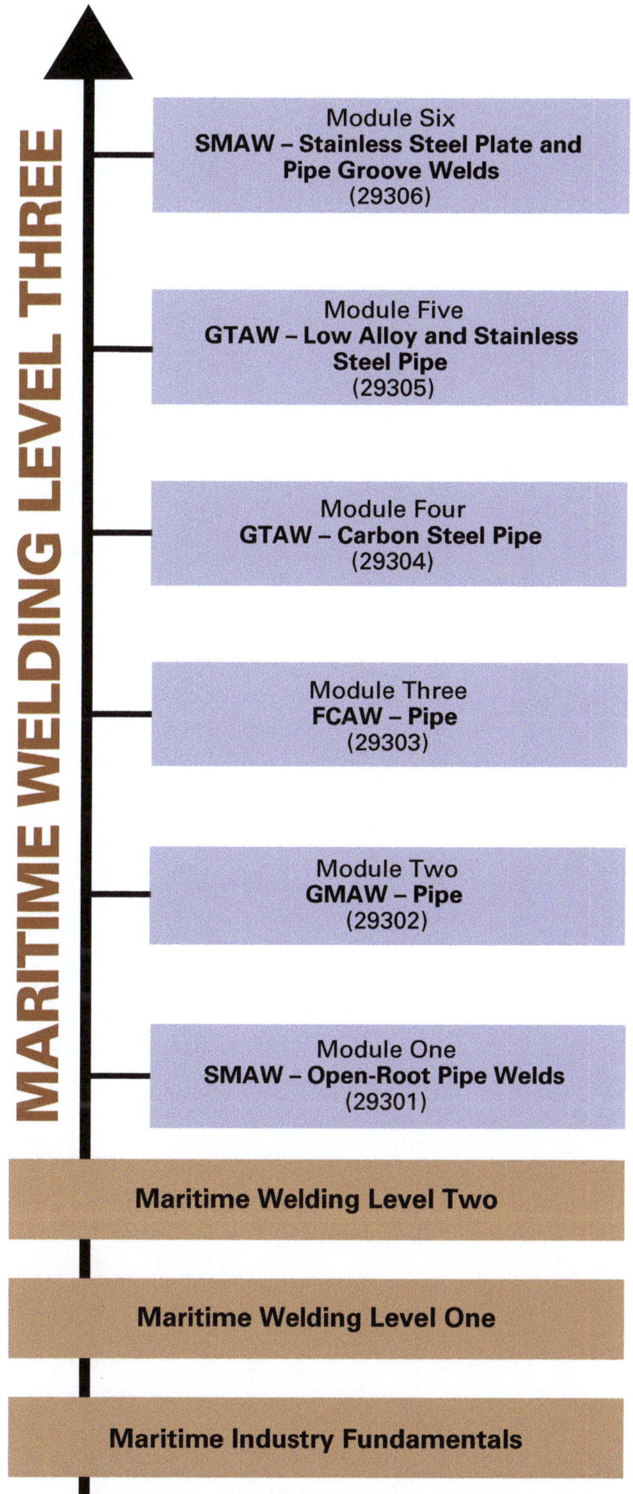

This course map shows all of the modules in *Maritime Welding Level Three.* The suggested training order begins at the bottom and proceeds up. Skill levels increase as you advance on the course map. The local Training Program Sponsor may adjust the training order.

SMAW – Open-Root Pipe Welds

Overview

Shielded metal arc welding (SMAW) is a well-established form of welding used extensively in commercial and industrial applications around the world. One routine use of SMAW is for joining pipe of various thicknesses and diameters. The most common joint used for these applications is the open-root V-groove joint. This module describes how to prepare and perform open-root V-groove welds on pipe in all positions using SMAW equipment and electrodes.

Module 29301

Trainees with successful module completions may be eligible for credentialing through the NCCER Registry. To learn more, go to **www.nccer.org** or contact us at 1.888.622.3720. Our website has information on the latest product releases and training, as well as online versions of our *Cornerstone* magazine and Pearson's product catalog.

Your feedback is welcome. You may email your comments to **curriculum@nccer.org**, send general comments and inquiries to **info@nccer.org**, or fill in the User Update form at the back of this module.

This information is general in nature and intended for training purposes only. Actual performance of activities described in this manual requires compliance with all applicable operating, service, maintenance, and safety procedures under the direction of qualified personnel. References in this manual to patented or proprietary devices do not constitute a recommendation of their use.

29301 V5

29301
SMAW – Open-Root Pipe Welds

Objectives

When you have completed this module, you will be able to do the following:

1. State the basic concepts of open-root SMAW and how to prepare for welding.
   a. State the basic concepts of open-root SMAW of pipe.
   b. Identify common welding safety practices.
   c. Explain how to prepare the area, materials, and equipment for SMAW of pipe.
2. Describe open-root pipe welding positions and related SMAW techniques.
   a. Identify and describe the four pipe welding positions and acceptable weld profiles.
   b. Describe the SMAW techniques for root and fill passes on pipe.
   c. Describe the techniques required to produce open-root SMAW pipe welds in all positions.

Performance Tasks

Under the supervision of your instructor, you should be able to do the following:

1. Make open-root pipe welds in the 1G-ROTATED position.
2. Make open-root pipe welds in the 2G position.
3. Make open-root pipe welds in the 5G position.
4. Make open-root pipe welds in the 6G position.

Trade Terms

Fast-freeze electrode
Feather
Hot pass
Keyhole
Pipe axis
Remaining passes
Root pass

Industry Recognized Credentials

If you are training through an NCCER-accredited sponsor, you may be eligible for credentials from NCCER's Registry. The ID number for this module is 29301. Note that this module may have been used in other NCCER curricula and may apply to other level completions. Contact NCCER's Registry at 888.622.3720 or go to **www.nccer.org** for more information.

Contents

Figures and Tables

[illegible]

Section One

1.0.0 SMAW Pipe Welding Preparations

Objective

State the basic concepts of open-root SMAW and how to prepare for welding.

a. State the basic concepts of open-root SMAW of pipe.
b. Identify common welding safety practices.
c. Explain how to prepare the area, materials, and equipment for SMAW of pipe.

Trade Terms

Fast-freeze electrode: An electrode that solidifies quickly, produces very little slag, and gives good penetration into the root of the joint.

Feather: To grind the start and/or stop areas of a weld to a tapered edge.

Keyhole: A small hole that is formed when the root faces of two pipe ends are melted away by the arc.

Pipe axis: An imaginary line that runs longitudinally (lengthwise) through the center of a pipe.

Remaining passes: A nonstandard term that is used to describe the fill and cover passes as a group.

Root pass: The initial weld pass made to produce a root bead.

Successfully performing open-root V-groove welds on pipe requires an understanding of SMAW pipe welding concepts, as well as a keen awareness of safety procedures and practices that must be observed before and during the welding operation. Although the welding process is similar to open-root plate welding, there are subtle differences and the round workpiece often requires constant changes in hand position. Another basic requirement is the ability to properly prepare the work area, materials, and welding equipment prior to any welding activity.

1.1.0 Basic SMAW Pipe Welding Concepts

The open-root V-groove weld (*Figure 1*) is the most common groove weld made on carbon, low-alloy, and stainless steel pipe. These welds are typically used for joining medium- and thick-walled pipe used in various piping systems. Piping containing or carrying material that can cause injury to people or damage to the environment, or those that operate at elevated pressures, may be categorized as critical piping. Injury or damage occurs when the piping system or other system components fail to contain the material as designed. Welds in critical piping may be required to meet stringent requirements. The welds must be expertly completed, thoroughly inspected, and at least as strong as the piping being welded. Piping that is used for hydronic heating and air conditioning applications, simple water systems, and similar applications may be assessed using less-stringent requirements.

Two common SMAW electrodes are used for open-root V-groove welds on carbon steel. E6010 electrodes are used for the **root pass** because a backing ring is not used. This type of electrode is a **fast-freeze electrode**. It solidifies quickly, produces very little slag, and gives good penetration into the root of the joint, even with a small root opening. However, welding a root pass on pipe takes some practice to perfect.

E7018 electrodes are the most common for the **remaining passes**. Base metal composition however, will determine which electrodes are acceptable for the remaining fill and cover passes.

For the purposes of this module, the dimensions shown are representative of codes and may not be specific to one specific code. Always refer to the applicable code or site-specific specifications for every weld.

1.2.0 Safety Practices

Safety considerations are a critical part of preparing for a welding operation. The following is a summary of safety procedures and practices that must be observed while cutting or welding.

Figure 1 Pipe welding.

Be advised that this is only a summary. Complete safety coverage is provided in the *Welding Level One* module, "Welding Safety." That module should be studied, if not already completed, before continuing. Above all, be sure to wear appropriate protective clothing and equipment at all times when welding or cutting.

1.2.1 Protective Clothing and Equipment

Protective clothing and equipment are a necessary part of every welding task. Follow the guidelines below as a general rule, and ensure that you are familiar with any unique requirements your employer may have regarding personal protective equipment (PPE).

- A welding hood equipped with a properly shaded lens is best for all forms of welding. The welding hood lens must be an appropriate shade for the application.
- Wear proper leather or flame-retardant clothing along with welding gloves for protection from flying sparks, molten metal, and heat.
- Wear high-top safety shoes or boots. Make sure that the tongue and lace area of the footwear is covered by a pant leg. If the tongue and lace area is exposed or if the footwear must be protected from burn marks, wear spats under the pant leg or leather chaps over the front and top of the footwear.
- Wear a 100-percent cotton cap without any mesh material. If a hard hat is required for the environment, use one that allows the attachment of rear deflector material and a face shield. A hard hat with a rear deflector is generally preferred when working overhead, and may be required by some employers and job sites.
- Wear earmuffs, or at least earplugs, to prevent sparks from entering the ear canals.

1.2.2 Housekeeping and Fall Protection

Welding is often performed outside in the weather and in the construction environment. Production work done inside large, open shops exposes a welder to equipment moving across the floor and overhead. Welders often have to crawl under or reach over objects to complete their work.

At times, welders must work at height. Fall protection equipment is required for all work at height (*Figure* 2). Always keep fall protection devices untangled and properly attached. Combined with the other equipment a welder must carry, there is a great deal to manage, but fall protection equipment must always be the priority. Fall protection equipment saves lives. Never

AWS Electrode Classifications

The four electrodes in the chart below are among those most commonly used in SMAW applications. As with all electrodes, they use the American Welding Society (AWS) electrode classification system. The E indicates Electrode; the first two numbers indicate the tensile strength of the welded joint in thousands of pounds per square inch; the third number indicates the positions in which the electrode was designed to be used; and the fourth number indicates the type of flux coating, the amount of penetration, and the acceptable type of current. Welders should be familiar with the characteristics of these electrodes when performing SMAW.

ELECTRODE NAME	TENSILE STRENGTH	WELDING POSITIONS	FLUX COATING	AMOUNT OF PENETRATION	CURRENT TYPE
E6010	60,000 PSI	ALL	CELLULOSE SODIUM	DEEP	DC ONLY
E6011	60,000 PSI	ALL	CELLULOSE POTASSIUM	DEEP	AC/DC
E6013	60,000 PSI	ALL	RUTILE POTASSIUM	MEDIUM	AC/DC
E7018	70,000 PSI	ALL	IRON POWDER LOW HYDROGEN	SHALLOW TO MEDIUM	AC/DC

Figure 2 Working at height.

attach any kind of welding gear to the harness that could interfere with its ability to perform its intended function.

Regardless of where you are working, always keep the work area as neat as possible. This makes the area safer and helps to reduce distractions. Mindful actions, such as making sure that welding cables are not tripping hazards, contribute to a safer environment. Electrode stubs and unused electrodes must not be left on the floor. They can cause someone to lose their footing and fall. Parts waiting to be welded or left over from previous work are a tripping hazard and should be kept out of the walking area. If the parts have sharp edges, they could cut welding cable insulation or torch gas hoses.

1.2.3 Fire and Explosion Prevention

Flammable gases are commonly found around welding and cutting operations, as are sparks and other significant sources of ignition. For that reason, and because many fire-related incidents over the years have been traced back to welding operations, it is crucial that fire and explosion prevention be given consistent attention. Follow these guidelines to maintain a safe environment:

- Never carry matches or gas-filled lighters. Sparks can cause the matches to ignite or the lighter to explode in your pocket, causing serious injury.
- Never use oxygen to blow dust or dirt from clothing. Traces of the oxygen can remain trapped in the fabric for some time. If a spark hits the fabric during this time, the clothing can burn rapidly and violently out of control.
- Make sure that any flammable material in the work area is removed or shielded by a fire-resistant covering. Approved fire extinguishers must be available before attempting any heating, welding, or cutting operations. If a hot-work permit and a fire watch are required, be sure these are in place before beginning.
- Ensure that all site safety requirements are being observed.
- Never release a large amount of oxygen or use oxygen in place of compressed air. Its presence around flammable materials or sparks can cause rapid and uncontrolled combustion. Keep oxygen away from oil, grease, and other petroleum products.
- Never release a large amount of fuel gas, especially acetylene. Propane tends to concentrate in and along low areas and can ignite at a considerable distance from the release point. Acetylene is lighter than air, but is even more dangerous than propane. When mixed with air or oxygen, acetylene will ignite at much lower concentrations than any other fuel gas.
- Even though shielding gases are not toxic, any gas released in a sufficient quantity can displace the oxygen in a small room or enclosed work area, creating an atmosphere with little or no oxygen. Avoid large releases of any type of gas, regardless of its characteristics.
- To prevent fires, maintain a neat and clean work area, and make sure that any metal scrap or slag is cold before disposal.
- Before cutting containers such as tanks or barrels, check to see if they have contained any explosive, hazardous, or flammable materials, including petroleum products, citrus products, or chemicals that decompose into toxic fumes when heated. As a standard practice, always clean and then fill any tanks or barrels with water, or purge them with a flow of inert gas such as nitrogen to displace any oxygen. Containers must be cleaned by steam cleaning, flushing with water, or washing with detergent until all traces of the material have been removed. Proper procedures for cutting or welding hazardous containers are described in the *American Welding Society (AWS) F4.1, Safe Practices for the Preparation of Containers and Piping for Welding and Cutting.*

> **WARNING!**
> Welding or cutting must never be performed on drums, barrels, tanks, vessels, or other containers until they have been emptied and cleaned thoroughly, eliminating all flammable materials and all substances (such as detergents, solvents, greases, tars, or acids) that might produce flammable, toxic, or explosive vapors when heated. Clean containers only in well-ventilated areas. Vapors can accumulate during cleaning, causing explosions and serious personal injury.

1.2.4 Work Area Ventilation

Many gases are introduced to the atmosphere through welding processes. Some of these gases are harmful, and all can displace or combine with atmospheric oxygen in a confined space. Follow these guidelines, as well as those of your employer, for work area ventilation:

- Make sure to follow all confined space procedures before conducting any welding or cutting in a confined space. Use powered extraction systems as needed.
- Never use oxygen to provide ventilation in confined spaces.
- Always perform cutting or welding operations in a well-ventilated area.

Cutting or welding operations involving zinc or cadmium materials or coatings result in toxic fumes. Another significant concern is Chromium 6, commonly referred to as *hexavalent chromium*, which is used in stainless steel. It releases toxic fumes when heated. Manganese is another hazardous metal used in stainless steel. It is also used in some welding rods, increasing the risk of exposure.

For long-term cutting or welding of materials containing these components, always wear an approved full-face, supplied-air respirator that uses breathing air supplied from outside the work area. For occasional short-term exposure, a HEPA-rated or metal-fume filter may be used on a standard respirator.

Powered fume-extraction equipment must be used when welding materials that contain zinc, cadmium, hexavalent chromium, manganese, and similar materials that produce hazardous fumes if the area is not naturally well-ventilated. *Figure 3* shows examples of both mobile and portable units.

1.3.0 Preparations for Welding

Before any welding procedure can take place, the work area must be made ready, the welding equipment must be properly set up, and the metal to be welded must be prepared.

1.3.1 Preparing the Welding Area

To practice welding, a welding table, bench, and/or appropriate pipe stands are needed. The welding surface must be made of steel. Provisions must be made for supporting the pipe welding coupons for out-of-position welding. *Figure 4* shows a typical welding station.

To set up the area for welding, follow these steps:

Step 1 Make sure that the area is properly ventilated. Make use of doors, windows, and fans.

Step 2 Check the area for fire hazards. Remove any flammable materials before proceeding.

Step 3 Know the location of the nearest fire extinguisher. Ensure the fire extinguisher is charged and that you and others in the immediate area know how to use it properly.

Step 4 Set up flash shields and curtains around the welding area as necessary.

1.3.2 Electrodes

Obtain a small number of the electrodes to be used. For the welding exercises in this module, ⅛" E6010, along with ⅛" and 3⁄32" E7018 electrodes, will be used. (Refer to *Table 1* for the metric equivalents

> ## Classes of Fire Extinguishers
> Fire extinguishers have different class ratings based on the types of fires they are used to extinguish. For example, Class A fire extinguishers are used for common materials like paper, wood, and plastic. Class B fire extinguishers are used for liquids such as gasoline and oil. Class C fire extinguishers are used for electrical equipment fires, and Class D extinguishers are used for combustible metals. Always make sure the fire extinguishers available on site are suitable for the types of fires that might occur.

(A) MOBILE UNIT

(B) PORTABLE UNIT

Figure 3 Mobile and portable extraction equipment.

Figure 4 Welding station.

of these rod sizes.) Keep these general guidelines in mind when running E6010 and E7018 electrodes on open-root V-groove pipe joints:

- When making the root pass with E6010 electrodes, use the **keyhole** technique to achieve proper root penetration.
- Use a whipping motion to control the puddle size with the E6010 electrode.
- Do not whip low-hydrogen E7018 electrodes.
- Maintain a short arc when running low-hydrogen electrodes.
- Set the amperage in the lower portion of the suggested range for better puddle control when welding vertically.
- Remove all slag between passes.

Take only enough E7018 electrodes from the oven to complete one weld joint. Store the electrodes in a proper pouch or rod holder to prevent the coating from being contaminated or damaged.

Never store electrodes loose on a table. They may end up on the floor where they can be contaminated, damaged, or create a tripping hazard. Some type of metal container, such as a pail, must also be available to discard hot electrode stubs.

Table 1 Electrode Dimensions and Metric Equivalents

DIAMETER		LENGTH	
Inches	mm	Inches	mm
3/32	2.4	12	300
1/8	3.2	14	350
5/32	4.0	14	350
3/16	4.8	14/18	350/450
7/32	5.6	18	450
1/4	6.4	18	450
5/16	8.0	18	450

WARNING

Do not throw electrode stubs on the floor. They roll easily and could become a tripping hazard.

1.3.3 Preparing the Welding Machine

Figure 5 shows the controls of common AC/DC and DC welding machines. Identify a welding machine to use and then follow these steps to set it up for welding:

Step 1 Verify that the welding machine can be used for DC welding.

Step 2 Know the location of the primary power disconnect. This could be a disconnect switch or circuit breaker for hard-wired equipment, or the power cord if the machine is simply plugged into a receptacle.

Step 3 Set the polarity to direct current electrode positive (DCEP).

Step 4 Set the amperage for the electrode type and size being used. Typical settings are shown in *Table 2*. Note that these settings are common, but the recommendations of the equipment manufacturer and/or your instructor should be followed in the absence of a WPS.

Step 5 Connect the clamp of the workpiece lead to the workpiece.

Beveling

The two ends of the pipe coupons must meet squarely. To accomplish this, the sharp, inside edge (root face) of the bevel must be ground flat, and square to the length of pipe. For the root opening to be uniform, final shaping should be carefully done with a grinder.

Step 6 Check to be sure that the electrode holder is not grounded.

Step 7 Turn on the welding machine.

CAUTION

When welding pipe, avoid making arc strikes outside of the weld groove on the surface of the pipe. An arc strike can cause a hardened spot that can crack as the pipe expands and contracts. An arc strike on the pipe surface is considered a defect, which will require repair or reworking.

1.3.4 Preparing Pipe Coupons

Pipe weld coupons should be cut from 3" to 12" (DN80 to 300) diameter Schedule 40 or Schedule 80 carbon steel pipe. Each welded joint requires two pipe coupons of the same diameter and schedule.

Figure 6 shows typical ASME, American Petroleum Institute (API), and AWS pipe bevel specifications for bevel angles, root faces, and root openings.

To prepare weld coupons for open-root V-groove weld joints, follow these steps:

Step 1 Clean heavy rust or mill scale from the inside and outside of the pipe with a grinder or wire brush.

Step 2 Bevel the end of the pipe to 30 to 37½ degrees by any acceptable beveling method, including manual, mechanical, and thermal methods.

Step 3 Cut off the end of the beveled pipe to the length specified by your instructor. Two coupons will be required for each weld.

NOTE

For 1G-ROTATED welding, the coupons may need to be cut longer to reach the rollers being used.

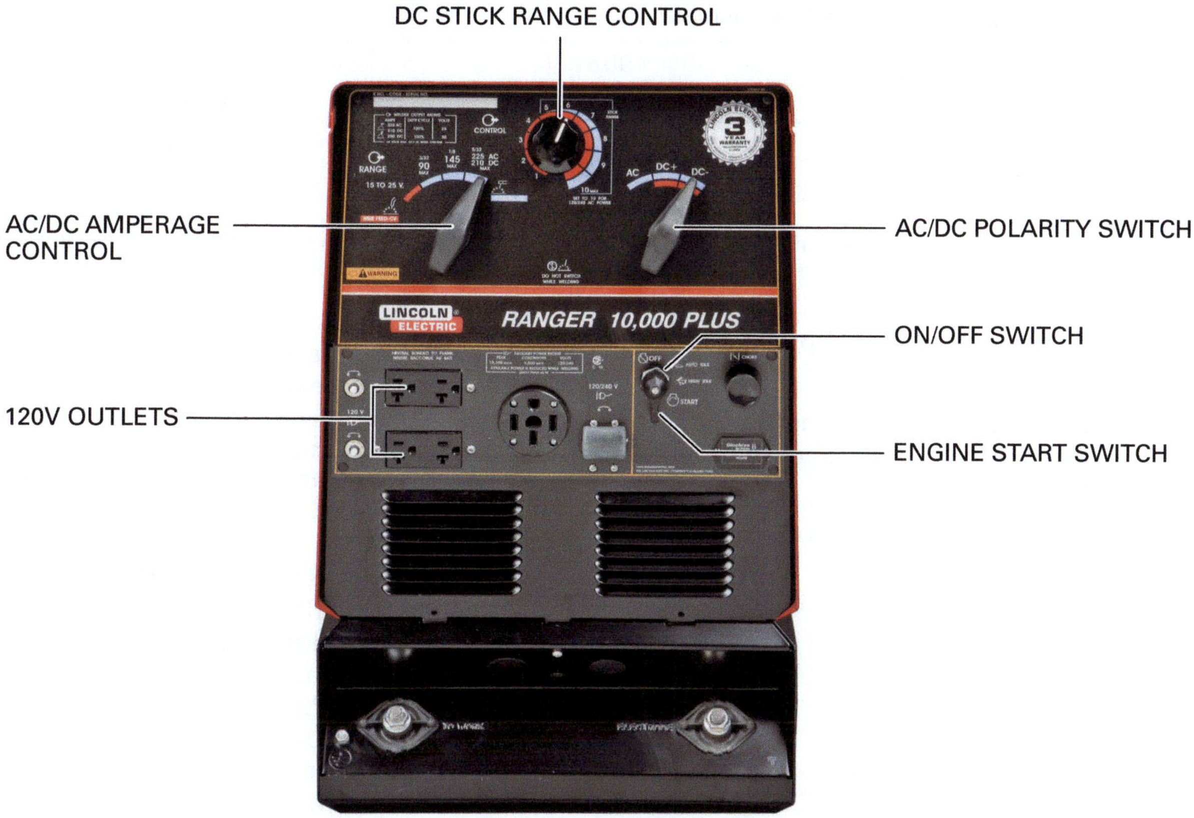

(A) AC/DC MACHINE

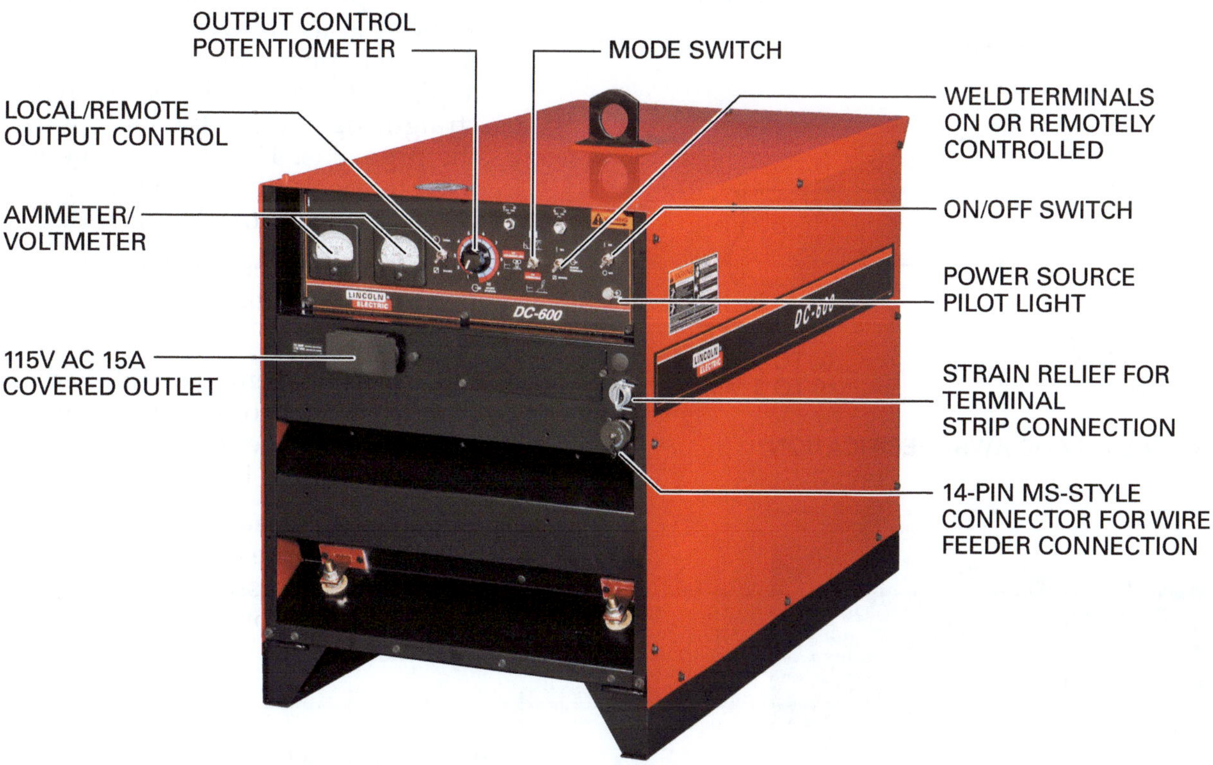

(B) DC MACHINE

Figure 5 Typical AC/DC and DC welding machines.

Table 2 Amperage Settings for Electrode Types and Sizes

Electrode	Size	Amperage (Approximate Range)
E6010	⅛"	65A to 130A
E7018	3/32"	70A to 120A
E7018	⅛"	110A to 150A

Note
Amperage recommendations vary by manufacturer, position, current type, and electrode brand. For specific advice, refer to the manufacturer's recommendations for the electrode being used, or to your site WPS.

The two coupons can now be prepared further for welding:

Step 1 Check the bevel. There should be no dross, and the bevel angle should be 30 to 37½ degrees.

Step 2 Grind or file a 0" to ⅛" (0 to 3 mm) root face on the bevel.

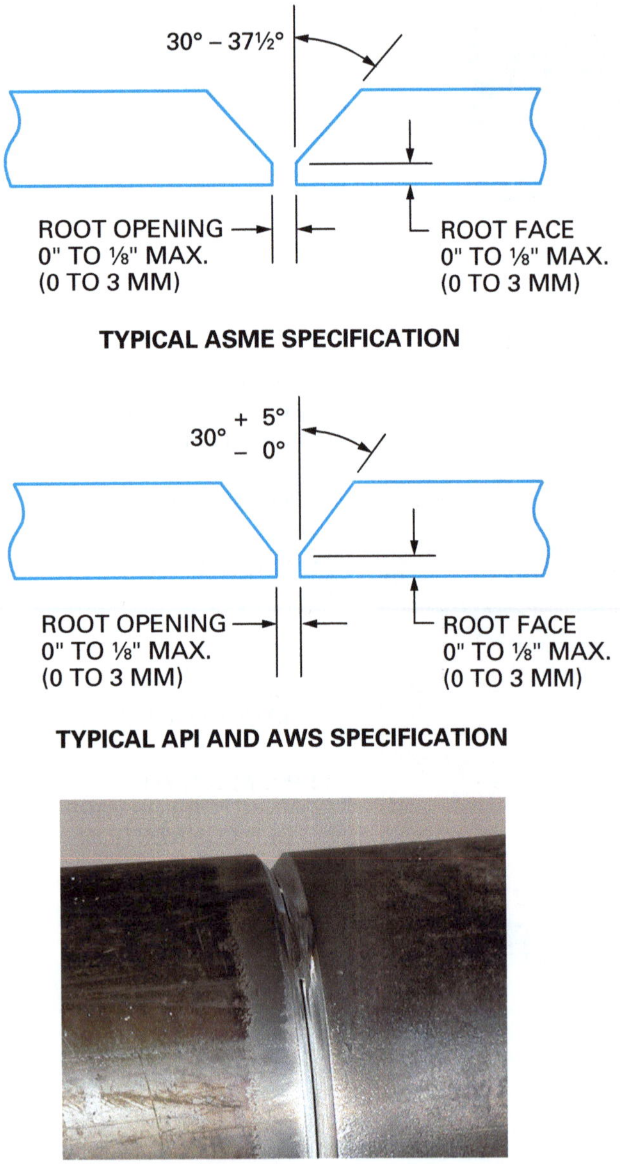

Figure 6 ASME, API, and AWS pipe bevel specifications.

NOTE

Welding codes allow the root face for open-root welds to be from 0" to ⅛" (0 to 3 mm) wide and the root opening to be from 0" to ⅛" (0 to 3 mm). Select and adjust the root face and opening as directed by the instructor as welding practice begins.

Step 3 Align the two pipe sections. The inside diameter (ID) surfaces of the two coupons should be even all around and evenly separated by the required root opening. Pieces of filler wire, metal shims, or pieces of bare welding electrode can be used to achieve the desired root opening between the two pipe sections until they are tack welded (*Figure 7*). Small-diameter pipe can be aligned by clamping both pieces to a piece of angle iron as shown in *Figure 8*. Large-diameter pipe can be aligned using a pipe alignment jig or by holding a straightedge across the joint, parallel to the **pipe axis**. The straightedge must be used to check all around the pipe in case one or both sections are distorted.

Step 4 Once aligned, secure the two pipe sections with a clamp, or with tack welds if necessary.

Step 5 With the root opening established and the pipe ends aligned, make the first of four tack welds. Longer tacks may be required on heavy-wall or large-diameter pipe.

Step 6 After the first tack weld, check the root opening opposite it. Adjust the root opening space if necessary, and make the second tack weld on the side opposite the first tack.

CAUTION

Remove the spacer material before making the second tack, so that it will not get caught as the heated metal expands.

Figure 7 Pipe coupons showing filler wire used as a spacer.

Figure 8 Pipe sections secured.

GOING GREEN

Conserve Materials

Pipe for practice welding is expensive and difficult to obtain. Make every effort to conserve the material available. Reuse weld coupons until all surfaces have been welded upon. Cut the coupon apart and reuse the pieces. Check with the instructor for the appropriate size of the coupons.

Step 7 Verify that the root opening remains correct before welding the third tack midway between the first and second tacks.

Step 8 Weld the fourth tack opposite the third tack and midway between the first and second tacks. There should now be four tack welds evenly spaced (90 degrees apart) around the pipe coupon (*Figure 9*).

Step 9 Clean and **feather** the tack welds with the edge of a grinding wheel. Feathering the ends of the tack welds with a grinder helps to fuse the root pass into the tack weld.

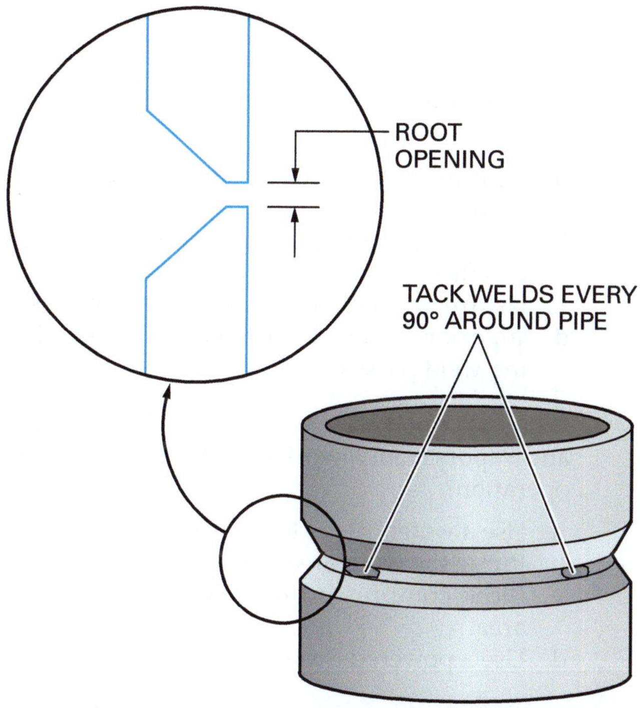

Figure 9 Tacked open-root V-groove weld coupon.

Welding Large-Diameter Pipe

The practice coupons for large-diameter pipe must be greater than 1½" (≈4 cm) in length. They need more metal to help absorb the increased heat used to make these welds.

Additional Resources

AWS F4.1: Recommended Safe Practices for Preparation for Welding and Cutting of Containers and Piping. Miami, FL: American Welding Society.

Modern Welding Technology. Current Edition. Howard B. Cary and Scott Helzer. Englewood Cliffs, NJ: Prentice Hall.

OSHA 1926.351, Arc Welding and Cutting. Current Edition. Washington, DC: Occupational Safety & Health Administration.

The Procedure Handbook of Arc Welding. Current Edition. Cleveland, OH: The James F. Lincoln Arc Welding Foundation.

Welding Handbook, Volumes 1 through 4. Current Edition. Miami, FL: American Welding Society.

Welding Pressure Pipelines and Piping Systems (PDF). The Lincoln Electric Company, Cleveland, OH, USA. **www.lincolnelectric.com**. Last accessed May 29, 2015.

1.0.0 Section Review

1. An E6010 electrode is best described as a(n) _____.
   a. light penetrating electrode that is slow to solidify
   b. E7018 electrode that has more metal and less flux
   c. fast-freeze electrode used for the root pass
   d. superior electrode for making the remaining weld passes

2. What is a good way to remove welding fumes and vapors from a work area during a welding operation?
   a. Use a water mister to spray the area.
   b. Purge the work area with zinc oxide.
   c. Blow compressed oxygen into the work area.
   d. Use a powered extraction system.

3. The surface of the welding table being used for practice welding should be made of _____.
   a. steel
   b. aluminum
   c. fiberglass
   d. wood

SECTION TWO

2.0.0 OPEN-ROOT V-GROOVE PIPE WELDING TECHNIQUES

Objective

Describe open-root pipe welding positions and related SMAW techniques.

a. Identify and describe the four pipe welding positions and acceptable weld profiles.
b. Describe the SMAW techniques for root and fill passes on pipe.
c. Describe the techniques required to produce open-root SMAW pipe welds in all positions.

Performance Tasks

1. Make open-root pipe welds in the 1G-ROTATED position.
2. Make open-root pipe welds in the 2G position.
3. Make open-root pipe welds in the 5G position.
4. Make open-root pipe welds in the 6G position.

Trade Terms

Hot pass: A nonstandard term for the weld pass that is made directly after the root pass.

The open-root V-groove weld begins with a welding pass referred to as the *root pass.* Additional welding passes, referred to as the *remaining passes*, are then added to complete the weld. Groove welds can be made in all positions on pipe.

2.1.0 Pipe Groove-Weld Test Positions

Groove-weld test positions are determined by the axis of the pipe (*Figure 10*). Four standard weld test positions are used with pipe:

- *Flat (1G-ROTATED)* – The pipe axis is horizontal, and the pipe is slowly rotated while it is being welded on the top. The weld beads are flat.
- *Horizontal (2G)* – The pipe axis is vertical and the pipe rotation is fixed. The weld beads are horizontal.
- *Multiple (5G)* – The pipe axis is horizontal and the pipe rotation is fixed. The weld beads are flat, vertical, and overhead.
- *Multiple inclined (6G)* – The pipe axis is inclined at approximately 45 degrees ±5 degrees. Pipe rotation is fixed. Weld beads are horizontal, vertical, and overhead.

The pipe axis positions for 1G-ROTATED, 2G, and 5G can vary ±15 degrees from the basic position. The pipe axis position for 6G can vary ±5 degrees from the nominal 45-degree angle. *Table 3* provides a quick reference for the four positions.

To determine the quality of a pipe weld, specific sections of pipe are removed for destructive testing. For positions 1G-ROTATED and 2G, two specimens are taken at random. They must be taken 180 degrees from each other, as shown on the left side of *Figure 11*. After removal, the specimens are put through bend tests to determine the quality of the welds. Alternatively, specimens may be tested by radiography, a non-destructive testing (NDT) method.

When 5G or 6G welds will be destructively tested, the test specimens must be cut from the four regions shown on the right side of *Figure 11*. These regions are midway between the twelve o'clock and three o'clock; three o'clock and six o'clock; six o'clock and nine o'clock; and nine o'clock and twelve o'clock positions. The twelve o'clock position is the top of the coupon groove; the six o'clock position is at the bottom.

2.1.1 Acceptable and Unacceptable Pipe Weld Profiles

Pipe groove welds should be made with slight reinforcement (not exceeding ⅛" [3 mm]) and a gradual transition to the base metal at each toe. The root pass should have complete penetration. The root reinforcement inside of the pipe may range in thickness from being flush to a maximum of ⅛" (3 mm) thick.

Pipe groove welds must not have excess reinforcement or any underfill at the face or root. They must be free from excessive undercut or any overlap. Excessively large face reinforcement reduces the pipe's strength. It causes the stresses in the pipe to be concentrated along the sides of the weld. As a result, the pipe will not expand and contract uniformly along its length. If a weld has any of the defects shown in *Figure 12*, it must be repaired.

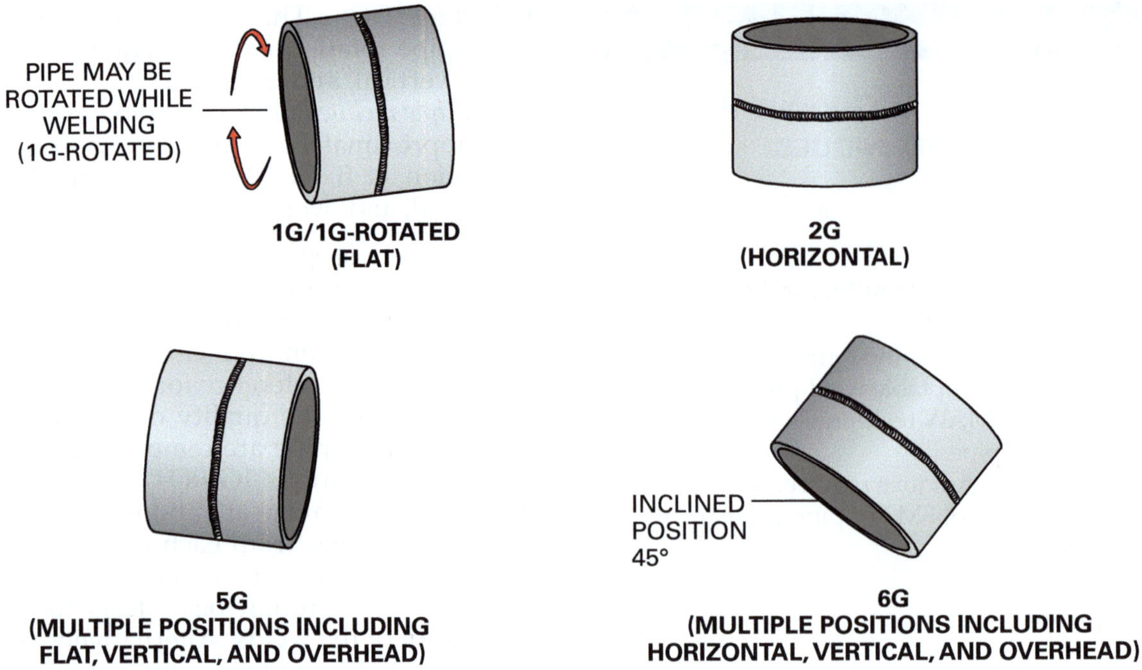

Figure 10 Four standard pipe groove-weld test positions.

Table 3 Pipe Groove-Weld Test Positions

Weld Test Position	Pipe Axis	Weld Beads	Pipe Position
1G-ROTATED	Horizontal	Flat	Slowly rotated
2G	Vertical	Horizontal	Fixed
5G	Horizontal	Flat, vertical, overhead	Fixed
6G	Inclined 45 degrees	Horizontal, vertical, overhead	Fixed

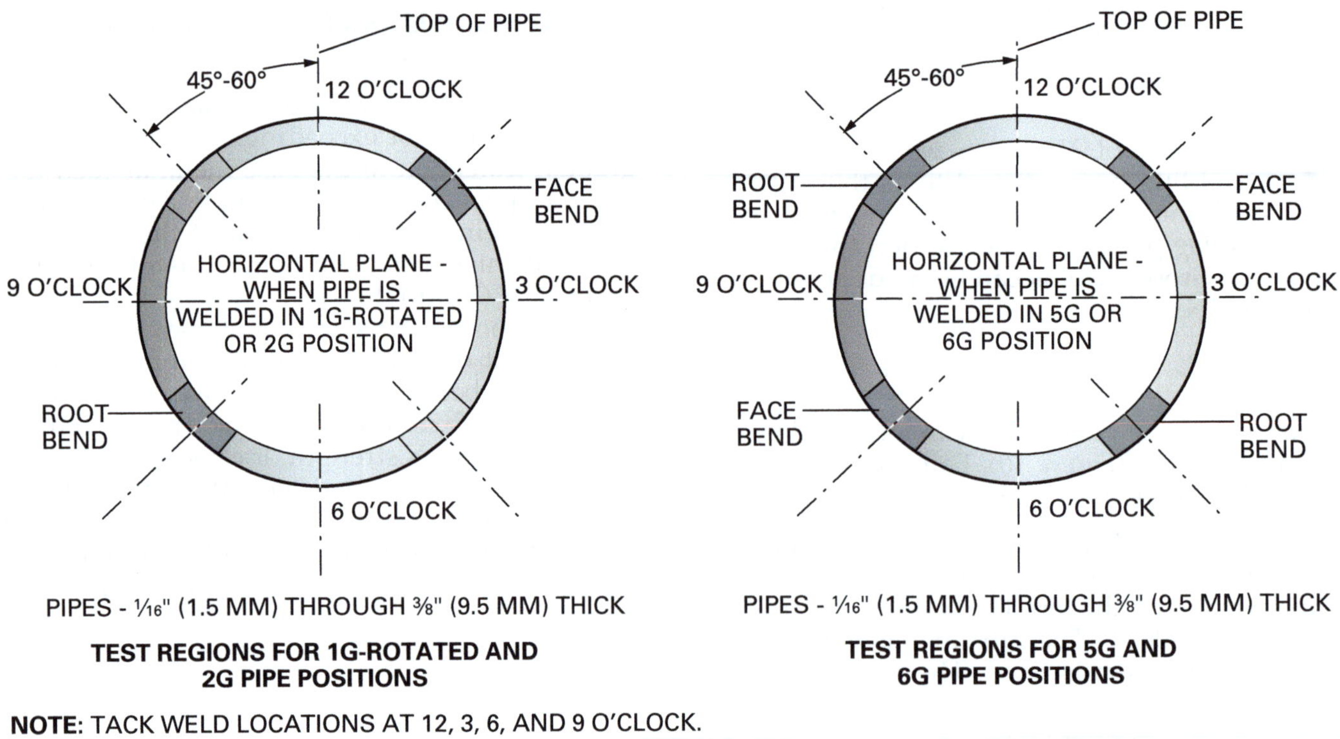

Figure 11 Test specimen regions for pipe positions 1G-ROTATED, 2G, 5G, and 6G.

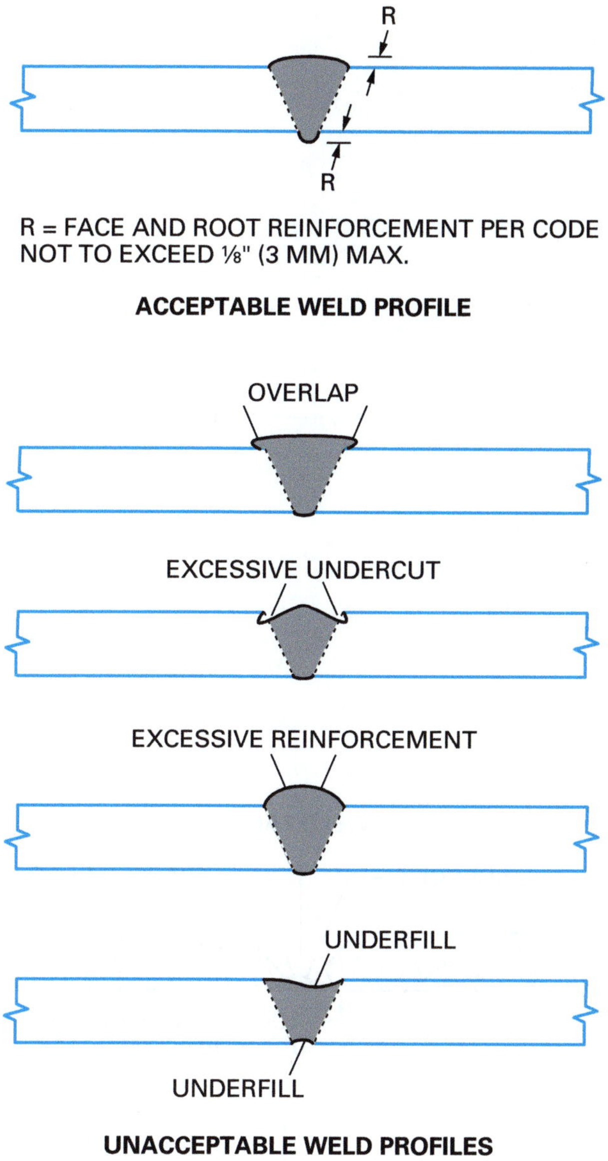

Figure 12 Acceptable and unacceptable pipe groove weld profiles.

2.2.0 The Root Pass

The greatest challenge of an open-root V-groove weld is making the root pass. The root pass must have complete penetration, but not an excessive amount of root reinforcement. The extent of penetration is very important. If the penetration is too little, it will create a weak joint. If there is too much penetration along with excessive root reinforcement, the inside diameter of the pipe will be reduced and the flow restricted.

The penetration is controlled by a technique called *running a keyhole.* A keyhole-shaped hole is formed when the root faces on the two pipe edges are melted away by the arc (*Figure 13*). The molten metal flows to the back side of the keyhole, forming the weld. The size of the keyhole should be maintained at $\frac{3}{16}$" to $\frac{1}{4}$" (5 mm to 6 mm) in diameter. If the keyhole is too large, there will be excessive melt-through with slag inclusions. If the keyhole closes up, a lack of penetration will result. The root reinforcement depth should be kept flush or at a maximum of $\frac{1}{8}$" (3 mm) deep.

One approach to running a keyhole that works well is to begin the root pass on top of the tack $\frac{1}{8}$" to $\frac{3}{16}$" (3 to 5 mm) behind the feathered portion of the tack. This helps preheat the joint for a better tie-in. When approaching the next tack, proceed past the feathered portion of the next tack to the same point. Ensure that the keyhole disappears before stopping the weld.

The keyhole is controlled by using fast-freeze electrodes, such as E6010, that solidify quickly. It is also controlled by slightly whipping the electrode forward and out of the joint, and then back into the puddle. Move the electrode forward about $\frac{3}{16}$" to $\frac{1}{4}$" (5 mm to 6 mm), then back about $\frac{1}{8}$" (3 mm), and then pause. Be careful not to increase the arc length when moving the electrode.

6GR Test Position

Welding pipe in the 6G position is a common welding test. A ring may be added to test for restricted accessibility (6GR).

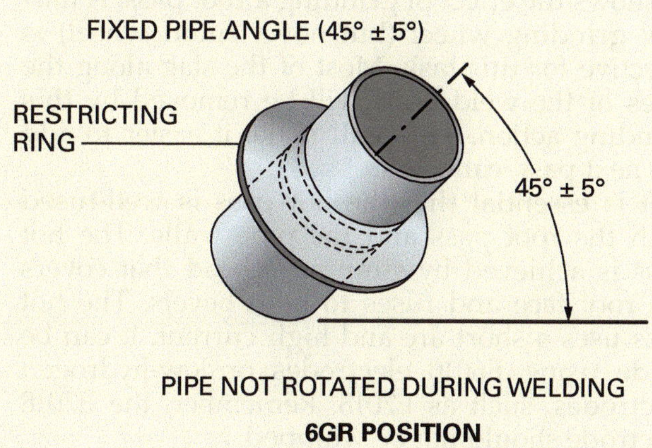

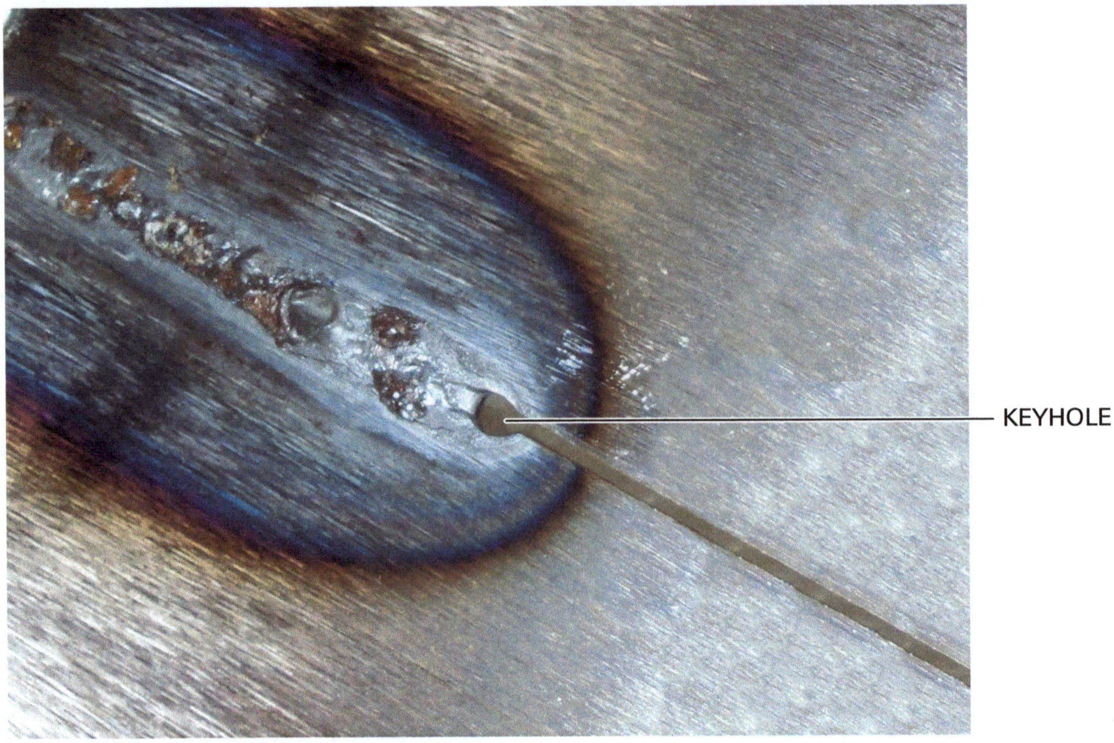

Figure 13 Root pass keyhole.

If the keyhole starts to grow larger, decrease the pause and increase the forward length of the whip. If the keyhole starts to close up, pause slightly longer and decrease the length of the whip. Other factors that affect the keyhole size are the amperage setting, root face size, and root opening. All these factors can be adjusted within the limits allowed by the welding codes or your site quality procedures to support this technique.

After the root pass has been run, it should be cleaned and inspected. Check the root face for excess buildup or undercut that could trap slag when the later passes are run. Remove excess buildup or undercut by grinding the face of the root pass with the edge of a grinding disk. *Figure 14* shows the effect of grinding a root pass. A narrow grinding wheel (but not a cutoff wheel) is effective for this task. Most of the slag along the sides of the weld bead will be removed by this grinding action. This will make it easier to add the next pass, called a **hot pass**.

It is essential that the hot pass is well-fused with the root pass and the pipe walls. The hot pass is achieved by running a bead that covers the root face and fuses to both bevels. The hot pass uses a short arc and high current. It can be made using E6010 electrodes or low-hydrogen electrodes, such as E7018. Remember, the E7018 electrode should not be whipped.

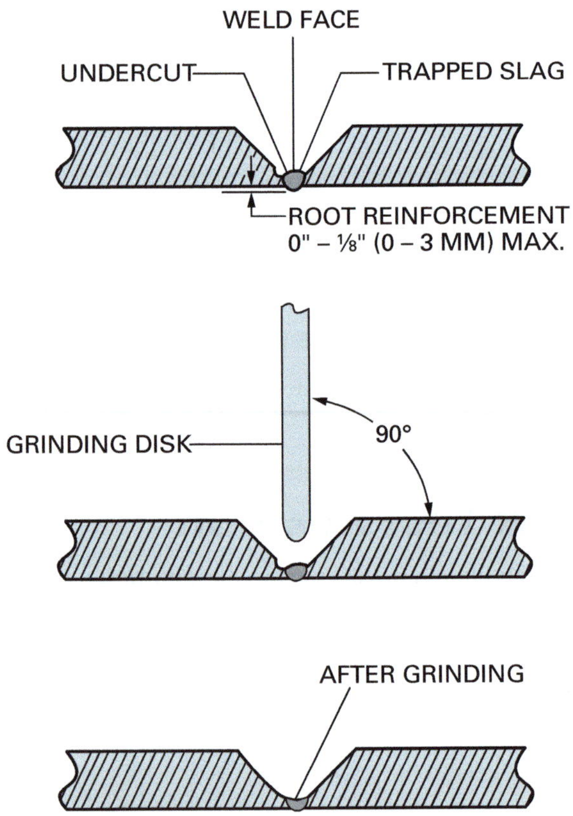

Figure 14 Grinding a root pass.

Use care not to grind through the root pass or widen the groove during the hot pass cleaning process. After the hot pass bead has been cleaned, the remaining passes are applied. Avoid starting and stopping each remaining pass in the same area; the start and stop location for each pass must be staggered.

2.3.0 SMAW of Open-Root V-Groove Pipe Joints

The following sections explain how to perform open-root V-groove welds on pipe in the following positions:

- Flat (1G-ROTATED) position
- Horizontal (2G) position
- Multiple (5G) position
- Multiple inclined (6G) position

2.3.1 Flat (1G-ROTATED) Position

Practice open-root V-groove pipe welds in the 1G-ROTATED position using the following electrodes:

- ⅛" E6010 electrodes for the root pass
- 3⁄32" E7018 electrodes for the hot pass (pass on top of the root)
- Either 3⁄32" or ⅛" E7018 electrodes for the remaining passes

> **NOTE**
> Electrode types and sizes may vary according to the related WPS. Do not assume that all open-root V-groove pipe welds will be completed using the electrodes used here.

Keep the electrode angle at 90 degrees (0-degree work angle) to the pipe axis with a 10- to 15-degree drag angle. Use a whipping motion to run a keyhole for the root pass. Be sure to fuse into the tack welds and restarts. Before running the hot pass, clean the face of the root pass with a file or grinder.

When running the remaining passes, use stringer beads with a slight circular or side-to-side motion. This ensures fusion at the toes of the weld bead. Take special care to fill the crater at the termination of the weld. Run all passes at or near the top of the pipe as the pipe is rotated.

Follow these steps to practice open-root V-groove pipe welds in the 1G-ROTATED position:

Step 1 Tack weld the practice pipe weld coupons together as explained earlier.

Step 2 Position the pipe weld coupon horizontally on two sets of rollers at a comfortable welding height. *Figure 15* shows roller supports commonly found in pipe welding shops. A welding positioner or pipe rotator is also required. The positioner shown in *Figure 16 (A)* is very popular. The pipe rotator shown in *Figure 16 (B)* is another option. This model can handle up to 16" (DN400) pipe and weights up to 1,200 pounds (544 kg). Adjustable roller stands can be used to help support longer stock.

Step 3 Make sure the workpiece clamp is attached directly onto the pipe coupon. This will prevent the welding current from passing through the roller bearings or arcing between the rollers and the pipe coupon.

LASH

To ensure good welding results, remember the LASH acronym:

- *Length of arc* (**L**) –The distance between the electrode and the base metal (usually one times the electrode diameter).
- *Angle* (**A**) –Two angles are critical:
  - *Travel angle* – The longitudinal angle of the electrode in relation to the axis of the weld joint
  - *Work angle* –The traverse angle of the electrode in relation to the axis of the weld joint
- *Speed* (**S**) –Travel speed is measured in inches per minute (IPM). The width of the weld will determine if the travel speed is correct.
- *Heat* (**H**) – Is controlled by the amperage setting. Heat is dependent upon the electrode diameter, base metal type, base metal thickness, welding position, and the interpass temperature.

Figure 15 Pipe roller supports.

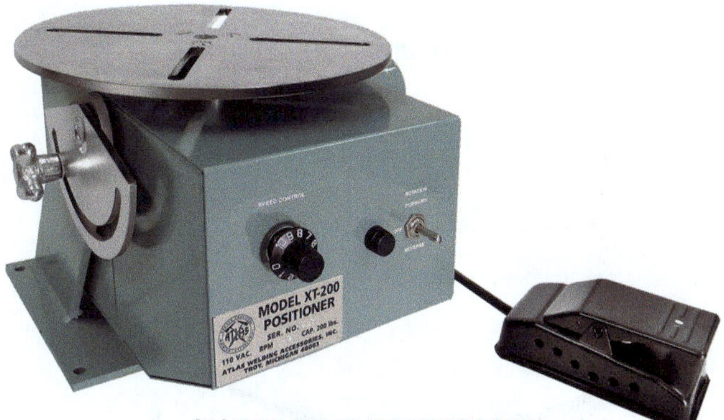

(A) WELDING POSITIONER

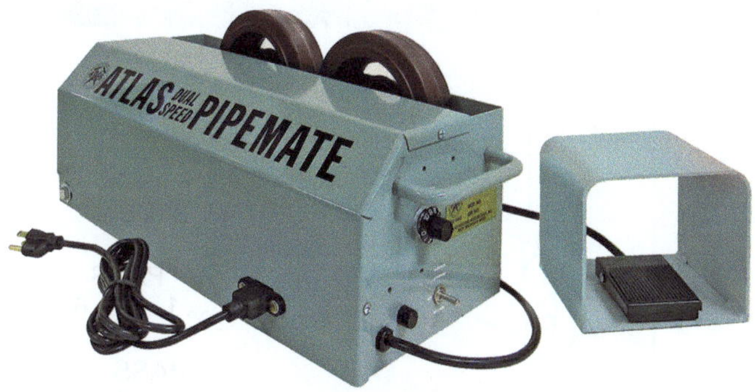

(B) PIPE ROTATOR

Figure 16 A welding positioner and a pipe rotator.

> **CAUTION**
>
> Failure to attach the workpiece clamp to the pipe coupon in the 1G-ROTATED position can result in variations in welding current, damage to the roller bearings, and arcing on the pipe rollers and coupons.

Step 4 Roll the pipe as necessary to keep the weld in the flat position.

Step 5 Run the root pass using 1/8" E6010 electrodes and a whipping motion. Position an existing tack weld at the eleven o'clock position, start the weld bead on the tack weld, and advance toward the twelve o'clock position.

Step 6 Clean the crater of slag before restarting with a fresh electrode.

Step 7 Grind the root pass with a disk grinder.

Step 8 Use the same rolling procedure to make the hot pass (the second bead). Use 3/32" E7018 electrodes.

Step 9 Use the same rolling procedure to make the remaining passes. Use 3/32" or 1/8" E7018 electrodes and stringer beads. Pay special attention at the fusions to prevent slag inclusions or excess buildup. The pipe rotation process and bead sequence are shown in *Figure 17*.

2.3.2 Horizontal (2G) Position

Practice 2G (horizontal) position open-root V-groove pipe welds using the following electrodes:

- 1/8" E6010 electrodes for the root pass
- 3/32" E7018 electrodes for the hot pass
- Either 3/32" or 1/8" E7018 electrodes for the remaining passes

For the root pass, the electrode should be held at a 10- to 15-degree drag angle and at a 0-degree work angle to the surface of the pipe. To prevent the weld puddle from sagging, the electrode work angle can be dropped slightly, but no more than 10 degrees. Use a whipping motion to run a keyhole for the root pass. Be sure to properly fuse and blend the tack welds and restarts.

Before running the hot pass, clean the face of the root pass with a file or grinder. When running the hot pass and remaining fill passes, use stringer beads and a slight circular or side-to-side motion. This ensures fusion at the toes of the weld bead. Take special care to fill the crater at the termination of the weld. *Figure 18* shows the 2G pipe weld coupon, work angles, and bead sequence.

Follow these steps to practice open-root V-groove pipe welds in the 2G position:

Step 1 Tack weld the practice pipe weld coupons together as explained earlier.

Step 2 Tack weld or fasten the pipe coupon to the positioning arm on the welding table with the axis of the pipe vertical.

Step 3 Position the arm to allow welding access all around the coupon.

Step 4 Run the root pass using 1/8" E6010 electrodes and a whipping motion. Start the root pass on a tack weld. Clean and chip the crater between restarts. Continue this procedure until the root pass is completed.

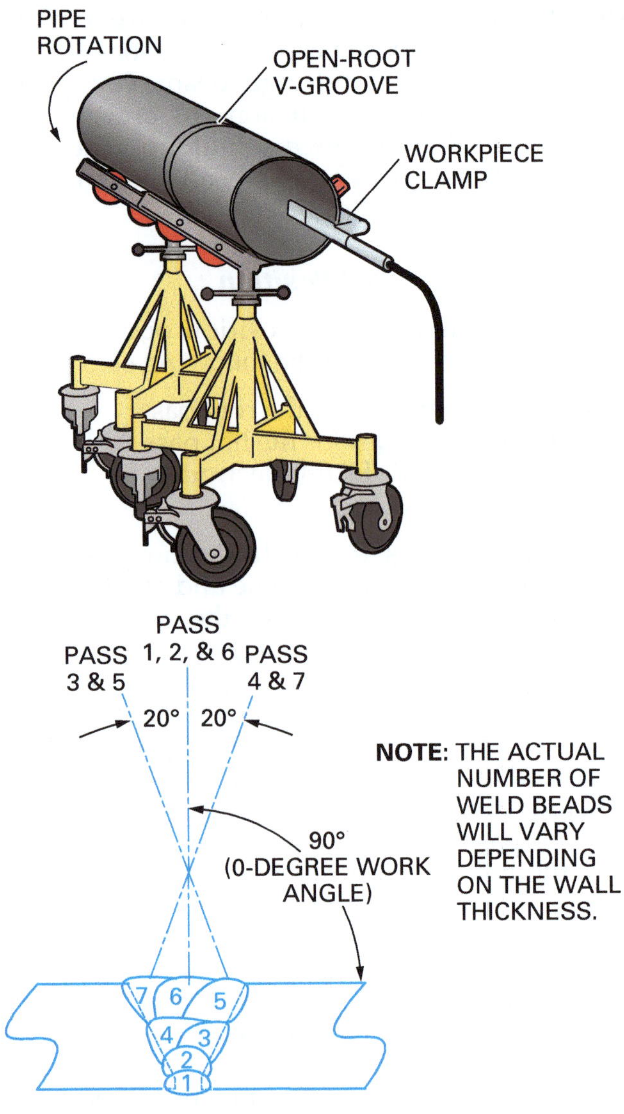

Figure 17 Pipe rotation and multiple-pass 1G-ROTATED bead sequence and work angles.

Step 5 Grind the root pass with a disk grinder to remove trapped slag.

Step 6 Run the hot pass using 3⁄32" E7018 electrodes.

Step 7 Run the remaining passes with either 3⁄32" or 1⁄8" E7018 electrodes and stringer beads.

2.3.3 Multiple (5G) Position

Practice the 5G multiple position open-root V-groove pipe welds using the following electrodes:

- 1⁄8" E6010 electrodes for the root pass
- 3⁄32" E7018 electrodes for the hot pass
- Either 3⁄32" or 1⁄8" E7018 electrodes and stringer beads for the remaining passes

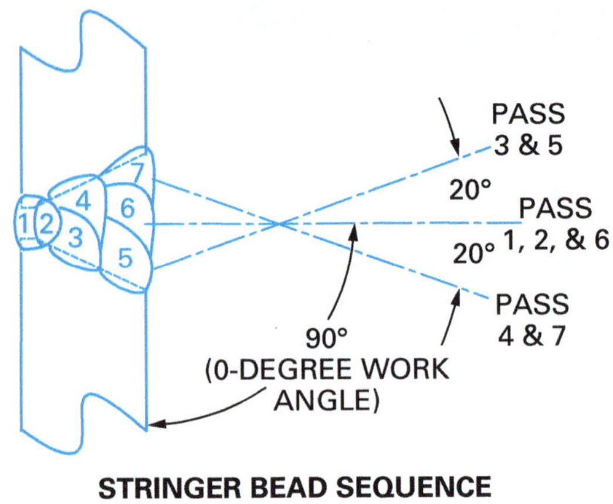

Figure 18 Multiple-pass 2G bead sequence and work angles.

Weld uphill by starting at the bottom of the pipe and welding toward the top. Welding downhill should also be practiced using fast-freeze electrodes. Use a whipping motion to run a keyhole for the root pass. Be sure to properly fuse and blend the tack welds and restarts. Before running the hot pass, clean the face of the root pass with a file or grinder.

When running the remaining passes, use a slight circular or side-to-side motion. This ensures fusion at the toes of the weld bead. Pay special attention to fill the crater at the termination of the weld. *Figure 19* shows the 5G pipe weld coupon, work angles, and bead sequence.

Follow these steps to practice open-root V-groove pipe welds in the 5G position:

Step 1 Tack weld the practice pipe weld coupons together as explained in the section related to pipe weld coupon preparation.

Step 2 Tack weld or fasten the pipe weld coupon to the positioning arm with the pipe axis horizontal and a tack weld at the bottom.

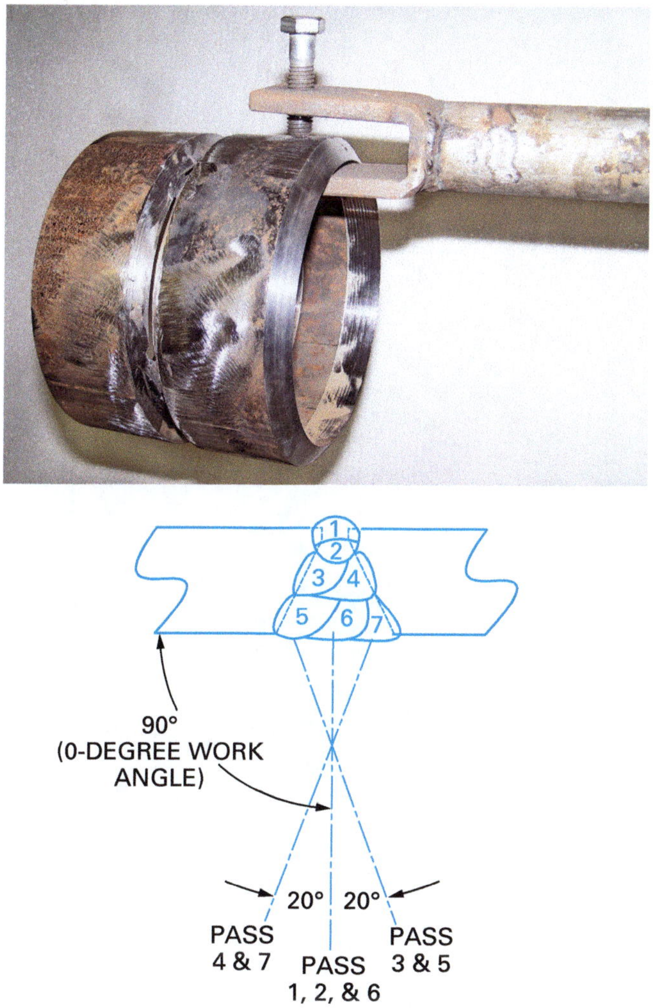

Figure 19 Multiple-pass 5G bead sequence and work angles.

Step 3 Position the arm to allow welding access all around the coupon.

Step 4 Run the root pass using ⅛" E6010 electrodes. Start the root pass on the tack weld at the bottom of the pipe and weld uphill. When welding at the top and bottom of the coupon (flat and overhead positions), the electrode typically points toward the pipe axis. When welding up on the coupon sides (vertical welding), a slight electrode push angle (maximum of 10 degrees) can be used to control the weld puddle.

Step 5 Grind the root pass with a disk grinder to remove trapped slag.

Step 6 Run the hot pass using 3⁄32" E7018 electrodes.

Step 7 Use stringer beads to run the remaining fill passes with either 3⁄32" or ⅛" E7018 electrodes.

2.3.4 Multiple Inclined (6G) Position

Practice the 6G multiple inclined (45-degree) position open-root V-groove pipe welds using the following electrodes:

- ⅛" E6010 electrodes for the root pass
- 3⁄32" E7018 electrodes for the hot pass
- Either 3⁄32" or ⅛" E7018 electrodes and stringer beads for the remaining passes

Weld uphill by starting at the bottom of the pipe and welding toward the top. Use a whipping motion to run a keyhole for the root pass.

Be sure to properly fuse and blend the tack welds and restarts. Before running the hot pass, clean the face of the root pass with a file or grinder.

When running the remaining fill passes, use a slight circular or side-to-side motion. This ensures fusion at the toes of the weld bead. Pay special attention to fill the crater at the termination of the weld. *Figure 20* shows the 6G pipe weld coupon, work angles, and bead sequence.

Follow these steps to practice open-root V-groove pipe welds in the 6G position:

> **NOTE**
> If required for training purposes, instructors may direct that a restricting ring be added to the 6G position coupon, forming a 6GR position coupon.

Step 1 Tack weld the two practice pipe coupons together as explained in the section related to pipe weld coupon preparation.

Step 2 Tack weld or fasten the pipe weld coupon to the positioning arm. Have the pipe axis inclined 45 degrees to the horizontal plane, with a tack weld at the bottom of the pipe.

Step 3 Position the arm to allow welding access all around the coupon.

Step 4 Run the root pass using ⅛" E6010 electrodes. Start the root pass on the tack weld at the bottom of the pipe and weld uphill. When welding at the top and bottom of the coupon (flat and overhead welds), the electrode should point toward the pipe axis. When welding up on the coupon sides (vertical welds), a slight electrode push angle (maximum of 10 degrees) can be used to control the weld puddle.

Step 5 Grind the root pass with a disk grinder to remove trapped slag.

Step 6 Run the hot pass using 3⁄32" E7018 electrodes.

Step 7 Use 3⁄32" or ⅛" E7018 electrodes and stringer beads to run the remaining passes.

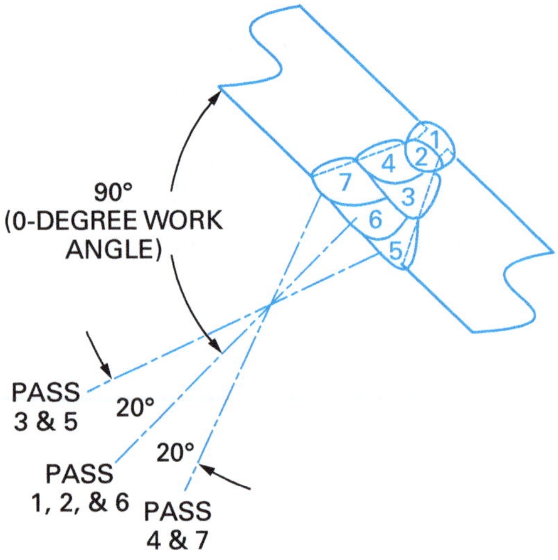

Figure 20 Multiple-pass 6G bead sequence and work angles.

Powerful Pipe Rotator

This pipe rotator is used for large, heavy pipe. Like most pipe rotators, the electric motor is activated by a foot switch, allowing hands-free operation. Although they are often called pipe rollers, using the term *pipe rotator* may avoid confusion with common rollers that are not powered.

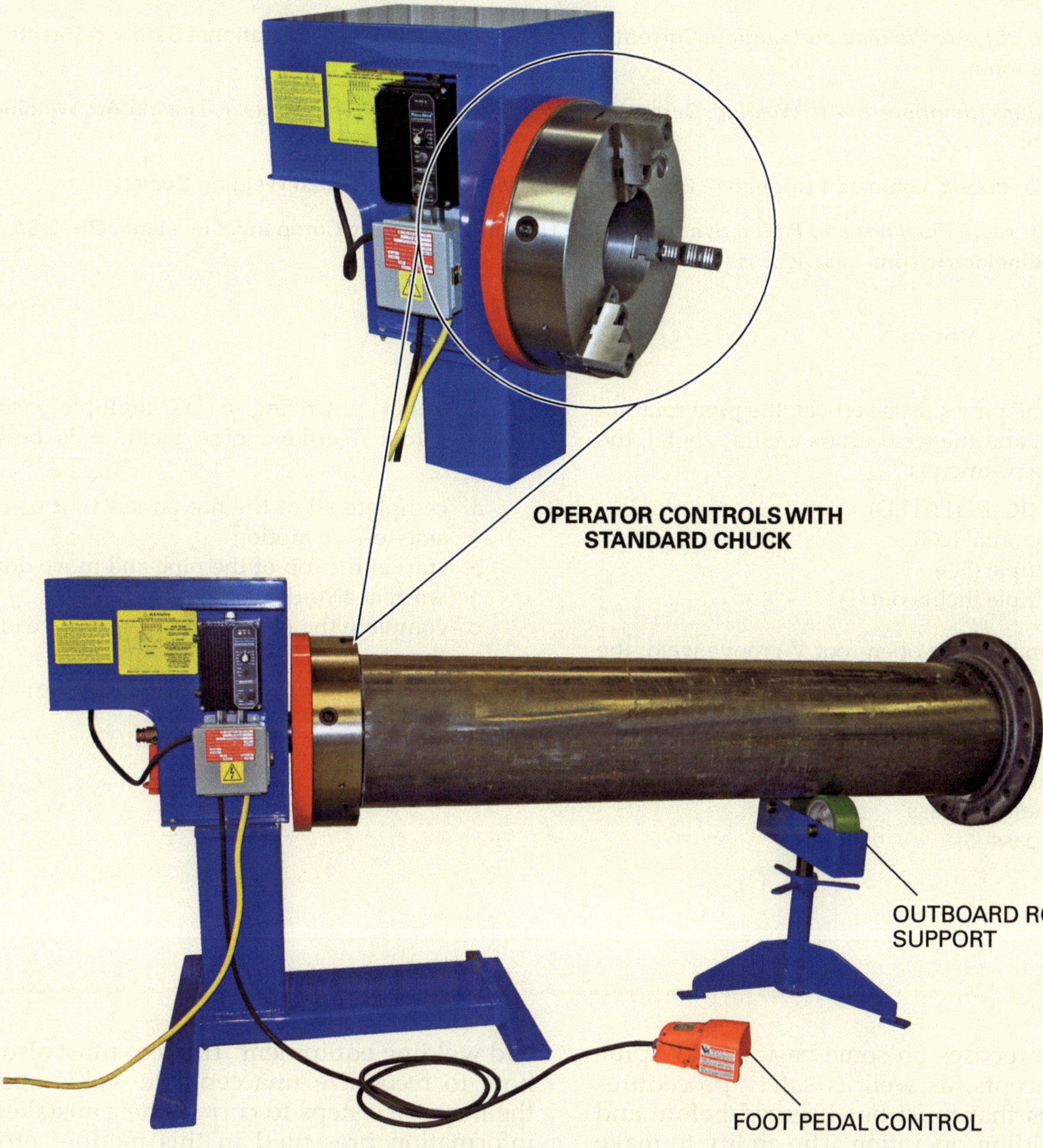

WELDING POSITIONER (PIPE ROTATOR) WITH PIPE INSTALLED

Additional Resources

AWS F4.1: Recommended Safe Practices for Preparation for Welding and Cutting of Containers and Piping. Miami, FL: American Welding Society.

Modern Welding Technology. Current Edition. Howard B. Cary and Scott Helzer. Englewood Cliffs, NJ: Prentice Hall.

OSHA 1926.351, Arc Welding and Cutting. Current Edition. Washington, DC: Occupational Safety & Health Administration.

The Procedure Handbook of Arc Welding. Current Edition. Cleveland, OH: The James F. Lincoln Arc Welding Foundation.

Welding Handbook, Volumes 1 through 4. Current Edition. Miami, FL: American Welding Society.

Welding Pressure Pipelines and Piping Systems (PDF). The Lincoln Electric Company, Cleveland, OH, USA. **www.lincolnelectric.com**. Last accessed May 29, 2015.

2.0.0 Section Review

1. When the pipe axis is vertical, the pipe rotation is fixed, and the weld beads are horizontal, the welding position is _____.
   a. flat (1G-ROTATED)
   b. horizontal (2G)
   c. multiple (5G)
   d. multiple inclined (6G)

2. When making an open-root V-groove weld, the weld pass that is done immediately after the root pass is known as the _____.
   a. hot pass
   b. secondary pass
   c. remainder pass
   d. fill pass

3. To begin practicing a 5G multiple-position open-root V-groove pipe weld, it is best to _____.
   a. complete all of the hot passes first using a side-to-side motion
   b. start at the top of the pipe and move downward at a 90-degree angle
   c. complete the overlapping stringer beads at the nine o'clock position
   d. start at the bottom of the pipe and weld toward the top

Summary

This module focuses on some basic SMAW pipe welding concepts, as well as safety procedures and practices that must be observed before and during a welding operation. The ability to make acceptable open-root V-groove welds on pipe in all positions is one of the more challenging skills a welder must develop. Good welds begin with properly preparing the work area, materials, and welding equipment. Trainees must also learn how to recognize unacceptable welds and take the necessary steps to correct their mistakes. The information presented in this module provides welders with the necessary knowledge to make open-root V-groove welds on medium- and thick-walled pipe using SMAW.

Review Questions

1. Welds made on critical piping must be expertly completed, thoroughly inspected, and _____.
   a. at least as strong as those made on non-critical piping
   b. made up of only one electrode material, such as E6010
   c. at least as strong as the piping being welded
   d. allowed time to slowly solidify to seat the backing ring

2. An E7018 electrode is typically used to make _____.
   a. the remaining passes to fill the groove
   b. welds in which strength is not a factor
   c. the initial root pass of the weld
   d. welds that are subject to fast-freeze conditions

3. If a hard hat is required during an SMAW job, it should be the type that allows for the attachment of _____.
   a. leather spats and a rear deflector mesh
   b. ear plugs and an oxygen breathing mask
   c. leather chaps and tinted goggle covers
   d. rear deflector material and a face shield

4. When preparing to practice open-root pipe welds, each welded joint requires _____.
   a. one small- and one large-diameter pipe coupon
   b. two pipe coupons of the same size and schedule
   c. one Schedule 40 and one Schedule 60 pipe coupon
   d. two 12" (31 cm) long sections of non-carbon pipe

5. To prepare an open-root V-groove weld coupon, bevel the end of the pipe to _____.
   a. 10 to 15 degrees
   b. 20 to 30½ degrees
   c. 30 to 37½ degrees
   d. 60 to 75 degrees

6. Positioning the pipe coupons for open-root V-groove welding practice requires tack welds to be made in _____.
   a. two places
   b. three places
   c. four places
   d. five places

7. When the pipe axis is horizontal and the pipe is slowly rotated while it is being welded on the top, the welding position is _____.
   a. Flat (1G-ROTATED)
   b. Horizontal (2G)
   c. Multiple (5G)
   d. Multiple inclined (6G)

8. An acceptable pipe groove weld profile is one in which the root pass has complete penetration and the root reinforcement on the inside of the pipe is no more than _____.
   a. ½" (13 mm)
   b. ⅜" (10 mm)
   c. ¼" (6 mm)
   d. ⅛" (3 mm)

9. The amount of penetration that is achieved during the root pass of an open-root V-groove pipe weld is controlled by a technique called _____.
   a. fusing the toes
   b. running a keyhole
   c. feathering the tack
   d. timing the arc

10. When practicing open-root V-groove pipe welds in the 5G position, which electrodes are used for the passes remaining after the root and hot passes?
    a. 3⁄32" or ⅛" E7018 electrodes
    b. 3⁄32" or ⅛" E6010 electrodes
    c. Only ⅛" E6010 electrodes
    d. Only 3⁄32" E6010 electrodes

Trade Terms Introduced in This Module

Fast-freeze electrode: An electrode that solidifies quickly, produces very little slag, and gives good penetration into the root of the joint.

Feather: To grind the start and/or stop areas of a weld to a tapered edge.

Hot pass: A nonstandard term for the weld pass that is made directly after the root pass.

Keyhole: A hole that is formed when the root faces on two pipe edges are melted away by the arc.

Pipe axis: An imaginary line that runs longitudinally (lengthwise) through the center of a pipe.

Remaining passes: A nonstandard term that is used to describe the fill and cover passes as a group.

Root pass: A weld pass that is made to produce a root bead.

Appendix

Performance Accreditation Tasks

PATs provide specific acceptable criteria for performance and help to ensure a true competency-based welding program for trainees.

The following tasks are designed to evaluate your ability to run open-root V-groove pipe welds with SMAW equipment in the four standard test positions using E6010 and E7018 electrodes. Perform each task when you are instructed to do so by your instructor. As you complete each task, take it to your instructor for evaluation. Do not proceed to the next task until instructed to do so.

For AWS 1G-ROTATED, 2G, and 5G certifications, refer to the current edition of *AWS EG3.0-96, Guide for the Training and Qualification of Welding Personnel; Level II—Advanced Welder* for bend test requirements. For AWS 6G certifications, refer to the current edition of *AWS EG4.0-96, Guide for the Training and Qualification of Welding Personnel; Level III—Expert Welder* for bend test requirements.

PATs 2 through 4 correspond to *AWS EG3.0-96, Section 3.3.6.1, Unit #1 SMAW, Learning Objective #10.* PAT 1 has no AWS correlation.

OPEN-ROOT V-GROOVE PIPE WELD IN THE 1G-ROTATED POSITION

Using $\frac{1}{8}$" (3.2 mm) E6010 electrodes for the root pass, $\frac{3}{32}$" (2.4 mm) E7018 electrodes for the hot pass, and $\frac{3}{32}$" or $\frac{1}{8}$" (2.4 or 3.2 mm) E7018 electrodes for the remaining passes, make an open-root V-groove weld on carbon steel pipe in the 1G-ROTATED position as shown.

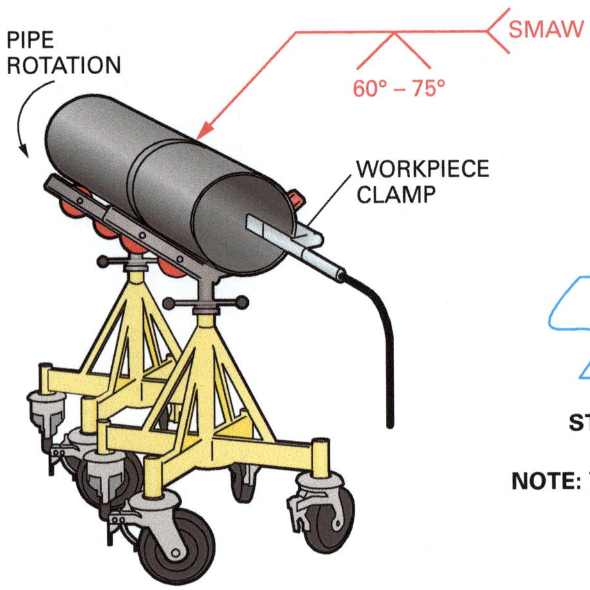

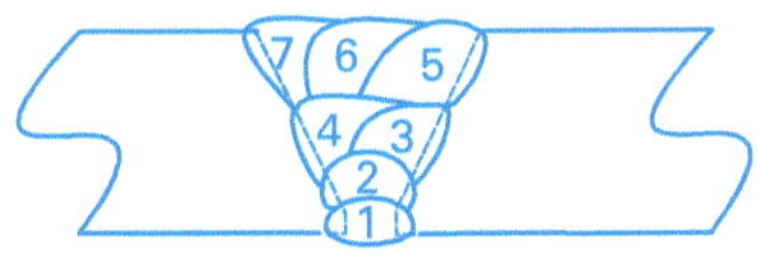

STRINGER BEAD SEQUENCE

NOTE: THE ACTUAL NUMBER OF WELD BEADS WILL VARY DEPENDING ON THE WALL THICKNESS.

Criteria for Acceptance:

- Uniform rippled appearance on the weld face ______
- Craters and restarts filled to the full cross section of the weld ______
- Uniform weld size ______
- Acceptable weld profile in accordance with the applicable codes and standards ______
- Smooth transition with complete fusion at the toes of the weld ______
- Complete joint penetration and face reinforcement with the root reinforcement at least flush with the inside of the pipe, to a maximum buildup of $\frac{1}{8}$" (3 mm) ______
- No porosity ______
- No undercut ______
- No inclusions ______
- No cracks ______

OPEN-ROOT V-GROOVE PIPE WELD IN THE 2G POSITION

Using 1/8" (3.2 mm) E6010 electrodes for the root pass, 3/32" (2.4 mm) E7018 electrodes for the hot pass, and 3/32" or 1/8" (2.4 or 3.2 mm) E7018 electrodes for the remaining passes, make an open-root V-groove weld on carbon steel pipe in the 2G position as shown.

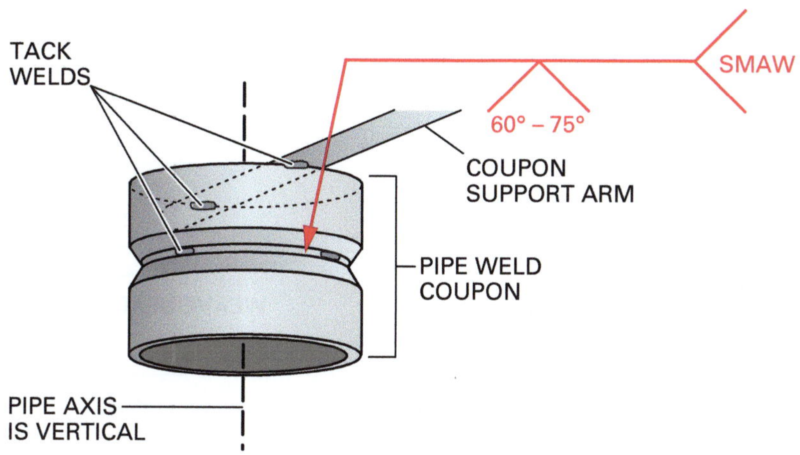

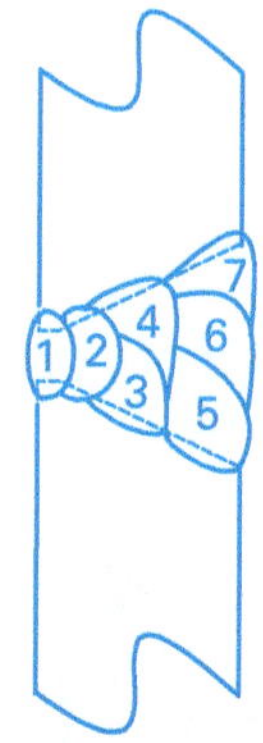

STRINGER BEAD SEQUENCE

Criteria for Acceptance:

- Uniform rippled appearance on the weld face ______
- Craters and restarts filled to the full cross section of the weld ______
- Uniform weld size ______
- Acceptable weld profile in accordance with the applicable codes and standards ______
- Smooth transition with complete fusion at the toes of the weld ______
- Complete joint penetration and face reinforcement with the root reinforcement at least flush with the inside of the pipe, to a maximum buildup of 1/8" (3 mm) ______
- No porosity ______
- No undercut ______
- No inclusions ______
- No cracks ______

OPEN-ROOT V-GROOVE PIPE WELD IN THE 5G POSITION

Using ⅛" (3.2 mm) E6010 electrodes for the root pass, 3⁄32" (2.4 mm) E7018 electrodes for the hot pass, and 3⁄32" or ⅛" (2.4 or 3.2 mm) E7018 electrodes for the remaining passes, make an open-root V-groove weld on carbon steel pipe in the 5G position as shown. For the root pass, the requirement is to weld uphill. The downhill progression can be used for additional practice.

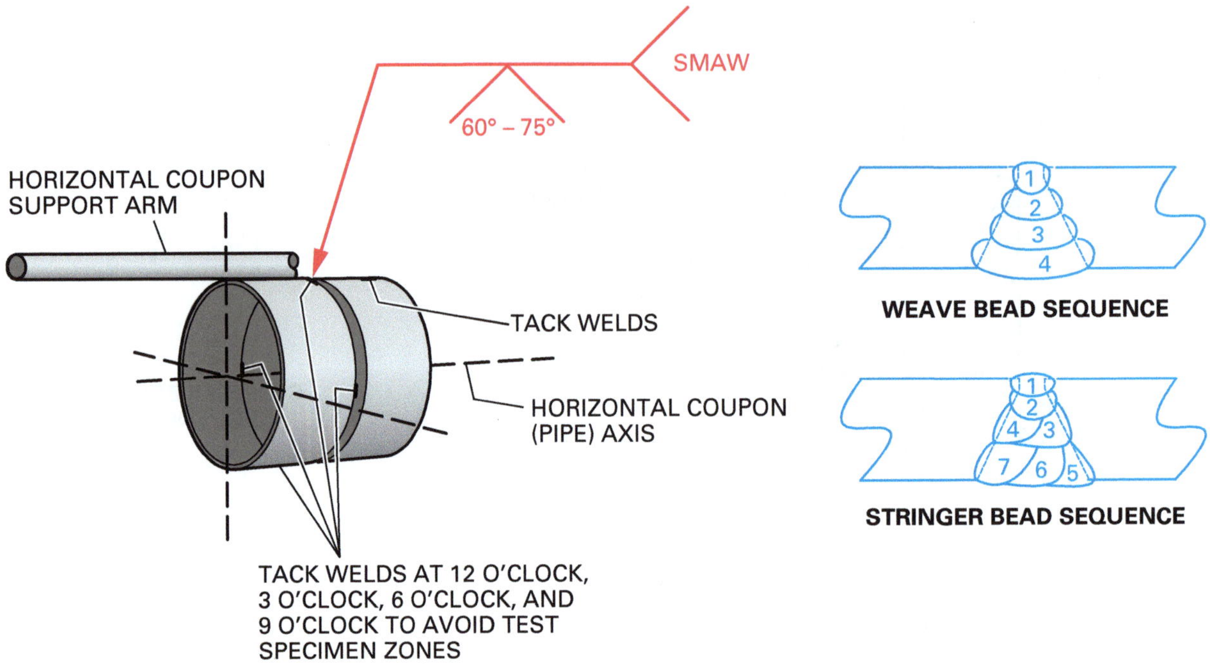

Criteria for Acceptance:

- Uniform rippled appearance on the weld face ______
- Craters and restarts filled to the full cross section of the weld ______
- Uniform weld size ______
- Acceptable weld profile in accordance with the applicable codes and standards ______
- Smooth transition with complete fusion at the toes of the weld ______
- Complete joint penetration and face reinforcement with the root reinforcement at least flush with the inside of the pipe, to a maximum buildup of ⅛" (3 mm) ______
- No porosity ______
- No undercut ______
- No inclusions ______
- No cracks ______

OPEN-ROOT V-GROOVE PIPE WELD IN THE 6G POSITION

Using 1/8" (3.2 mm) E6010 electrodes for the root pass, 3/32" (2.4 mm) E7018 electrodes for the hot pass, and 3/32" or 1/8" (2.4 or 3.2 mm) E7018 electrodes for the remaining passes, make an open-root V-groove weld on carbon steel pipe in the 6G position as shown. If required for qualification test purposes, a restricting ring may be added to the 6G position coupon to form a 6GR position coupon. For the root pass, the requirement is to weld uphill. The downhill progression can be used for additional practice.

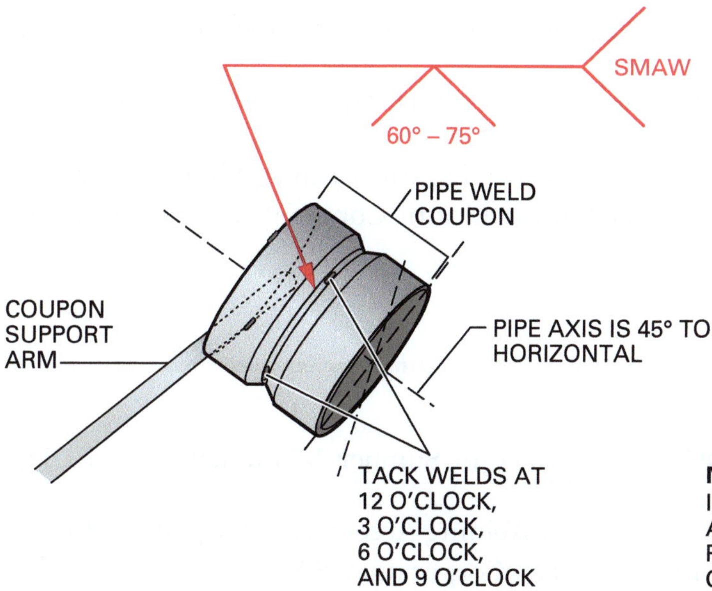

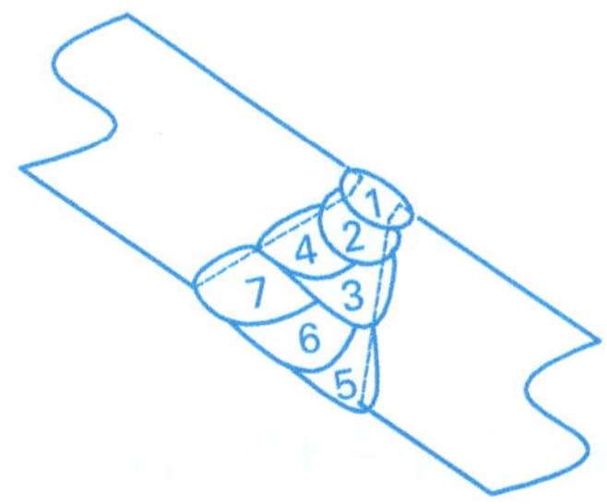

STRINGER BEAD SEQUENCE

NOTE:
IF REQUIRED FOR QUALIFICATION PURPOSES, A RESTRICTING RING MAY BE ADDED TO THE 6G POSITION COUPON TO FORM A 6GR POSITION COUPON.

Criteria for Acceptance:

- Uniform rippled appearance on the weld face ____________
- Craters and restarts filled to the full cross section of the weld ____________
- Uniform weld size ____________
- Acceptable weld profile in accordance with the applicable codes and standards ____________
- Smooth transition with complete fusion at the toes of the weld ____________
- Complete joint penetration and face reinforcement with the root reinforcement at least flush with the inside of the pipe, to a maximum buildup of 1/8" (3 mm) ____________
- No porosity ____________
- No undercut ____________
- No inclusions ____________
- No cracks ____________

Additional Resources

This module presents thorough resources for task training. The following resource material is recommended for further study.

AWS F4.1: Recommended Safe Practices for Preparation for Welding and Cutting of Containers and Piping. Miami, FL: American Welding Society.

Modern Welding Technology. Current Edition. Howard B. Cary and Scott Helzer. Englewood Cliffs, NJ: Prentice Hall.

OSHA 1926.351, Arc Welding and Cutting. Current Edition. Washington, DC: Occupational Safety & Health Administration.

The Procedure Handbook of Arc Welding. Current Edition. Cleveland, OH: The James F. Lincoln Arc Welding Foundation.

Welding Handbook, Volumes 1 through 4. Current Edition. Miami, FL: American Welding Society.

Welding Pressure Pipelines and Piping Systems (PDF). The Lincoln Electric Company, Cleveland, OH, USA. **www.lincolnelectric.com**. Last accessed May 29, 2015.

Figure Credits

The Lincoln Electric Company, Cleveland, OH, USA, Module Opener, Figures 1–3, 5

Topaz Publications, Inc., Figures 4, 6 (photo)

Terry Lowe, Figures 7, 8, 9 (photo), 13, 18-19 (photos)

Courtesy of Sumner Manufacturing Co., Figure 15

Atlas Welding Accessories, Inc., Figure 16

All-Fab Corporation, SA03

Section Review Answer Key

Answer	Section Reference	Objective
Section One		
1. c	1.1.0	1a
2. d	1.2.4	1b
3. a	1.3.1	1c
Section Two		
1. b	2.1.0	2a
2. a	2.2.0	2b
3. d	2.3.3	2c

NCCER CURRICULA — USER UPDATE

NCCER makes every effort to keep its textbooks up-to-date and free of technical errors. We appreciate your help in this process. If you find an error, a typographical mistake, or an inaccuracy in NCCER's curricula, please fill out this form (or a photocopy), or complete the online form at **www.nccer.org/olf**. Be sure to include the exact module ID number, page number, a detailed description, and your recommended correction. Your input will be brought to the attention of the Authoring Team. Thank you for your assistance.

Instructors – If you have an idea for improving this textbook, or have found that additional materials were necessary to teach this module effectively, please let us know so that we may present your suggestions to the Authoring Team.

NCCER Product Development and Revision

13614 Progress Blvd., Alachua, FL 32615

Email: curriculum@nccer.org
Online: www.nccer.org/olf

❑ Trainee Guide ❑ Lesson Plans ❑ Exam ❑ PowerPoints Other ______________________

Craft / Level: Copyright Date:

Module ID Number / Title:

Section Number(s):

Description:

Recommended Correction:

Your Name:

Address:

Email: Phone:

GMAW – Pipe

Overview

Gas metal arc welding (GMAW) is a form of welding that joins two metals together using a filler metal wire electrode and a shielding gas. GMAW is an effective method for making high-quality, open-root V-groove welds on pipe. This module covers some basic concepts of open-root GMAW of pipe and describes how to prepare and perform open-root V-groove welds on medium- and thick-walled pipe in all positions.

Module 29302

Trainees with successful module completions may be eligible for credentialing through the NCCER Registry. To learn more, go to **www.nccer.org** or contact us at 1.888.622.3720. Our website has information on the latest product releases and training, as well as online versions of our *Cornerstone* magazine and Pearson's product catalog.

Your feedback is welcome. You may email your comments to **curriculum@nccer.org**, send general comments and inquiries to **info@nccer.org**, or fill in the User Update form at the back of this module.

This information is general in nature and intended for training purposes only. Actual performance of activities described in this manual requires compliance with all applicable operating, service, maintenance, and safety procedures under the direction of qualified personnel. References in this manual to patented or proprietary devices do not constitute a recommendation of their use.

29302 V5

29302
GMAW – Pipe

Objectives

When you have completed this module, you will be able to do the following:

1. State the basic concepts of open-root GMAW of pipe and how to prepare for welding.
   a. State the basic concepts of GMAW.
   b. Identify common welding safety practices.
   c. Explain how to prepare the area and materials for GMAW pipe welding.
   d. Explain how to prepare the equipment for GMAW pipe welding.
2. Describe open-root V-groove pipe welding positions and GMAW techniques.
   a. Identify and describe the four pipe welding positions and acceptable weld profiles.
   b. Describe the technique for GMAW of the root and fill passes.
   c. Describe the techniques required to produce open-root GMAW pipe welds in all positions.

Performance Tasks

Under the supervision of your instructor, you should be able to do the following:

1. Make GMAW open-root V-groove pipe welds in the 1G-ROTATED position using the correct filler metal and shielding gas.
2. Make GMAW open-root V-groove pipe welds in the 2G position using the correct filler metal and shielding gas.
3. Make GMAW open-root V-groove pipe welds in the 5G position using the correct filler metal and shielding gas.
4. Make GMAW open-root V-groove pipe welds in the 6G position using the correct filler metal and shielding gas.

Trade Terms

Axis
Composite wire electrode
Electrode extension
Feather
Keyhole
Microwire
Pipe axis
Remaining passes
Root pass
Standoff distance
Stickout

Industry Recognized Credentials

If you are training through an NCCER-accredited sponsor, you may be eligible for credentials from NCCER's Registry. The ID number for this module is 29302. Note that this module may have been used in other NCCER curricula and may apply to other level completions. Contact NCCER's Registry at 888.622.3720 or go to **www.nccer.org** for more information.

Contents

Figures and Tables

[illegible]

Section One

1.0.0 GMAW Concepts and Preparations

Objective

State the basic concepts of open-root GMAW of pipe and how to prepare for welding.

a. State the basic concepts of GMAW.
b. Identify common welding safety practices.
c. Explain how to prepare the area and materials for GMAW pipe welding.
d. Explain how to prepare the equipment for GMAW pipe welding.

Trade Terms

Axis: The center line running the length of an object. In welding, the term axis is most often used in relation to an electrode or a section of pipe.

Composite wire electrode: A generic term for multicomponent filler metal electrodes. For GMAW, the tubular electrode contains a powdered metal inside a metal sheath. A flux-cored electrode contains flux inside a metal sheath.

Electrode extension: The length of the wire that extends beyond the tip of the welding gun contact tube.

Feather: To grind the start and/or stop areas of a weld to a tapered edge.

Microwire: Generally considered to be wire electrode that is 0.045" (1.1 mm) in diameter or smaller. However, due to the small size of most GMAW wire electrodes in comparison to SMAW electrodes and similar filler metals, some refer to all GMAW-related wire electrodes as microwire.

Pipe axis: An imaginary line that runs longitudinally (lengthwise) through the center of a pipe.

Remaining passes: A nonstandard term that is used to describe the fill and cover passes as a group.

Root pass: The initial weld pass made to produce a root bead.

Standoff distance: The distance from the gas nozzle or insulating nozzle to the workpiece.

Stickout: The distance from the gas nozzle or insulating nozzle to the end of the electrode.

Being able to effectively perform open-root V-groove welds on pipe requires an understanding of GMAW pipe welding concepts, as well as safety procedures and practices that must be observed before and during the welding operation. It also requires a basic understanding of how to properly prepare the work area, materials, and welding equipment prior to any welding activity.

1.1.0 Basic GMAW Pipe Welding Concepts

GMAW is an arc welding process that uses a continuously fed solid or **composite wire electrode** to provide the filler metal. A shielding gas is added to protect the weld zone by surrounding the arc and displacing oxygen-laden air (*Figure 1*). The GMAW process is commonly used to weld carbon, low-alloy, and stainless steel, as well as aluminum and other metals.

GMAW equipment includes a gun equipped with a gas nozzle, a gas hose, a gas control solenoid valve, a gas regulator/flowmeter, and a shielding gas cylinder. A wire feeder is used to consistently feed the electrode, which also provides the filler metal, to the gun. *Figure 2* shows typical GMAW equipment.

Welding pipe with the GMAW process is a fast and effective method of producing high-quality welds. Welding can be continuous due to the wire

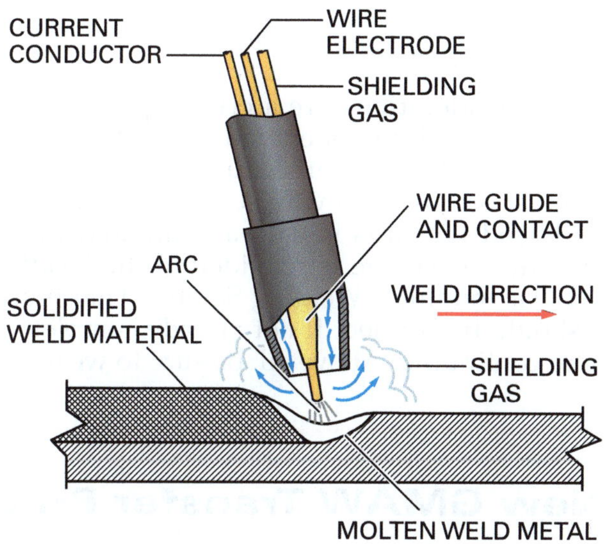

NOTE: Other gun configurations and styles are available.

Figure 1 GMAW process.

Figure 2 GMAW equipment.

electrode feeder, which reduces discontinuities and restarts. With some materials, such as stainless steel, it is common field practice to use gas tungsten arc welding (GTAW) for the **root pass** and GMAW for the **remaining passes**.

For the purposes of this module, the dimensions used are representative of various codes and standards and may not be specific to an exact code or standard. Always refer to the applicable code, standard, or site WPS.

1.2.0 Safety Practices

Safety considerations are a critical part of preparing for a welding operation. The following is a summary of safety procedures and practices that must be observed while cutting or welding. Be advised that this is only a summary. Complete safety coverage is provided in the *Welding Level One* module, "Welding Safety." That module should be studied, if not already completed, before continuing. Above all, be sure to wear appropriate protective clothing and equipment at all times when welding or cutting.

1.2.1 Protective Clothing and Equipment

Protective clothing and equipment are a necessary part of every welding task. Follow the guidelines below as a general rule, and ensure that you are familiar with any unique requirements your employer may have regarding personal protective equipment (PPE).

- A welding hood equipped with a properly shaded lens is best for all forms of welding. The welding hood lens must be an appropriate shade for the application.
- Wear proper leather or flame-retardant clothing along with welding gloves for protection from flying sparks, molten metal, and heat.
- Wear high-top safety shoes or boots. Make sure that the tongue and lace area of the footwear is covered by a pant leg. If the tongue and lace area is exposed or if the footwear must be

New GMAW Transfer Processes

A number of manufacturers have introduced proprietary versions of short-circuiting transfer that rely on intelligent inverter power sources. These units adjust the voltage and current moment-by-moment throughout each stage of the short-circuit transfer process so the droplets form and then transfer to the weld puddle optimally. The result is minimal spatter, excellent fusion, and higher-quality welds. These new technologies are especially popular in settings involving stainless steel pipe welds. The process is referred to as Surface Tension Transfer® (STT®) by Lincoln Electric, and as RMD Pro® or Pro-Pulse® by Miller.

protected from burn marks, wear spats under the pant leg or leather chaps over the front and top of the footwear.

- Wear a 100-percent cotton cap without any mesh material. If a hard hat is required for the environment, use one that allows the attachment of rear deflector material and a face shield. A hard hat with a rear deflector is generally preferred when working overhead, and may be required by some employers and job sites.
- Wear earmuffs, or at least earplugs, to prevent sparks from entering the ear canals.

1.2.2 Housekeeping and Fall Protection

Welding is often performed outside in the weather and in the construction environment. Production work done inside large, open shops exposes a welder to equipment moving across the floor and overhead. Welders often have to crawl under or reach over objects to complete their work.

At times, welders must work at height. Fall protection equipment is required for all work at height (*Figure 3*). Always keep fall protection devices untangled and properly attached. Combined with the other equipment a welder must carry, there is a great deal to manage, but fall protection equipment must always be the priority. Fall protection equipment saves lives. Never attach any kind of welding gear to the harness that could interfere with its ability to perform its intended function.

Regardless of where you are working, always keep the work area as neat as possible. This makes the area safer and helps to reduce distractions. Mindful actions, such as making sure that welding cables are not tripping hazards, contribute to a safer environment. Electrode stubs and unused electrodes must not be left on the floor.

Figure 3 Working at height.

They can cause someone to lose their footing and fall. Parts waiting to be welded or left over from previous work are a tripping hazard and should be kept out of the walking area. If the parts have sharp edges, they could cut welding cable insulation or torch gas hoses.

1.2.3 Fire and Explosion Prevention

Flammable gases are commonly found around welding and cutting operations, as are sparks and other significant sources of ignition. For that reason, and because many fire-related incidents over the years have been traced back to welding operations, it is crucial that fire and explosion prevention be given consistent attention. Follow these guidelines to maintain a safe environment:

- Never carry matches or gas-filled lighters. Sparks can cause the matches to ignite or the lighter to explode in your pocket, causing serious injury.
- Never use oxygen to blow dust or dirt from clothing. Traces of the oxygen can remain trapped in the fabric for some time. If a spark hits the fabric during this time, the clothing can burn rapidly and violently out of control.
- Make sure that any flammable material in the work area is removed or shielded by a fire-resistant covering. Approved fire extinguishers must be available before attempting any heating, welding, or cutting operations. If a hot-work permit and a fire watch are required, be sure these are in place before beginning.
- Ensure that all site safety requirements are being observed.
- Never release a large amount of oxygen or use oxygen in place of compressed air. Its presence around flammable materials or sparks can cause rapid and uncontrolled combustion. Keep oxygen away from oil, grease, and other petroleum products.
- Never release a large amount of fuel gas, especially acetylene. Propane tends to concentrate in and along low areas and can ignite at a considerable distance from the release point. Acetylene is lighter than air, but is even more dangerous than propane. When mixed with air or oxygen, acetylene will ignite at much lower concentrations than any other fuel gas.
- Even though shielding gases are not toxic, any gas released in a sufficient quantity can displace the oxygen in a small room or enclosed work area, creating an atmosphere with little or no oxygen. Avoid large releases of any type of gas, regardless of its characteristics.

- To prevent fires, maintain a neat and clean work area, and make sure that any metal scrap or slag is cold before disposal.
- Before cutting containers such as tanks or barrels, check to see if they have contained any explosive, hazardous, or flammable materials, including petroleum products, citrus products, or chemicals that decompose into toxic fumes when heated. As a standard practice, always clean and then fill any tanks or barrels with water, or purge them with a flow of inert gas such as nitrogen to displace any oxygen. Containers must be cleaned by steam cleaning, flushing with water, or washing with detergent until all traces of the material have been removed. Proper procedures for cutting or welding hazardous containers are described in the *American Welding Society (AWS) F4.1, Safe Practices for the Preparation of Containers and Piping for Welding and Cutting.*

WARNING!

Welding or cutting must never be performed on drums, barrels, tanks, vessels, or other containers until they have been emptied and cleaned thoroughly, eliminating all flammable materials and all substances (such as detergents, solvents, greases, tars, or acids) that might produce flammable, toxic, or explosive vapors when heated. Clean containers only in well-ventilated areas. Vapors can accumulate during cleaning, causing explosions and serious personal injury.

1.2.4 Work Area Ventilation

Many gases are introduced to the atmosphere through welding processes. Some of these gases are harmful, and all can displace or combine with atmospheric oxygen in a confined space. Follow these guidelines, as well as those of your employer, for work area ventilation:

- Make sure to follow all confined space procedures before conducting any welding or cutting in a confined space. Use powered extraction systems as needed.
- Never use oxygen to provide ventilation in confined spaces.
- Always perform cutting or welding operations in a well-ventilated area.

Cutting or welding operations involving zinc or cadmium materials or coatings result in toxic fumes. Another significant concern is Chromium 6, commonly referred to as *hexavalent chromium,* which is used in stainless steel. It releases toxic fumes when heated. Manganese is another hazardous metal used in stainless steel. It is also used in some welding rods, increasing the risk of exposure.

For long-term cutting or welding of materials containing these components, always wear an approved full-face, supplied-air respirator that uses breathing air supplied from outside the work area. For occasional short-term exposure, a HEPA-rated or metal-fume filter may be used on a standard respirator.

Powered fume-extraction equipment must be used when welding materials that contain zinc, cadmium, hexavalent chromium, manganese, and similar materials that produce hazardous fumes if the area is not naturally well-ventilated. *Figure 4* shows examples of both mobile and portable units.

1.3.0 Preparations for Welding

Before any welding procedure can take place, the work area must be made ready, the welding equipment must be properly set up, and the metal to be welded must be prepared.

1.3.1 Preparing the Welding Area

To practice welding, a welding table, bench, or stand is needed. The welding surface must be made of steel. Provisions must be made for supporting the pipe welding coupons for out-of-position welding. *Figure 5* shows a typical welding station.

Classes of Fire Extinguishers

Fire extinguishers have different class ratings based on the types of fires they are used to extinguish. For example, Class A fire extinguishers are used for common materials like paper, wood, and plastic. Class B fire extinguishers are used for liquids such as gasoline and oil. Class C fire extinguishers are used for electrical equipment fires, and Class D extinguishers are used for combustible metals. Always make sure the fire extinguishers available on site are suitable for the types of fires that might occur.

(A) MOBILE UNIT

(B) PORTABLE UNIT

Figure 4 Mobile and portable extraction equipment.

To set up the area for welding, follow these steps:

Step 1 Make sure that the area is properly ventilated. Make use of doors, windows, and fans.

Step 2 Check the area for fire hazards. Remove any flammable materials before proceeding.

Step 3 Know the location of the nearest fire extinguisher. Ensure the fire extinguisher is charged and that you and others in the immediate area know how to use it properly.

Step 4 Set up flash shields and curtains around the welding area as necessary.

1.3.2 Preparing Pipe Weld Coupons

Pipe weld coupons should be cut from 3" to 12" (DN80 to 300) diameter Schedule 40 or Schedule 80 carbon steel pipe. Each welded joint requires two pipe coupons of the same diameter and schedule.

Figure 6 shows typical ASME, American Petroleum Institute (API), and AWS pipe bevel specifications for bevel angles, root faces, and root openings.

Figure 5 Welding station.

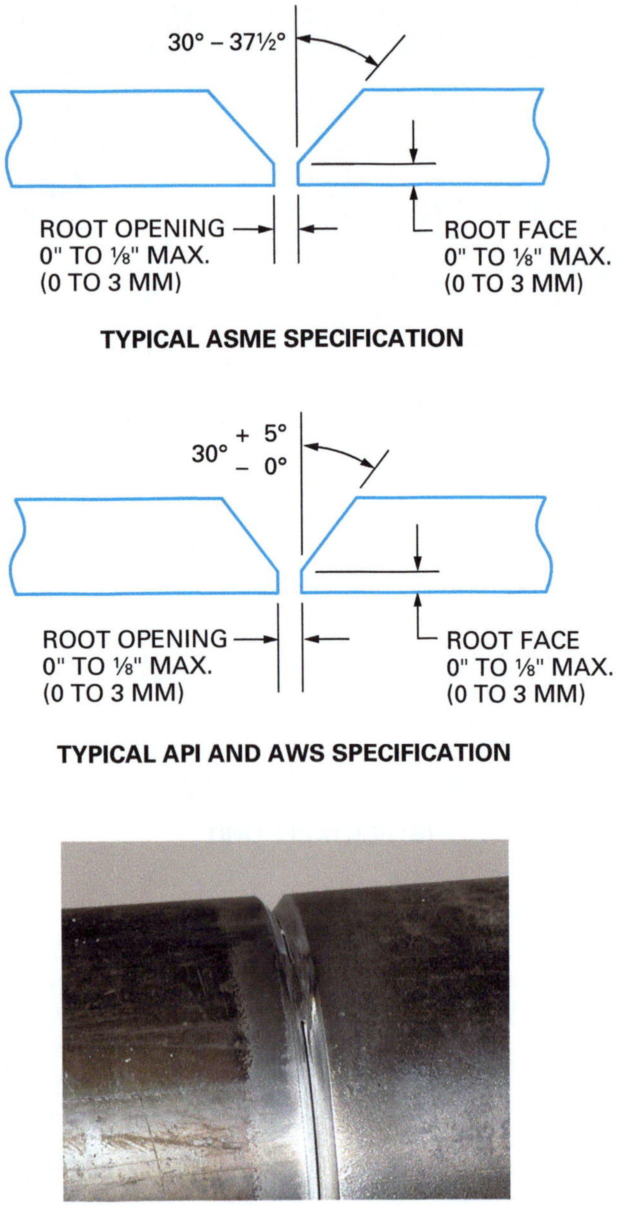

Figure 6 ASME, API, and AWS pipe bevel specifications.

To prepare weld coupons for open-root V-groove weld joints, follow these steps:

Step 1 Clean all rust or mill scale from the inside and outside of the pipe with a grinder, wire brush, or other common method.

Step 2 Bevel the end of the pipe to 30 to 37½ degrees by any acceptable beveling method, including manual, mechanical, and thermal methods.

Step 3 Cut off the end of the beveled pipe to the length specified by your instructor. Two coupons will be required for each weld.

> **NOTE**
> For 1G-ROTATED welding, the coupons may need to be cut longer to reach the rollers being used.

The two coupons can now be prepared further for welding:

Step 1 Check the bevel. There should be no dross, and the bevel angle should be 30 to 37½ degrees.

Step 2 Grind or file a 0" to ⅛" (0 to 3 mm) root face on the bevel.

> **NOTE**
> Welding codes allow the root face for open-root welds to be from 0" to ⅛" (0 to 3 mm) wide and the root opening to be from 0" to ⅛" (0 to 3 mm). Select and adjust the root face and opening as directed by the instructor as welding practice begins.

Step 3 Align the two pipe sections. The inside diameter (ID) surfaces of the two coupons should be even all around and evenly separated by the required root opening. Pieces of filler wire, metal shims, or pieces of bare welding electrode can be used to achieve the desired root opening between the two pipe sections until they are tack welded. Small-diameter pipe can be aligned by clamping both pieces to a piece of angle iron. Large-diameter pipe can be aligned using a pipe alignment jig or by holding a straightedge across the joint, parallel to the **pipe axis**. The straightedge must be used to check all around the pipe in case one or both sections are distorted.

> **GOING GREEN**
> ### Water Disposal
> To protect the environment and save resources, make sure to properly dispose of any water used in the cutting of tanks, barrels, or metals. If the water can be reused, save it and use it again for the next cutting.

Step 4 Once aligned, secure the two pipe sections with a clamp, or with tack welds if necessary.

Step 5 With the root opening established and the pipe ends aligned, make the first of four tack welds. Longer tacks may be required on heavy-wall or large-diameter pipe.

Step 6 After the first tack weld, check the root opening opposite it. Adjust the root opening space if necessary, and make the second tack weld on the side opposite the first tack.

> **CAUTION**
> Remove the spacer material before making the second tack, so that it will not get caught as the heated metal expands.

Step 7 Verify that the root opening remains correct before welding the third tack midway between the first and second tacks.

Step 8 Weld the fourth tack opposite the third tack and midway between the first and second tacks. There should now be four tack welds evenly spaced (90 degrees apart) around the pipe coupon (*Figure 7*).

Step 9 Clean and **feather** the tack welds with the edge of a grinding wheel. Feathering the ends of the tack welds with a grinder helps to fuse the root pass into the tack weld.

1.4.0 Preparing GMAW Equipment

Identify a proper welding machine for GMAW, and follow these steps to set it up for use:

Step 1 Verify that the welding machine can be used for GMAW.

Step 2 Make sure that the area is properly ventilated.

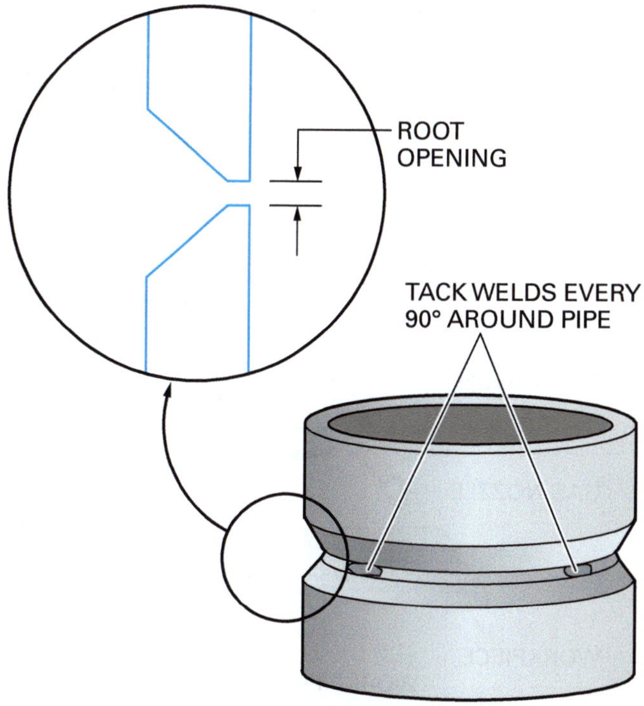

Figure 7 Tacked open-root V-groove weld coupon.

Step 3 Know the location of the primary power disconnect. This could be a disconnect switch or circuit breaker for hard-wired equipment, or the power cord if the machine is simply plugged into a receptacle.

Step 4 Configure the welding equipment for GMAW as directed by the instructor. The typical arrangement is shown in *Figure 8*. Configure the gun polarity. Equip the gun with the correct liner material, contact tube for the diameter of the electrode wire being used, and the desired nozzle for the application.

Machined Bevels and Root Faces for GMAW

For V-groove joints, machined bevels and root faces provide the cleanest and most uniform way to prepare materials for GMAW welds. This is especially true for stainless steel or aluminum from which all contaminants of thermal cutting must be removed. SMAW, FCAW, and GTAW can tolerate some poor fit-up for root passes. Controlling root pass penetration and obtaining good fusion with GMAW is more difficult if the fit-up is poor. In practice, SMAW or GTAW is usually used for the root pass.

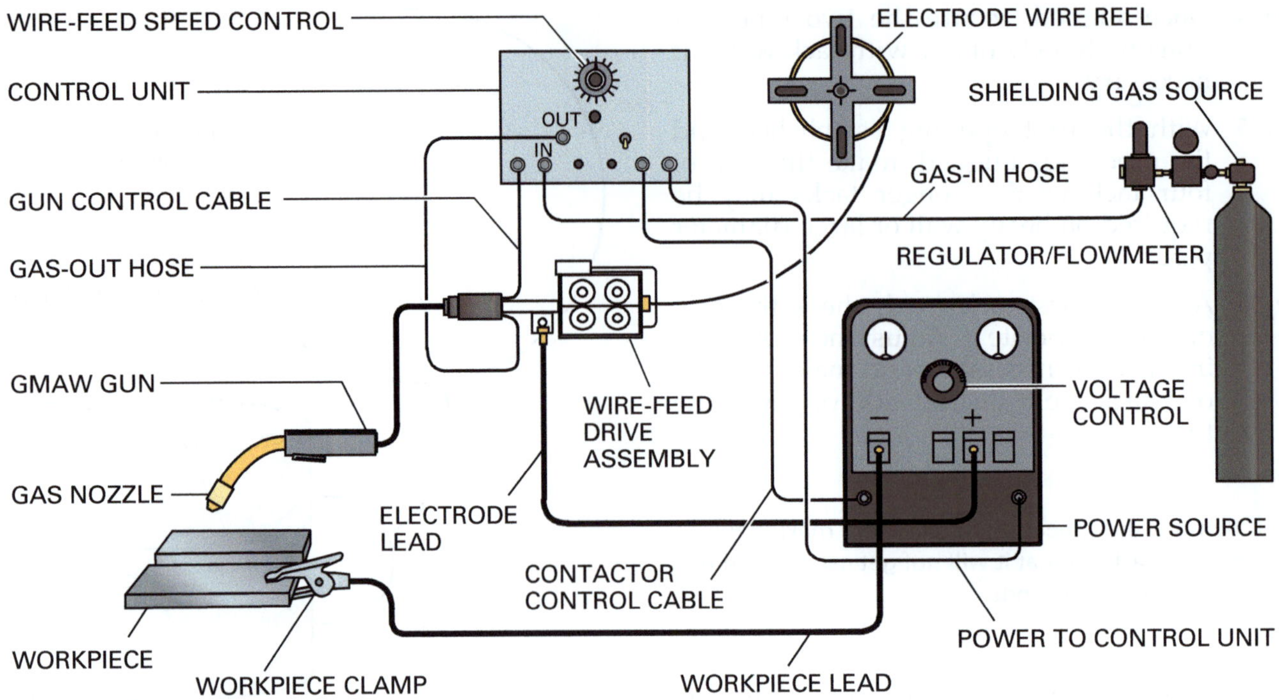

NOTE: The polarity of the gun and workpiece lead is determined by the type of filler metal and application.

Figure 8 Typical GMAW equipment arrangement.

Step 5 Configure and load the wire feeder and gun with the proper type and diameter of electrode wire, in accordance with the manufacturer's instructions.

Step 6 Connect the proper shielding gas for the application, as specified by the wire electrode manufacturer, WPS, site quality standards, and/or your instructor.

Step 7 Connect the clamp of the workpiece lead to the table or workpiece.

Step 8 Turn on the welding machine and purge the gun as directed by the gun manufacturer's instructions.

Step 9 Set the initial welding voltage and wire feed speed for the type and size electrode wire, as recommended by the manufacturer.

1.4.1 Welding Voltage

Arc length is determined by the voltage, which is set at the power source. Arc length is the distance from the wire electrode tip to the base metal or to the molten pool at the base metal. If the voltage is set too high, the arc will be too long. This could cause the wire to melt and fuse to the contact tube. A voltage that is too high also causes porosity and excessive spatter. Remember that the voltage must be increased or decreased as the wire feed speed is increased or decreased. Set the voltage according to the manufacturer's recommendations.

1.4.2 Welding Amperage

When using a standard constant-voltage welding machine, the wire feed speed determines the welding current (amperage) after the initial recommended setting has been made by the welder. The welding machine provides the amperage to melt the wire electrode while maintaining the selected welding voltage. Within limits, when the wire feed speed is increased, the needed welding amperage and deposition rate also increase. This results in higher welding heat, deeper penetration, and higher beads. When the wire feed speed is decreased, the welding amperage automatically decreases. With lower welding amperage and less heat, the deposition rate drops, and the weld beads are smaller with less penetration.

Note that some constant-voltage welding machines used for GMAW provide different degrees of current modification. Examples are slope and inductance adjustments. Welding machines with a slope adjustment allow you to vary the amount of amperage change in relation to the voltage range of the unit. This allows the correct short-circuiting current to be set up for certain welding uses and/or wire electrode requirements. Standard constant-voltage GMAW units have a current slope that is fixed by the manufacturer for general welding applications and conditions.

Other units allow adjustment of the response time for short-circuit current using an induction control. Instead of an instant short-circuit current draw, the current draw can ramp up over a varying period of time. This is done by using adjustable inductance in the welding machine circuits. This evens out the arc characteristics during short-circuiting transfer. Pulse-transfer units allow the peak current for the pulse, and the background current between pulses, to be adjusted to match specific welding applications and/or wire electrode requirements.

Hi-Lo Gauge

A gauge, such as the one shown here, can be used to check that internal surfaces are even all around.

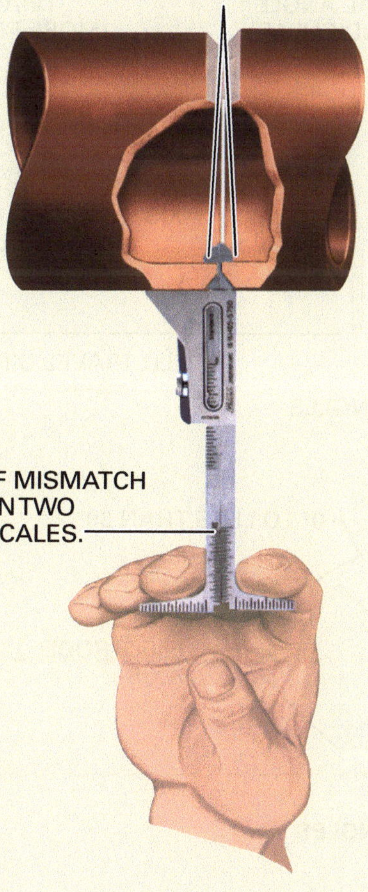

Tack Welds

When welding pipe with GMAW, it may be necessary to remove, instead of weld over, a tack weld that shows signs of defects. Welding over a tack weld that has defects can cause penetration and fusion problems.

1.4.3 Weld Travel Speed

The travel speed must be considered before beginning a weld. Weld travel speed is the speed at which the electrode tip passes across the base metal in the forward direction of the weld. It is measured in inches per minute (ipm) or millimeters per second (mm/s). Travel speed has a significant effect on penetration and bead size. Slower travel speeds build higher beads with deeper penetration. Faster travel speeds build smaller beads with less penetration. Ideally, the electrode tip should be positioned at the leading edge of the weld puddle during travel.

1.4.4 Gun Position

In GMAW, the gun position has a greater effect on weld penetration than either voltage or travel speed. The gun position for GMAW, in relation to the direction of the weld, is defined the same as it is for SMAW welding.

The gun travel angle (*Figure 9*) is a less-than-90-degree angle between the electrode axis and a line perpendicular to the weld axis at the point of electrode contact in a plane determined by the electrode axis and the weld axis. For pipe, the plane is determined by the electrode axis and a line tangent to the pipe surface at the same point.

- *Neutral angle* – Describes the travel angle when the center line of the gun electrode is exactly perpendicular (90 degrees) to the welding surface. The term is nonstandard for a 0-degree travel angle. Weld penetration is moderate when the gun electrode is used in this position.
- *Push angle* – Generally used for welding thin materials or when shallow penetration is required. A push angle is the travel angle created when the welding gun is tilted back, and the electrode tip precedes the gun in the direction of the weld. In this position, the electrode tip and shielding gas are directed ahead of the weld bead. Welding with this angle yields a smaller weld bead with less penetration. Push angles greater than 15 degrees are not advised for manual welding.

Wire Electrode Manufacturer's Recommendations

Always follow the manufacturer's recommendations for shielding gas and for the initial setting of the welding voltage and wire feed speed parameters. These balanced parameters are critical and are based on the welding position, size, and composition of the solid or composite wire being used.

- *Drag angle* – Describes the angle created when the gun is tilted forward, and the electrode tip trails the gun in the direction of the weld. In this position, the electrode tip and shielding gas are directed back toward the weld bead. Welding with this angle yields a higher and wider weld bead with more penetration. The greatest penetration is achieved using a drag angle from 15 to 25 degrees.

The gun work angle, also shown in *Figure 9*, is a less-than-90-degree angle between a line perpendicular to the major workpiece surface at the point of electrode contact and a plane determined by the electrode axis and the weld axis. For a T-joint or corner joint, the line is perpendicular to the non-butting member. For pipe, the plane is determined by the electrode axis and a line tangent to the pipe surface at the same point.

1.4.5 Electrode Extension, Stickout, and Standoff Distance

Electrode extension is the length of the wire that extends beyond the end of the welding gun contact tip. Changing the extension affects the welding current for low-conductivity metal wire electrodes. This is because the extension distance changes the preheating of the wire caused by its resistance to current flow. As the extension increases, wire preheating increases. Because the welding power source is self-regulating, it does

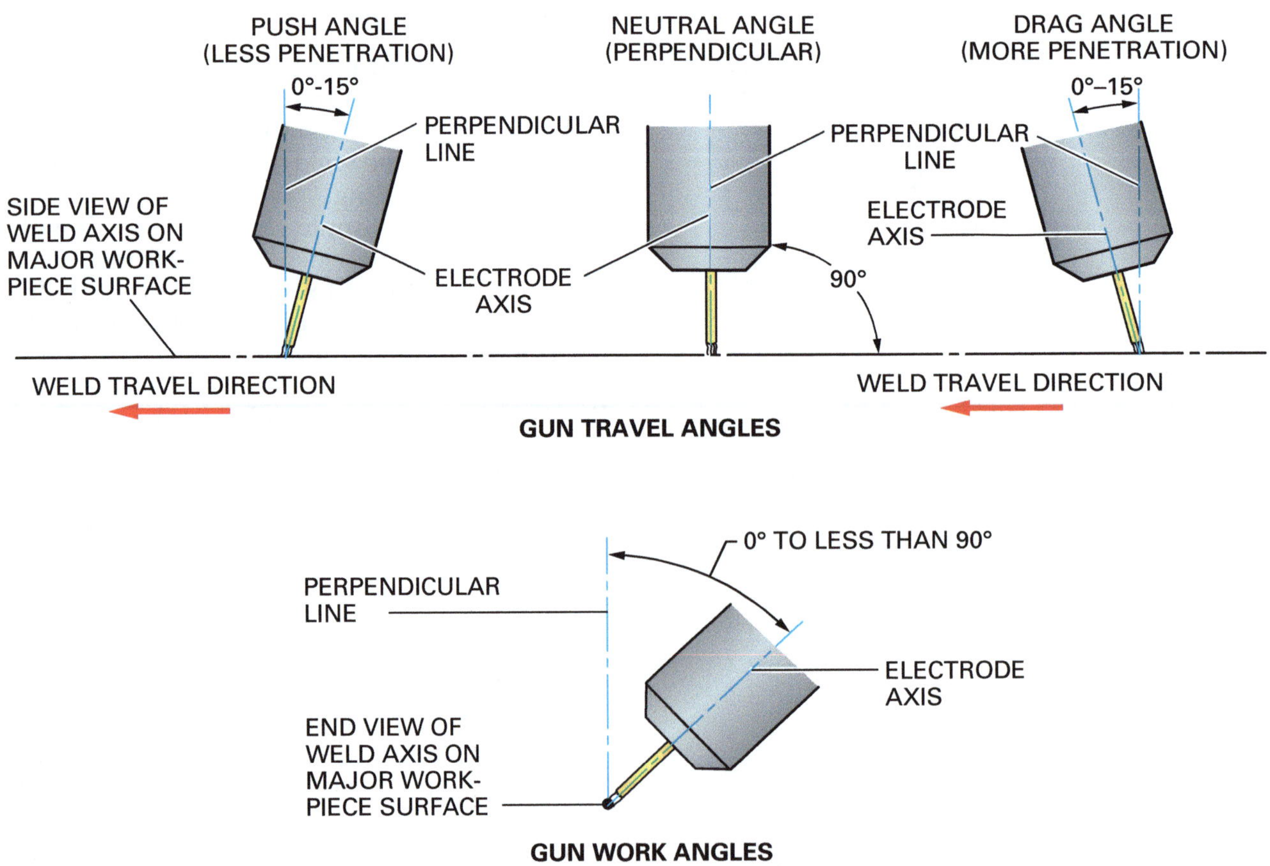

Figure 9 GMAW gun travel and work angles.

not have to furnish as much welding current to melt the wire, so the current output automatically decreases. This results in less penetration. Increasing extension is useful for bridging gaps and compensating for mismatch. However, it can cause overlap or a lack of fusion.

When the extension is decreased, the power source is forced to increase its current output to burn off the wire. Too little extension can cause the wire to weld itself to the contact tip, and porosity can develop in the weld. For high-conductivity metal wires, the preheating effect of wire resistance is minimal. Wire speed, current (if variable), and voltage settings have a more direct effect on penetration and deposit rate.

The GMAW extension for **microwire** (0.030" to 0.045" or 0.8 to 1.1 mm) varies between ¼" to 1" (≈6 to 25 mm), depending on the transfer mode. *Figure 10* shows electrode extensions for two GMAW gun configurations. It also shows other gun terms and parts. **Stickout** is the distance from the gas nozzle or insulating nozzle to the end of the electrode. **Standoff distance** is the distance from the gas nozzle or insulating nozzle to the workpiece. Contact tip extension, or setback, usually depends on the transfer mode for the GMAW application.

Arc Voltage

A minimum arc voltage is needed to maintain spray transfer. However, penetration is not directly related to voltage. Penetration increases with voltage for a time, but it decreases if the voltage is increased above the optimum level.

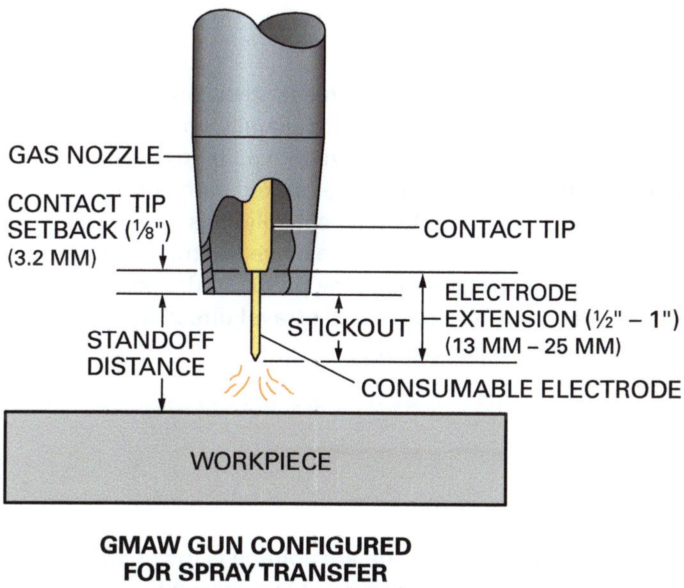

Figure 10 GMAW gun transfer configurations.

Travel Speed and Wire Feed Speed

New welders are tempted to turn down the wire feed speed if they have difficulty controlling the weld puddle. Wire electrodes must be run at certain balanced parameters that cannot be changed individually. Voltage, wire feed speed, and travel speed must be adjusted and balanced together to control the weld puddle.

1.4.6 Gas Nozzle Cleaning

As welding progresses, weld spatter builds up on the gas nozzle and contact tube. If the gas nozzle is not cleaned occasionally, it will limit or interrupt the shielding gas flow. This will cause porosity in the weld.

Clean the gas nozzle with a reamer, round file, or the tang of a standard file. After cleaning, the nozzle can be sprayed with, or dipped in, a special anti-spatter compound (*Figure 11*). The anti-spatter compound helps prevent the spatter from sticking to the nozzle. Use only anti-spatter compounds specifically designed for GMAW gas nozzles. Always refer to the SDS for specific safety concerns.

Figure 11 Anti-spatter compound in spray form.

GMAW of Stainless Steel

For stainless steel, a push angle is normally used for square or grooved butt joints. Stainless steel is usually welded in the flat or horizontal positions using spray transfer. Vertical and overhead welds usually employ short-circuit transfer for at least the root passes, if not the remaining passes. Pulsed spray transfer can be used for the remaining passes. Remember that all open joints in stainless steel materials must be backed with copper, stainless steel, or a shielding gas. This prevents air from reaching the weld puddle before it solidifies.

When practicing welding using stainless steel wire electrodes and carbon steel coupons, use the travel angle and transfer methods just outlined. Also use carbon steel backing rings, backing-gas pipe plugs, or dams, as required. When welding thick-walled pipe, the gun should be moved back-and-forth in the travel direction, and slightly side-to-side. On thinner materials, only the back-and-forth motion is needed.

Spatter Control

When there is excessive stickout, more voltage is lost in the electrode length and the arc voltage is lower. Irregular arc action and spatter may also occur. Remember that penetration is decreased as stickout is increased.

With shorter stickout and less voltage drop in the wire, higher arc voltage is available. However, with spray transfer, if the stickout is too short, spatter will increase and rapidly build up in the nozzle and on the contact tube. When using short-circuit transfer with contact tube extension, spatter buildup in the nozzle and on the contact tube is reduced.

Additional Resources

AWS F4.1: Recommended Safe Practices for Preparation for Welding and Cutting of Containers and Piping. Miami, FL: American Welding Society.

Modern Welding Technology. Current Edition. Howard B. Cary and Scott Helzer. Englewood Cliffs, NJ: Prentice Hall.

OSHA 1926.351, Arc Welding and Cutting. Current Edition. Washington, DC: Occupational Safety & Health Administration.

The Procedure Handbook of Arc Welding. Current Edition. Cleveland, OH: The James F. Lincoln Arc Welding Foundation.

Welding Handbook, Volumes 1 through 4. Current Edition. Miami, FL: American Welding Society.

Welding Pressure Pipelines and Piping Systems (PDF). The Lincoln Electric Company, Cleveland, OH, USA. **www.lincolnelectric.com**. Last accessed May 29, 2015.

Gas Metal Arc Welding – Product and Procedures Selection (PDF). The Lincoln Electric Company, Cleveland, OH, USA. **www.lincolnelectric.com**. Last accessed May 29, 2015.

1.0.0 Section Review

1. The shielding gas used in GMAW is intended to _____.
   a. provide breathing air for the welder
   b. reduce toxic fumes from the workpiece
   c. preheat the electrode before the arc starts
   d. move oxygen away from the weld zone

2. The fuel gas that is heavier than air and tends to collect in low areas is _____.
   a. oxygen
   b. nitrogen
   c. acetylene
   d. propane

3. Pipe weld coupons to be used for GMAW practice should be cut from carbon steel pipe that is Schedule _____.
   a. 10 or 30
   b. 20 or 60
   c. 40 or 80
   d. 60 or 90

4. When using a standard constant-voltage power source, after the initial recommended settings have been made by the welder, the welding amperage is usually controlled by the _____.
   a. wire electrode speed feed
   b. electrode-to-puddle distance
   c. GMAW gun work angle
   d. power source arc length

Section Two

2.0.0 Open-Root V-Groove Pipe Welds

Objective

Describe open-root V-groove pipe welding positions and GMAW techniques.

a. Identify and describe the four pipe welding positions and acceptable weld profiles.
b. Describe the technique for GMAW of the root and fill passes.
c. Describe the techniques required to produce open-root GMAW pipe welds in all positions.

Performance Tasks

1. Make GMAW open-root V-groove pipe welds in the 1G-ROTATED position using the correct filler metal and shielding gas.
2. Make GMAW open-root V-groove pipe welds in the 2G position using the correct filler metal and shielding gas.
3. Make GMAW open-root V-groove pipe welds in the 5G position using the correct filler metal and shielding gas.
4. Make GMAW open-root V-groove pipe welds in the 6G position using the correct filler metal and shielding gas.

Trade Terms

Keyhole: A hole that is formed when the root faces on two pipe ends are melted away by the arc.

The open-root V-groove weld is the most common groove weld made on carbon, low-alloy, and stainless steel pipe. These welds are typically used for joining medium- and thick-walled pipe used in various piping systems.

Piping containing or carrying material that can cause injury to people or damage to the environment, or those that operate at elevated pressures, may be categorized as critical piping. Injury or damage occurs when the piping system or other system components fail to contain the material as designed. Welds in critical piping may be required to meet stringent requirements. The welds must be expertly completed, thoroughly inspected, and at least as strong as the piping being welded. Piping that is used for hydronic heating and air conditioning applications, simple water systems, and similar applications may be assessed using less stringent requirements.

2.1.0 Pipe Groove-Weld Test Positions

Groove welds may be made in all positions on pipe. The weld position is determined by the axis of the pipe (*Figure 12*). Four standard weld test positions are used with pipe:

- *Flat (1G-ROTATED)* – The pipe axis is horizontal, and the pipe is slowly rotated while it is being welded on the top. The weld beads are flat.
- *Horizontal (2G)* – The pipe axis is vertical and the pipe rotation is fixed. The weld beads are horizontal.
- *Multiple (5G)* – The pipe axis is horizontal and the pipe rotation is fixed. The weld beads are flat, vertical, and overhead.
- *Multiple inclined (6G)* – The pipe axis is inclined at approximately 45 degrees ±5 degrees. Pipe rotation is fixed. Weld beads are horizontal, vertical, and overhead.

The pipe axis positions for 1G-ROTATED, 2G, and 5G can vary ±15 degrees from the basic position. The pipe axis position for 6G can vary ±5 degrees from the nominal 45-degree angle. *Table 1* provides a quick reference for the four positions. To determine the quality of a pipe weld, specific sections of pipe are removed for destructive testing. For positions 1G-ROTATED and 2G, two specimens are removed. They must be taken from opposite sides of the pipe, as shown in *Figure 13*. Once removed, the specimens are put through different bend tests, which reveal the quality of the welds. Alternatively, specimens may be tested by radiography, a non-destructive testing (NDT) method.

When 5G or 6G welds will be destructively tested, the test specimens must be cut from the four regions shown in *Figure 13*. These regions are midway between the twelve o'clock and three o'clock; three o'clock and six o'clock; six o'clock and nine o'clock; and nine o'clock and twelve o'clock positions. The twelve o'clock position is the top of the coupon groove; the six o'clock position is at the bottom.

2.1.1 Acceptable and Unacceptable Pipe Weld Profiles

Pipe groove welds should be made with slight reinforcement (not exceeding ⅛" or 3 mm) and a gradual transition to the base metal at each toe. The root pass should have complete penetration.

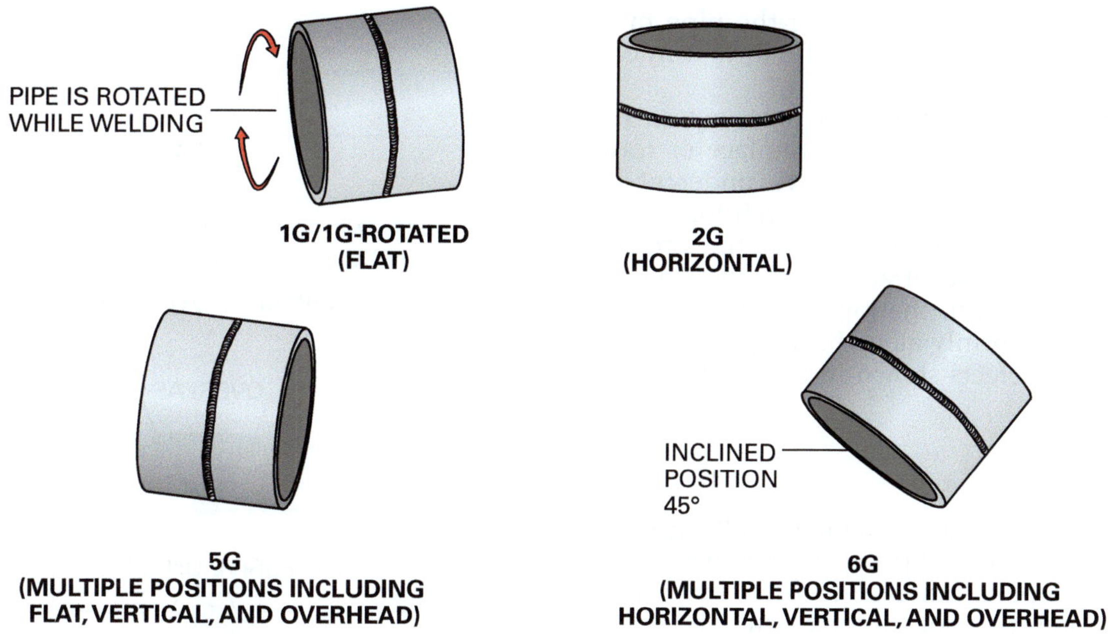

Figure 12 Four standard pipe groove-weld test positions.

Table 1 Pipe Groove-Weld Test Positions

Weld Test Position	Pipe Axis	Weld Beads	Pipe Position
1G-ROTATED	Horizontal	Flat	Slowly rotated
2G	Vertical	Horizontal	Fixed
5G	Horizontal	Flat, vertical, overhead	Fixed
6G	Inclined 45 degrees	Horizontal, vertical, overhead	Fixed

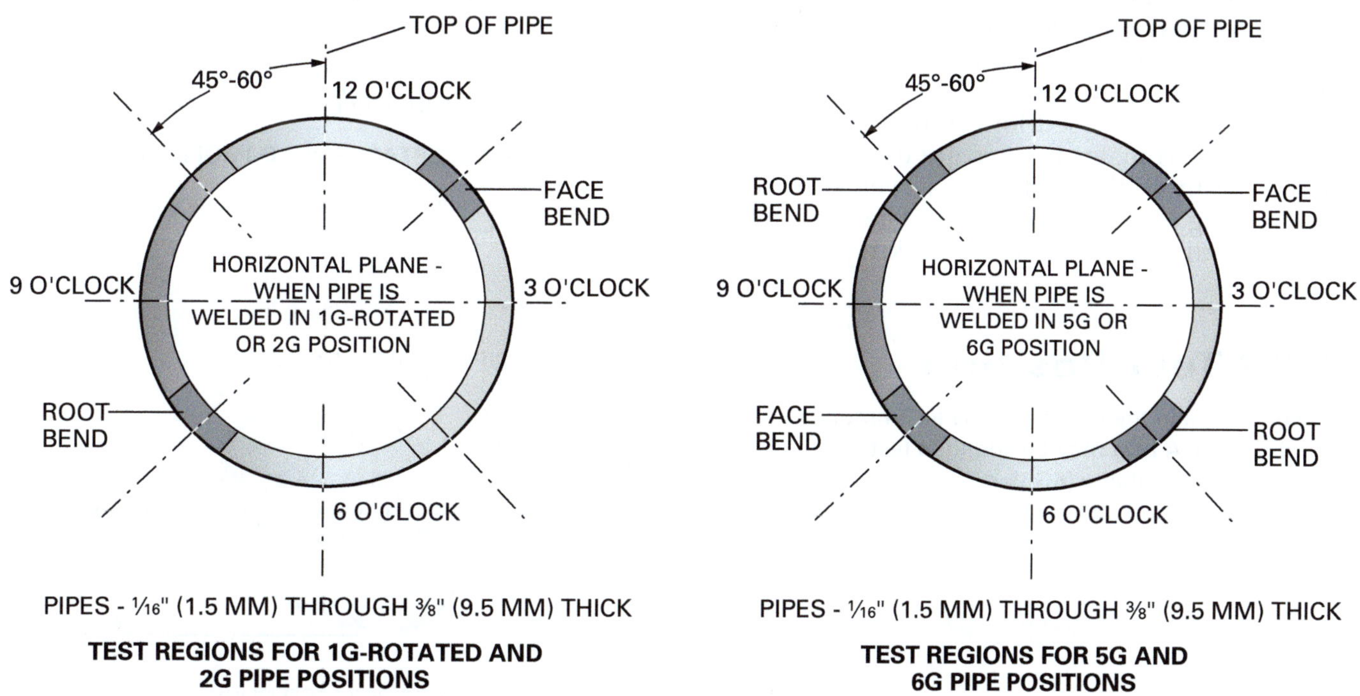

NOTE: Tack weld locations at 12, 3, 6, and 9 o'clock.

Figure 13 Test specimen regions for pipe positions 1G-ROTATED, 2G, 5G, and 6G.

The root reinforcement inside of the pipe may range in thickness from being flush to a maximum of ⅛" (3 mm) thick.

Pipe groove welds must not have excess reinforcement or any underfill at the face or root. They must be free from excessive undercut or any overlap. Excessively large face reinforcement reduces the pipe's strength. It causes the stresses in the pipe to be concentrated along the sides of the weld. As a result, the pipe will not expand and contract uniformly along its length. If a weld has any of the defects shown in *Figure 14*, it must be repaired.

2.2.0 The Root Pass

The greatest challenge of an open-root V-groove weld is making the root pass. The root pass must have complete penetration and fusion, but not an excessive amount of root reinforcement. The extent of penetration and fusion is very important. If the penetration is not sufficient, it will create a weak joint. If there is too much penetration along with excessive root reinforcement, the inside diameter of the pipe will be reduced and the flow restricted. Incomplete fusion will result in a defective weld.

One approach to running a **keyhole** that works well is to begin the root pass on top of the tack ⅛" to $^3/_{16}$" (3 to 5 mm) behind the feathered portion of the tack. This helps preheat the joint for a better tie-in. When approaching the next tack, proceed past the feathered portion of the next tack to the same point. Ensure that the keyhole disappears before stopping the weld.

When GMAW is used for the root pass, short-circuiting arc transfer is typically used. The progression of multiple-position (5G) welds for root pass welding is usually downhill. The remaining passes usually progress uphill. Backing gas must

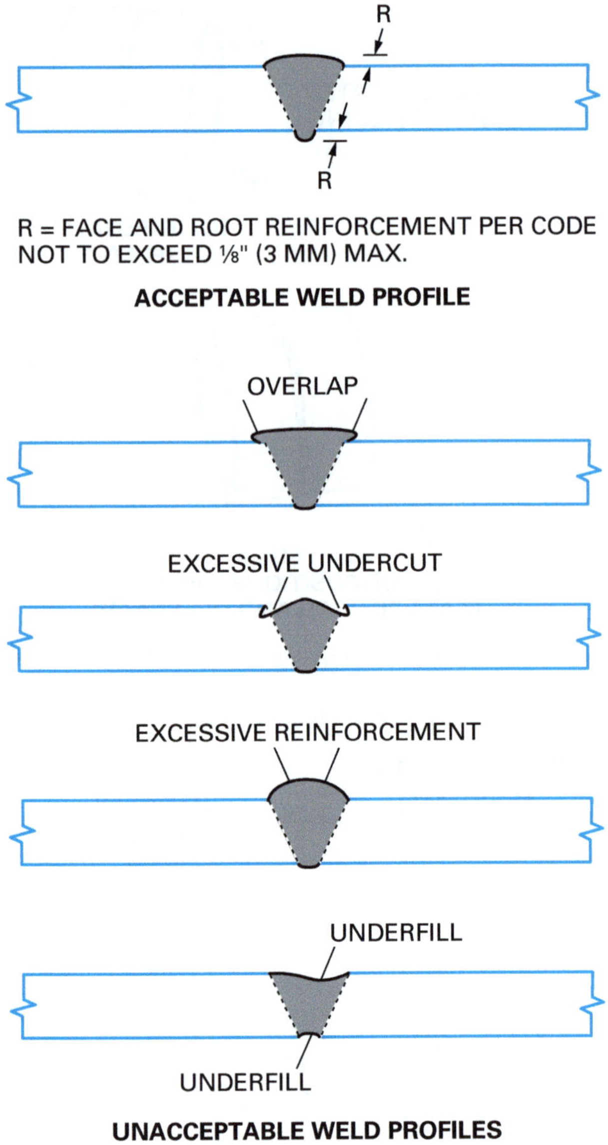

Figure 14 Acceptable and unacceptable pipe groove-weld profiles.

6GR Test Position

Welding pipe in the 6G position is a common welding test. A ring may be added to test for restricted accessibility (6GR).

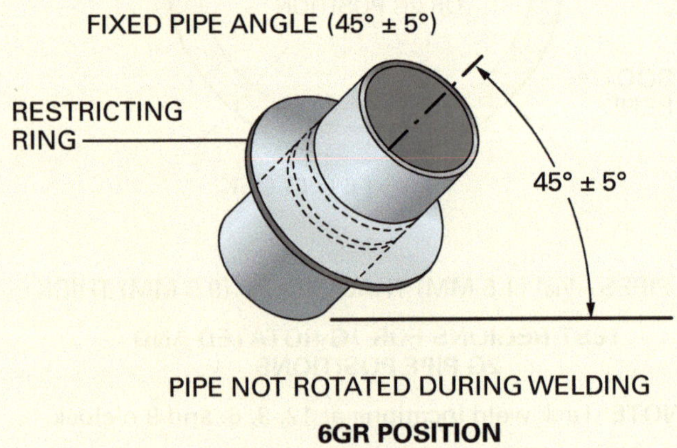

be used for stainless, low-alloy, and nonferrous alloys. *Figure 15* shows changes to the travel position that may be needed for multiple (5G) position pipe welds when welding uphill or downhill with GMAW.

After the root pass has been placed, it should be cleaned and inspected. Check the weld for incomplete fusion and penetration. Root reinforcement should be maintained at ⅛" (≈3mm) or less.

Remove any excess buildup or undercut by grinding the face of the root pass with the edge of a grinding disk. Be careful not to grind through the root pass or widen the groove. A narrow grinding wheel (not to be confused with a cutoff wheel) is effective for this purpose.

CAUTION

When welding pipe, avoid making arc strikes outside of the weld groove on the surface of the pipe. An arc strike can cause a hardened spot that can crack as the pipe expands and contracts. An arc strike on the pipe surface is considered a defect, which will require repair or reworking.

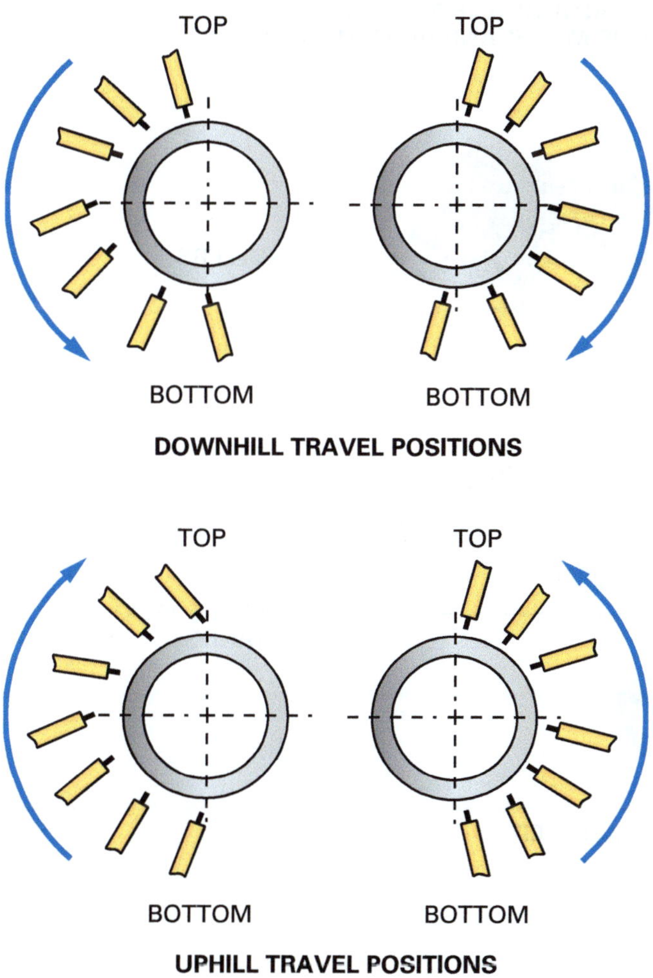

Figure 15 Approximate travel position progressions.

2.3.0 Practicing Open-Root V-Groove Welds

The following sections explain how to perform open-root V-groove GMAW in the following positions:

- Flat (1G-ROTATED)
- Horizontal (2G)
- Multiple (5G)
- Multiple inclined (6G)

2.3.1 Flat (1G-ROTATED) Position

Practice the 1G-ROTATED position for open-root V-groove pipe welds. Use a shielding gas and a solid wire electrode with a wire diameter specified by the instructor. For the root pass, keep the gun work angle at 90 degrees (0-degree work angle) to the pipe axis with a 10- to 15-degree drag angle. Use a slight forward and backward motion with a slight side-to-side oscillation to run the root pass and to control penetration. Be sure to fuse into the tack welds. Clean the face of the root pass with a brush.

When running the remaining passes, use stringer or weave beads with a slight side-to-side motion. This ensures fusion at the toes of the weld bead. Try to keep the wire at the leading edge of the weld puddle to ensure proper penetration. Note that the use of weave beads should be avoided on stainless steel pipe. Take special care to fill the crater at the termination of the weld. Run all passes at or near the top of the pipe as the pipe is rotated.

Follow these steps to practice open-root V-groove pipe welds in the 1G-ROTATED position:

Step 1 Tack weld the practice pipe weld coupons together as explained earlier.

Step 2 Position the pipe coupon horizontally on two sets of rollers at a comfortable welding height. *Figure 16* shows roller supports commonly found in pipe welding shops. A welding positioner or pipe rotator is also required. The positioner shown

Fill Passes

Follow these guidelines when running the fill passes:

- Fill craters to prevent cracking.
- Ensure pipe cleanliness by slightly grinding or filing the crater.
- Stagger the location of the starts and stops for each weld.

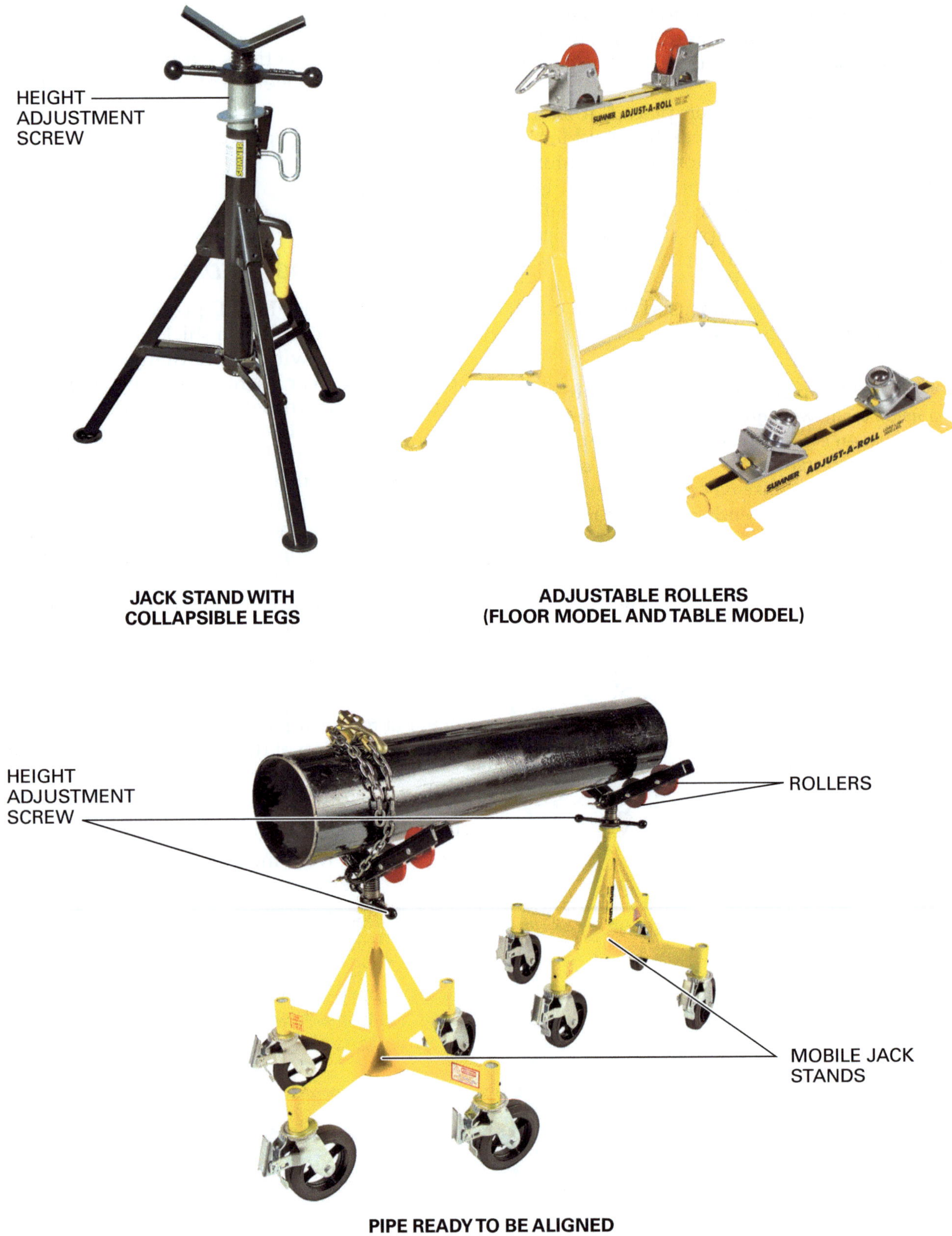

Figure 16 Pipe roller supports.

in *Figure 17 (A)* is very popular. The pipe rotator, shown in *Figure 17 (B)*, is another option. This model can handle up to 16" (DN400) pipe and weights up to 1,200 pounds (544 kg). Adjustable roller stands can be used to help support longer stock.

Step 3 Make sure that the workpiece clamp is attached directly onto the pipe coupon. This will prevent the welding current from passing through the roller bearings or arcing between the rollers and the pipe coupon.

CAUTION: Failure to attach the workpiece clamp to the pipe coupon in the 1G-ROTATED position can result in variations in welding current, damage to the roller bearings, and arcing on the rollers and pipe coupon.

Step 4 Run the root pass using a slight forward and backward motion and a slight side-to-side oscillation. Position a tack weld at the eleven o'clock position. Start the weld bead on the tack weld, and advance toward the twelve o'clock position.

Step 5 Roll the pipe as necessary to keep the weld in the flat position.

Step 6 Clean the crater and root pass to remove silica deposits and any potential discontinuities before restarting.

Step 7 Use the same rolling procedure to make the remaining passes. Use stringer or weave beads as applicable. Pay special attention at the fusions to prevent slag inclusions or excess buildup. Overlap the passes, and clean the weld after each pass. *Figure 18* shows the related bead sequences and work angles.

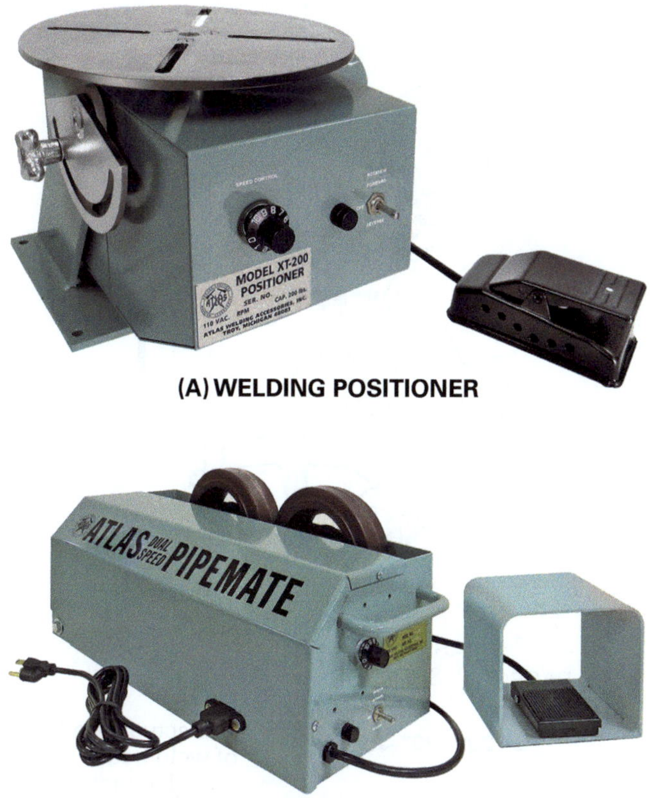

Figure 17 A welding positioner and a pipe rotator.

2.3.2 Horizontal (2G) Position

Practice open-root V-groove pipe welds in the 2G position using a shielding gas and the type and diameter of wire electrode specified by your instructor. For the root pass, the gun should be at a 10- to 15-degree drag angle and at 90 degrees (0-degree work angle) to the surface of the pipe. To prevent the weld puddle from sagging, the gun work angle can be dropped slightly, but not more than 10 degrees. For the root pass, use a slight side-to-side oscillation to control penetration using a zig-zag or crescent-shaped pattern to control penetration, as shown in *Figure 19*. Be sure to properly fuse and blend the tack welds.

Open-Root V-Groove Stainless-Steel Pipe Welds

When practicing open-root V-groove stainless-steel pipe welds, always use push angles in place of drag angles. Also, practice with a backing gas using backing-gas pipe plugs or dams.

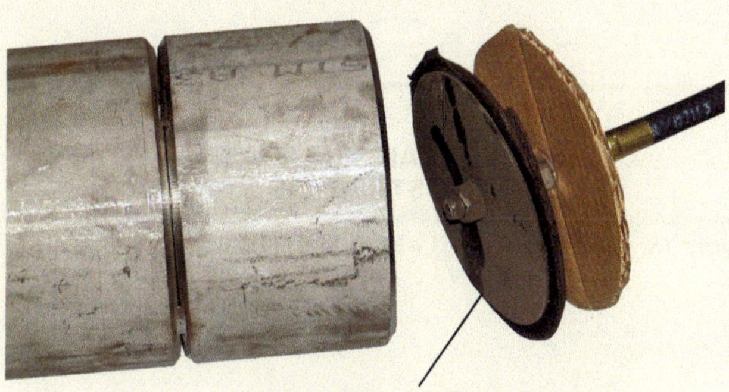

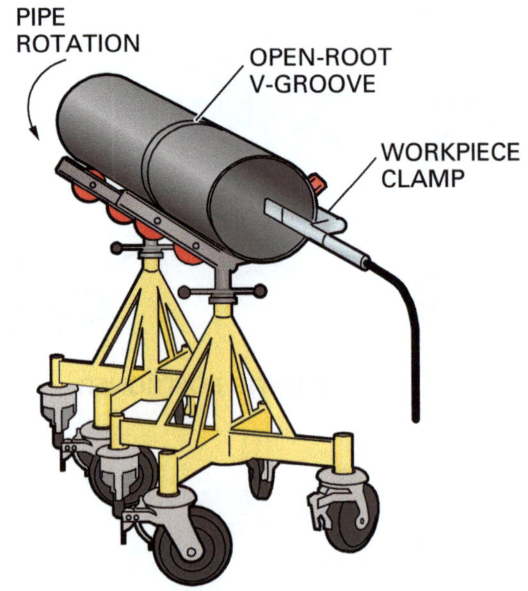

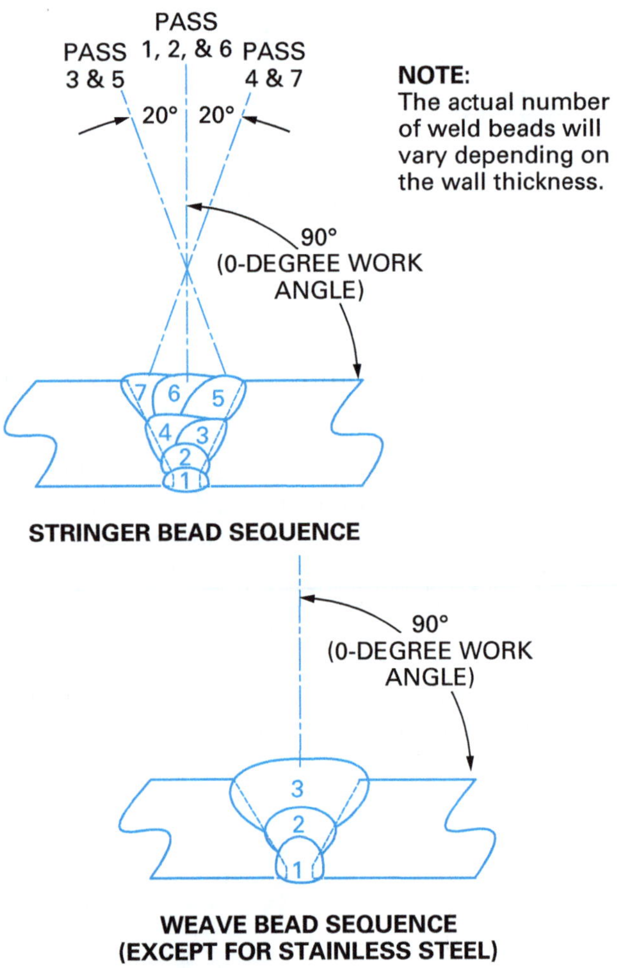

Figure 18 Multiple-pass 1G-ROTATED bead sequences and work angles.

When running the remaining passes, use stringer beads and a slight side-to-side motion to ensure fusion at the toes of the weld bead. Try to keep the wire electrode at the leading edge of the weld puddle. Take special care to fill the crater at the termination of the weld. *Figure 20* shows the 2G pipe bead sequence and work angles.

Follow these steps to practice open-root V-groove pipe welds in the 2G position:

Step 1 Tack weld the practice pipe weld coupons together as explained earlier.

Step 2 Clamp or tack weld the pipe coupon with the axis of the pipe vertical.

Step 3 Run the root pass using the motions shown in *Figure 19*. Start the root pass on a tack weld. Clean the crater between restarts, if restarts are necessary. Continue this procedure until the root pass has been completed.

Step 4 Clean the root pass to remove silica deposits and any potential discontinuities before restarting.

Step 5 Run the remaining passes at the proper work angles. Overlap the passes, and clean the weld after each pass.

2.3.3 Multiple (5G) Position

Practice the multiple (5G) position open-root V-groove pipe welds using a shielding gas and the type and diameter of electrode wire specified by your instructor. The gun should be at approximately a 10- to 15-degree drag or push angle for the root and remaining passes. The type of angle depends on whether your instructor directs you to run downhill or uphill passes. In some cases,

Welding Thick-Walled Pipe

When welding a V-groove joint in thick-walled pipe, it is sometimes difficult to maintain the proper electrode extension. This may be because of the gun size or the depth/angle of the groove. In these cases, extend the contact tube beyond the nozzle as necessary to keep a ⅜" (9–10 mm) electrode extension.

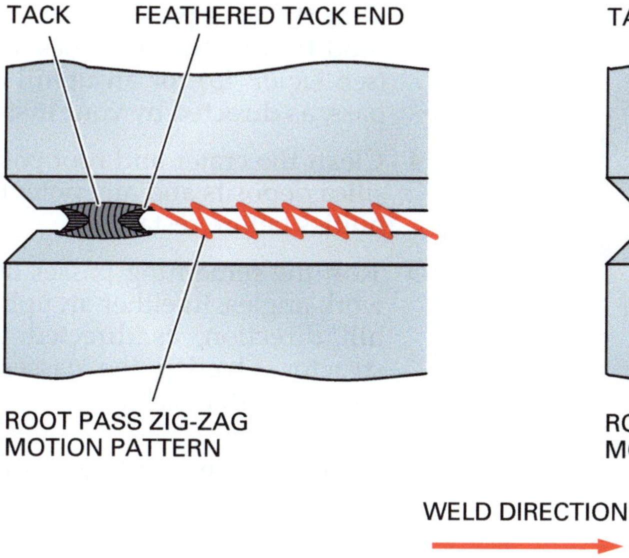

Figure 19 Root-pass patterns.

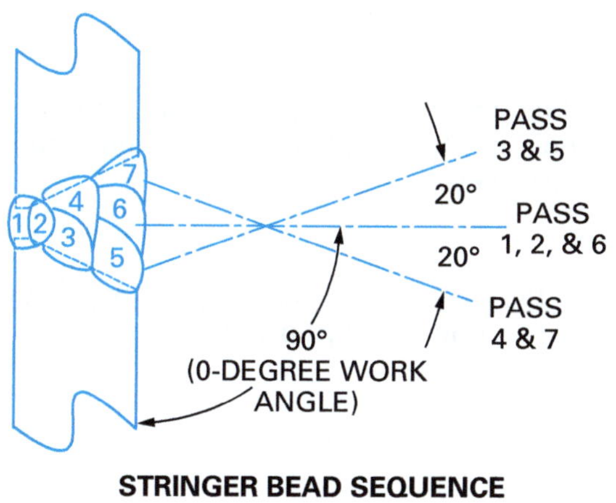

Figure 20 Multiple-pass 2G bead sequence and work angles.

the root pass is made downhill and the remaining passes are made uphill. Refer to *Figure 15* for the travel position changes that may be necessary as the passes are made on both sides of the pipe.

When running the remaining passes, use stringer or weave beads with a slight side-to-side motion to ensure fusion at the toes of the weld bead or weave bead. Try to keep the wire electrode at the leading edge of the weld puddle. Pay special attention to fill the crater at the termination of the weld. *Figure 21* shows the 5G pipe bead sequences and work angles.

Follow these steps to practice open-root V-groove pipe welds in the 5G position:

Step 1 Tack weld the practice pipe weld coupons together as explained earlier.

Step 2 Clamp or tack weld the pipe weld coupon into position with the pipe axis horizontal. Be sure to position the tack welds away from the areas from which the test coupons will be cut.

Use of GMAW for a Root Pass

In practice, the WPS or site standards may require the root pass of critical pipe welds be applied using GTAW or SMAW. For noncritical pipe welds, GMAW may be allowed for the root pass.

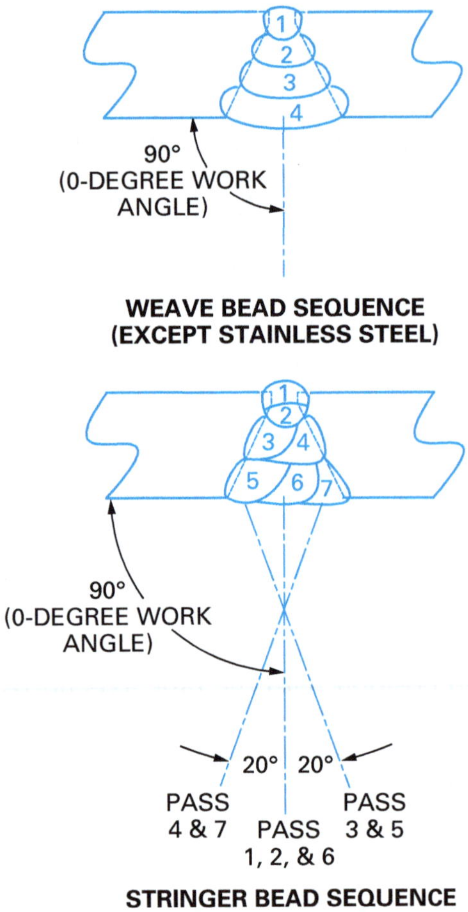

Figure 21 Multiple-pass 5G bead sequences and work angles.

Step 3 Run the root pass with a stringer bead. Modify the travel angles as necessary (see *Figure 15*) for an uphill or downhill pass, as directed by your instructor.

Step 4 Clean the crater and root pass to remove silica deposits and any potential discontinuities before restarting.

Step 5 Run the remaining passes at the proper work angles, in either an uphill or downhill direction, as directed by your instructor. Overlap the passes, and clean the weld after each pass.

2.3.4 Multiple Inclined (6G) Position

Practice the 6G multiple inclined (45 degrees) position open-root V-groove pipe welds. Use a shielding gas and a solid or composite wire electrode with a wire diameter as specified by your instructor. The gun should be at approximately a 10- to 15-degree drag or push angle for the root and the remaining passes. The type of angle depends on whether the instructor directs downhill or uphill passes to be run. In some cases, the root pass is made downhill, and the remaining passes are made uphill. Refer to *Figure 15* for the travel position changes that may be needed as the passes are made on both sides of the pipe.

When running the remaining passes, use weave beads or stringer beads and a slight side-to-side motion. This ensures fusion at the toes of the weld bead. Try to keep the wire electrode at the leading edge of the weld puddle. Pay special attention to fill the crater at the termination of the weld. *Figure 22* shows the 6G pipe bead sequence and work angles.

Follow these steps to practice open-root V-groove pipe welds in the 6G position:

NOTE

If required for training purposes, a restricting ring may be added to the 6G position coupon to form a 6GR position coupon.

Step 1 Tack weld the practice pipe weld coupons together as explained earlier.

Starting Pipe Welds

When starting a pipe weld, briefly use a short electrode extension to create a hot weld that will ensure proper penetration and flatten the weld face. This will make terminating the weld at the starting point easier.

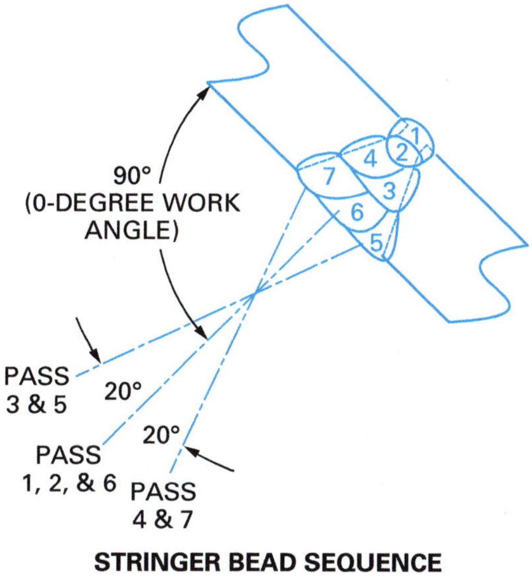

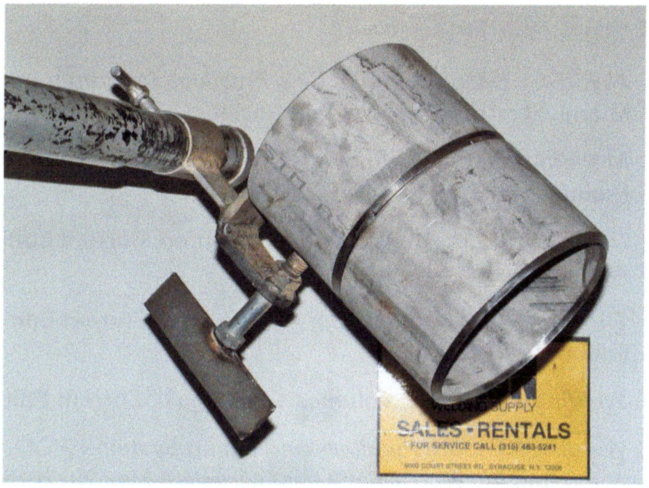

Figure 22 Multiple-pass 6G bead sequence and work angles.

Step 2 Clamp or tack weld the pipe weld coupon into position with the pipe axis inclined 45 degrees to the horizontal plane. Be sure to position the tack welds so that they are away from the areas from which the test specimen will be cut.

Step 3 Run the root pass with a stringer bead. Modify the travel angles as needed (see *Figure 15*) for an uphill or downhill pass, as directed by your instructor.

Step 4 Clean the crater and root pass to remove silica deposits and any potential discontinuities before restarting.

Step 5 Run the remaining passes at the proper work angles. Use either an uphill or downhill direction, as directed by your instructor. Overlap the passes, and clean the weld after each pass.

Additional Resources

AWS F4.1: Recommended Safe Practices for Preparation for Welding and Cutting of Containers and Piping. Miami, FL: American Welding Society.

Modern Welding Technology. Current Edition. Howard B. Cary and Scott Helzer. Englewood Cliffs, NJ: Prentice Hall.

OSHA 1926.351, Arc Welding and Cutting. Current Edition. Washington, DC: Occupational Safety & Health Administration.

The Procedure Handbook of Arc Welding. Current Edition. Cleveland, OH: The James F. Lincoln Arc Welding Foundation.

Welding Handbook, Volumes 1 through 4. Current Edition. Miami, FL: American Welding Society.

Welding Pressure Pipelines and Piping Systems (PDF). The Lincoln Electric Company, Cleveland, OH, USA. **www.lincolnelectric.com**. Last accessed May 29, 2015.

Gas Metal Arc Welding – Product and Procedures Selection (PDF). The Lincoln Electric Company, Cleveland, OH, USA. **www.lincolnelectric.com**. Last accessed May 29, 2015.

2.0.0 Section Review

1. In the multiple inclined (6G) pipe groove-welding position, the pipe axis is inclined approximately _____.
   a. 10 degrees
   b. 30 degrees
   c. 45 degrees
   d. 60 degrees

2. What occurs if there is excessive root penetration and root reinforcement during a root pass?
   a. The inside diameter of the pipe is reduced and the flow will be restricted.
   b. The welding arc will short circuit between the base metal and the toe.
   c. The root joint will become excessively brittle due to high arc temperatures.
   d. The shielding gas will no longer protect the weld puddle from impurities.

3. The two types of welding beads that are typically used when running the remaining passes during a GMAW pipe welding procedure are _____.
   a. weave or toe
   b. inclined or stagger
   c. root or stringer
   d. stringer or weave

Summary

Becoming an accomplished welder requires skill development, which is acquired through practice. One of the common skills that welders need is the ability to make acceptable GMAW open-root V-groove welds on pipe in all positions. This module presents basic GMAW pipe welding concepts and covered safety procedures and practices that must be followed before and during GMAW. It also describes how to properly prepare for GMAW and how to recognize unacceptable welds. Techniques for successful open-root V-groove welds in the 1G-ROTATED, 2G, 5G, and 6G positions are presented. The information in this module will provide welders with a thorough understanding of how to complete sound GMAW open-root V-groove pipe welds.

Review Questions

1. The filler metal that is used in GMAW comes from a _____.
   a. manually fed wire alloy
   b. single rod-shaped electrode
   c. sample welding coupon
   d. continuously fed wire electrode

2. During GMAW, a welder should wear a cap made out of 100-percent _____.
   a. polyester
   b. cotton
   c. wool
   d. rayon

3. Before GMAW is performed on any tank or barrel, the container should be filled with water or purged with an inert gas in order to _____.
   a. displace any oxygen
   b. establish an ambient temperature
   c. remove shielding gases
   d. offset any vacuum condition

4. Pipe weld coupons used for practice should be cut from carbon steel pipe with a diameter of _____.
   a. 1" (DN25)
   b. 2" to 3" (DN50 to 80)
   c. 3" to 12" (DN80 to 300)
   d. 14" or larger (DN350)

5. Once the two pipe weld coupon halves are aligned and secured, it is time to _____.
   a. bevel the two pipe ID surfaces
   b. make the first of the tack welds
   c. measure the root reinforcement
   d. remove any rust or mill scale

6. Arc length is determined by _____.
   a. gun position
   b. amperage
   c. voltage
   d. travel speed

7. Within limits, when the wire electrode speed is increased, the _____.
   a. voltage and deposition rate decrease
   b. deposition rate and penetration decrease
   c. welding amperage automatically decreases
   d. welding amperage and deposition rate increase

8. With GMAW, what has a greater effect on weld penetration than either voltage or travel speed?
   a. Gun position
   b. Amperage
   c. Arc length
   d. Gas flow

9. The contact tip extension, or setback, for a GMAW application is usually dependent on the _____.
   a. travel angle
   b. transfer mode
   c. voltage level
   d. power source

10. When the pipe axis is horizontal and the weld beads are flat, vertical, and overhead, the welding position is _____.
    a. flat (1G-ROTATED)
    b. horizontal (2G)
    c. multiple (5G)
    d. multiple inclined (6G)

11. Four test specimens cut from four separate regions around the pipe are used to destructively test _____.
    a. 5G or 6G welds
    b. 2G or 3G welds
    c. 1G or 2G welds
    d. 1G-ROTATED welds

12. One technique that works well when running a keyhole in an open-root V-groove weld is to begin the root pass on top of the _____.
    a. leg reinforcement
    b. hot pass weld
    c. keyhole joint
    d. tack weld

13. When GMAW is used for the root pass in an open-root V-groove weld, the transfer mode selected is typically _____.
    a. spray arc transfer
    b. short-circuiting arc transfer
    c. globular arc transfer
    d. thermal arc transfer

14. When making the remaining weld passes on stainless steel pipe, avoid using _____.
    a. tack welds
    b. push angles
    c. weave beads
    d. stringer beads

15. To prevent the weld puddle from sagging during a weld in the 2G position, the gun work angle can be dropped a maximum of _____.
    a. 10 degrees
    b. 15 degrees
    c. 20 degrees
    d. 25 degrees

Trade Terms Introduced in This Module

Axis: The center line running the length of an object. In welding, the term axis is most often used in relation to an electrode or a section of pipe.

Composite wire electrode: A generic term for multicomponent filler metal electrodes. For GMAW, the tubular electrode contains a powdered metal inside a metal sheath. A flux-cored electrode contains flux inside a metal sheath.

Electrode extension: The length of the wire that extends beyond the tip of the welding gun contact tube.

Feather: To grind the start and/or stop areas of a weld to a tapered edge.

Keyhole: A hole that is formed when the root faces on two pipe ends are melted away by the arc.

Microwire: Generally considered to be wire electrode that is 0.045" (1.1 mm) in diameter or smaller. However, due to the small size of most GMAW wire electrodes in comparison to SMAW electrodes and similar filler metals, some refer to all GMAW-related wire electrodes as microwire.

Pipe axis: An imaginary line that runs longitudinally (lengthwise) through the center of a pipe.

Remaining passes: A nonstandard term that is used to describe the fill and cover passes as a group.

Root pass: The initial weld pass made to produce a root bead.

Standoff distance: The distance from the gas nozzle or insulating nozzle to the workpiece.

Stickout: The distance from the gas nozzle or insulating nozzle to the end of the electrode.

Appendix

Performance Accreditation Tasks

PATs provide specific acceptable criteria for performance and help to ensure a true competency-based welding program for students.

The following tasks are designed to evaluate your ability to run open-root V-groove welds with GMAW equipment in the four standard test positions, using carbon steel wire of the appropriate diameter, and shielding gas. Perform each task when you are instructed to do so by your instructor. As you complete each task, show it to your instructor for evaluation. Do not proceed to the next task until told to do so by your instructor.

For AWS 1G-ROTATED, 2G, and 5G certifications, refer to *AWS EG3.0:1996 Guide for the Training and Qualification of Welding Personnel; Level II—Advanced Welder* for bend test requirements. For AWS 6G certifications, refer to *AWS EG4.0:1996 Guide for the Training and Qualification of Welding Personnel; Level III—Expert Welder* for bend test requirements.

PATs 2 and 3 correspond to *AWS EG3.0-96, Section 3.3.6.2, Unit #2 GMAW, Learning Objective #11*. PAT 4 corresponds to *AWS EG4.0-96, Section 3.3.6.3, Unit #3 GMAW, Learning Objective #7*. PAT 1 has no AWS correlation.

OPEN-ROOT V-GROOVE PIPE WELD IN THE 1G-ROTATED POSITION

Using carbon steel wire of the appropriate diameter, shielding gas, and stringer beads, make an open-root V-groove weld on carbon steel pipe in the 1G-ROTATED position.

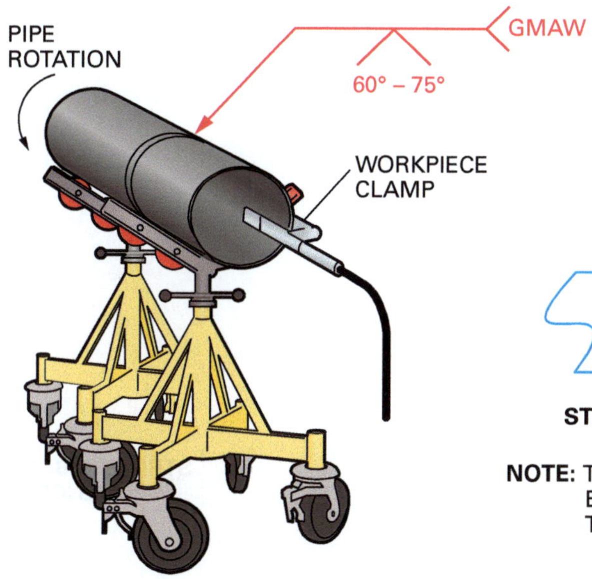

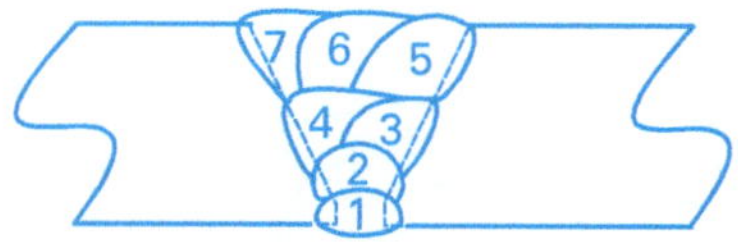

STRINGER BEAD SEQUENCE

NOTE: THE ACTUAL NUMBER OF WELD BEADS WILL VARY DEPENDING ON THE WALL THICKNESS.

Criteria for Acceptance:

- Uniform appearance on the bead face ______
- Craters and restarts filled to the full cross section of the weld ______
- Acceptable weld profile in accordance with the applicable codes and standards ______
- Smooth transition with complete fusion at the toes of the weld ______
- Complete joint penetration and face reinforcement with the root reinforcement at least flush with the inside of the pipe, to a maximum buildup of ⅛" (3 mm) ______
- No porosity ______
- No undercut ______
- No cracks ______
- No overlap ______
- No incomplete fusion ______

OPEN-ROOT V-GROOVE PIPE WELD IN THE 2G POSITION

Using carbon steel wire of the appropriate diameter, shielding gas, and stringer beads, make an open-root V-groove weld on carbon steel pipe in the 2G position.

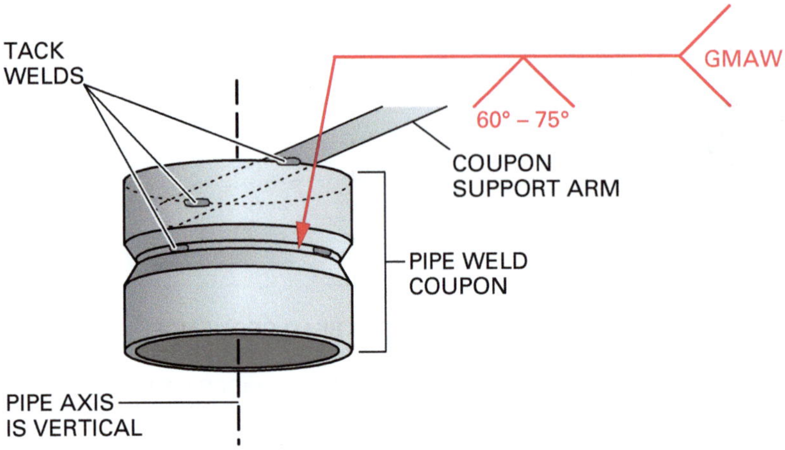

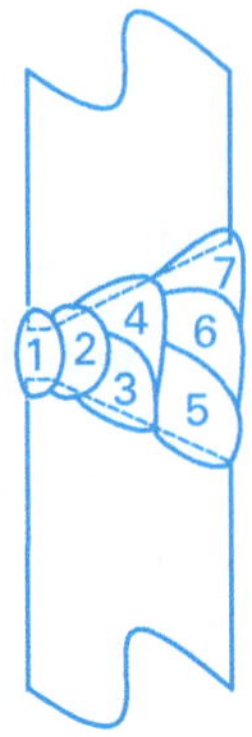

STRINGER BEAD SEQUENCE

Criteria for Acceptance:

- Uniform appearance on the bead face ______
- Craters and restarts filled to the full cross section of the weld ______
- Acceptable weld profile in accordance with the applicable codes and standards ______
- Smooth transition with complete fusion at the toes of the weld ______
- Complete joint penetration and face reinforcement with the root reinforcement at least flush with the inside of the pipe, to a maximum buildup of ⅛" (3 mm) ______
- No porosity ______
- No undercut ______
- No cracks ______
- No overlap ______
- No incomplete fusion ______

OPEN-ROOT V-GROOVE PIPE WELD IN THE 5G POSITION

Using carbon steel wire of the appropriate diameter, shielding gas, and stringer or weave beads, make an open-root V-groove weld on carbon steel pipe in the 5G position. The root pass can be accomplished uphill or downhill at the discretion of the instructor. Fill and cap passes shall be done uphill.

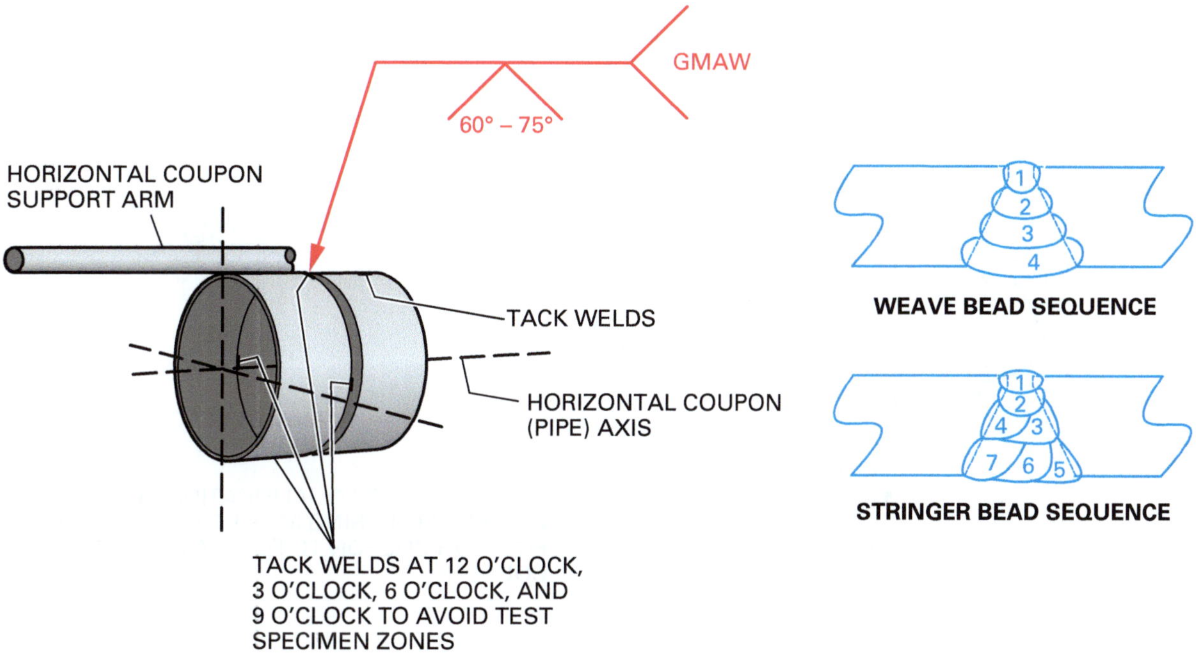

Criteria for Acceptance:

- Uniform appearance on the bead face ________
- Craters and restarts filled to the full cross section of the weld ________
- Acceptable weld profile in accordance with the applicable codes and standards ________
- Smooth transition with complete fusion at the toes of the weld ________
- Complete joint penetration and face reinforcement with the root reinforcement at least flush with the inside of the pipe, to a maximum buildup of ⅛" (3 mm) ________
- No porosity ________
- No undercut ________
- No cracks ________
- No overlap ________
- No incomplete fusion ________

OPEN-ROOT V-GROOVE PIPE WELD IN THE 6G (OR 6 GR) POSITION

Using carbon steel wire of the appropriate diameter, shielding gas, and stringer beads, make an open-root V-groove weld on carbon steel pipe in the 6G (or 6GR) position. The root pass can be accomplished uphill or downhill at the discretion of the instructor. Fill and cap passes shall be done uphill.

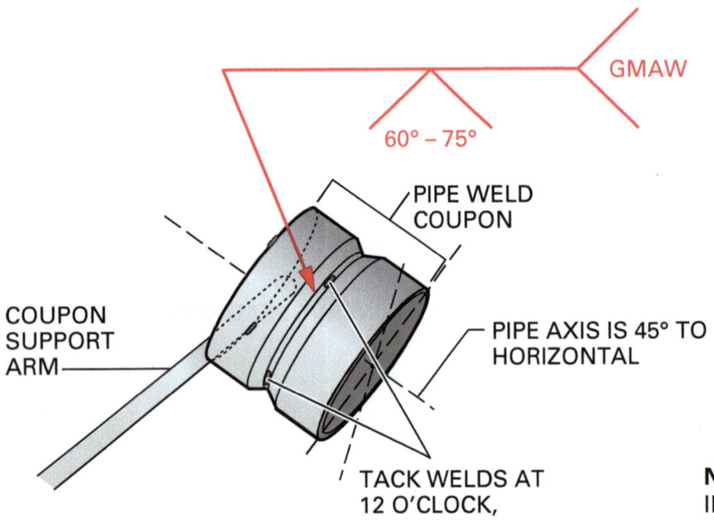

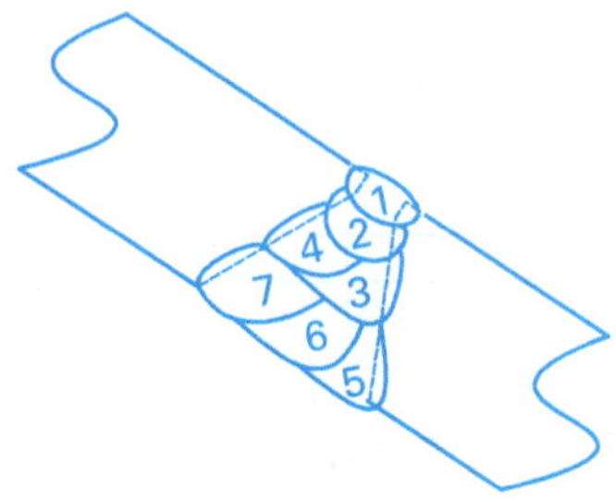

STRINGER BEAD SEQUENCE

NOTE:
IF REQUIRED FOR QUALIFICATION PURPOSES, A RESTRICTING RING MAY BE ADDED TO THE 6G POSITION COUPON TO FORM A 6GR POSITION COUPON.

Criteria for Acceptance:

- Uniform appearance on the bead face ____________
- Craters and restarts filled to the full cross section of the weld ____________
- Acceptable weld profile in accordance with the applicable codes and standards ____________
- Smooth transition with complete fusion at the toes of the weld ____________
- Complete joint penetration and face reinforcement with the root reinforcement at least flush with the inside of the pipe, to a maximum buildup of ⅛" (3 mm) ____________
- No porosity ____________
- No undercut ____________
- No cracks ____________
- No overlap ____________
- No incomplete fusion ____________

Additional Resources

This module presents thorough resources for task training. The following resource material is recommended for further study.

AWS F4.1: Recommended Safe Practices for Preparation for Welding and Cutting of Containers and Piping. Miami, FL: American Welding Society.

Modern Welding Technology. Current Edition. Howard B. Cary and Scott Helzer. Englewood Cliffs, NJ: Prentice Hall.

OSHA 1926.351, Arc Welding and Cutting. Current Edition. Washington, DC: Occupational Safety & Health Administration.

The Procedure Handbook of Arc Welding. Current Edition. Cleveland, OH: The James F. Lincoln Arc Welding Foundation.

Welding Handbook, Volumes 1 through 4. Current Edition. Miami, FL: American Welding Society.

Welding Pressure Pipelines and Piping Systems (PDF). The Lincoln Electric Company, Cleveland, OH, USA. **www.lincolnelectric.com**. Last accessed May 29, 2015.

Gas Metal Arc Welding – Product and Procedures Selection (PDF). The Lincoln Electric Company, Cleveland, OH, USA. **www.lincolnelectric.com**. Last accessed May 29, 2015.

Figure Credits

The Lincoln Electric Company, Cleveland, OH, USA, Module Opener, Figures 2–4, 11

Topaz Publications, Inc., Figures 5, 6 (photo), 20 (photo), 21 (photo), 22 (photo), SA03

Terry Lowe, Figure 7 (photo)

G.A.L. Gage Co., SA01

Courtesy of Sumner Manufacturing Co., Figure 16

Atlas Welding Accessories, Inc., Figure 17

Section Review Answer Key

Answer	Section Reference	Objective
Section One		
1. d	1.1.0	1a
2. d	1.2.3	1b
3. c	1.3.2	1c
4. a	1.4.2	1d
Section Two		
1. c	2.1.0	2a
2. a	2.2.0	2b
3. d	2.3.1	2c

NCCER CURRICULA — USER UPDATE

NCCER makes every effort to keep its textbooks up-to-date and free of technical errors. We appreciate your help in this process. If you find an error, a typographical mistake, or an inaccuracy in NCCER's curricula, please fill out this form (or a photocopy), or complete the online form at **www.nccer.org/olf**. Be sure to include the exact module ID number, page number, a detailed description, and your recommended correction. Your input will be brought to the attention of the Authoring Team. Thank you for your assistance.

Instructors – If you have an idea for improving this textbook, or have found that additional materials were necessary to teach this module effectively, please let us know so that we may present your suggestions to the Authoring Team.

NCCER Product Development and Revision

13614 Progress Blvd., Alachua, FL 32615

Email: curriculum@nccer.org

Online: www.nccer.org/olf

❑ Trainee Guide ❑ Lesson Plans ❑ Exam ❑ PowerPoints Other ____________

Craft / Level: Copyright Date:

Module ID Number / Title:

Section Number(s):

Description:

Recommended Correction:

Your Name:

Address:

Email: Phone:

FCAW – Pipe

Overview

The flux-core arc welding (FCAW) process is excellent for pipe welding, although the root pass is seldom applied using FCAW. FCAW is a versatile process than can be done with or without a shielding gas, depending upon the wire electrode used. This module describes the two primary approaches to FCAW and provides guidance to practice the techniques.

Module 29303

Trainees with successful module completions may be eligible for credentialing through the NCCER Registry. To learn more, go to **www.nccer.org** or contact us at 1.888.622.3720. Our website has information on the latest product releases and training, as well as online versions of our *Cornerstone* magazine and Pearson's product catalog.

Your feedback is welcome. You may email your comments to **curriculum@nccer.org**, send general comments and inquiries to **info@nccer.org**, or fill in the User Update form at the back of this module.

This information is general in nature and intended for training purposes only. Actual performance of activities described in this manual requires compliance with all applicable operating, service, maintenance, and safety procedures under the direction of qualified personnel. References in this manual to patented or proprietary devices do not constitute a recommendation of their use.

29303 V5

29303
FCAW – Pipe

Objectives

When you have completed this module, you will be able to do the following:

1. State the basic concepts of FCAW of pipe and how to prepare for welding.
   a. State the basic concepts of FCAW.
   b. Identify common welding safety practices.
   c. Explain how to prepare the area and materials for FCAW pipe welding.
   d. Explain how to prepare the equipment for FCAW pipe welding.
2. Describe V-groove pipe welding positions and FCAW techniques.
   a. Identify and describe the four pipe welding positions and acceptable weld profiles.
   b. Describe the technique for FCAW of the root and fill passes.
   c. Describe the techniques required to produce FCAW pipe welds in all positions.

Performance Tasks

Under the supervision of your instructor, you should be able to do the following:

1. Make FCAW open-root V-groove pipe welds in the 1G-ROTATED position.
2. Make FCAW open-root V-groove pipe welds in the 2G position.
3. Make FCAW open-root V-groove pipe welds in the 5G position.
4. Make FCAW open-root V-groove pipe welds in the 6G position.

Trade Terms

Flux-cored wire electrode

Industry Recognized Credentials

If you are training through an NCCER-accredited sponsor, you may be eligible for credentials from NCCER's Registry. The ID number for this module is 29303. Note that this module may have been used in other NCCER curricula and may apply to other level completions. Contact NCCER's Registry at 888.622.3720 or go to **www.nccer.org** for more information.

Contents

Figures and Tables

SECTION ONE

1.0.0 FCAW PIPE WELDING AND PREPARATIONS

Objective

State the basic concepts of FCAW of pipe and how to prepare for welding.

a. State the basic concepts of FCAW.
b. Identify common welding safety practices.
c. Explain how to prepare the area and materials for FCAW pipe welding.
d. Explain how to prepare the equipment for FCAW pipe welding.

Trade Terms

Flux-cored wire electrode: A wire electrode that is tube-like and filled with flux and/or degassing components with other elements to improve the chemical and mechanical characteristics of the weld.

Flux-cored arc welding (FCAW) is an arc-welding process that uses a continuously fed **flux-cored wire electrode**. FCAW wire electrodes provide the filler metal for the weld while also providing varying levels of protection during the welding process. This section introduces and compares the two primary FCAW processes and describes how to prepare for welding.

1.1.0 Basic FCAW Pipe Welding Concepts

Common welding processes require that the weld zone be protected from the atmosphere and the oxygen it contains while the weld is in progress. In SMAW for example, the coating of flux on the outside of the rod provides the needed protection. For GMAW, protection is provided by a consistent flow of shielding gas that displaces air, pushing it away from the weld zone. The GMAW wire electrode may be made from a composite material, but flux is not integrated into the wire.

FCAW, like the other processes, requires weld-zone protection. The protection can be provided one of two ways. In the self-shielded FCAW process (referred to as *FCAW-S*), the wire electrode provides a sufficient amount and type of flux to completely protect the weld zone. Like an SMAW electrode, the required shielding can be generated entirely from the flux portion of the electrode.

It is important to note that an external shielding gas should never be used with the FCAW-S process. The materials that comprise the flux core actually rely on the surrounding atmosphere for the needed chemical reactions. Using a shielding gas displaces the atmosphere, interrupting the chemical reactions and causing an inferior weld. Think of the flux-cored wire electrode as an SMAW electrode turned inside-out. The FCAW-S process is especially well-suited to outdoor environments, where wind can disturb the protective cloud of a shielding gas.

A second FCAW process does require a shielding gas, but uses a different type of flux-cored wire electrode. This process is referred to as *gas-shielded FCAW*, or *FCAW-G*. Flux is still a component of the electrode, but it has a different composition. FCAW-G flux-cored wire electrodes do require a shielding gas to be used. Both processes are shown in *Figure 1*.

In all types of flux-cored wires, the flux contains agents for degassing the weld, stabilizing the arc, and forming a protective slag over the weld bead. Other elements are often added to the flux-cored wire to alloy with the base metal and improve its chemical and/or mechanical characteristics. The proper electrode diameter is related to metal thickness and the welding machine in use. A smaller wire makes it easier to weld thinner metal.

Flux-Cored Wire Electrodes

The FCAW-S and FCAW-G processes require different types of flux-cored wire electrodes as described in this module. Electrode manufacturers typically use different product names to help users easily identify the proper products. The Lincoln Electric Company, for example, has three primary flux-cored wire electrode products. The Innershield® product line is designed for FCAW-S, while the Outershield® and UltraCore® products are designed for FCAW-G. ESAB provides the Core-Bright® and Coreshield® product lines for FCAW-S, and the Dual-Shield® product line for FCAW-G. It is essential that welders become familiar with the available products and ensure the correct electrode is chosen for the FCAW process in use.

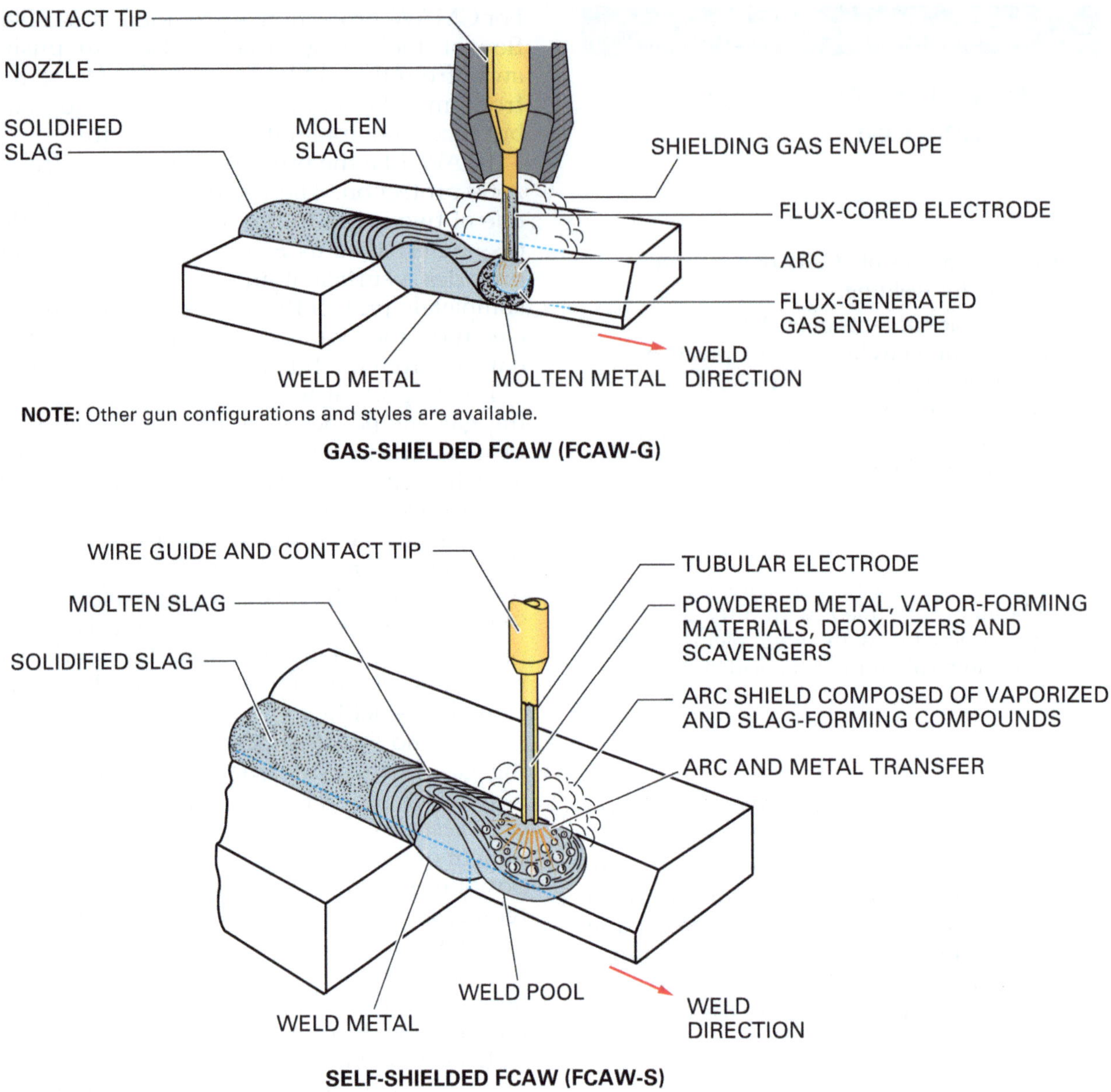

Figure 1 FCAW process.

The equipment used for FCAW-G and FCAW-S processes is basically the same, except for the components needed to apply the shielding gas in FCAW-G. Like GMAW, this additional equipment includes guns equipped with a gas nozzle, gas hoses, a gas control solenoid valve, a gas regulator/flowmeter, and a shielding gas source. *Figure 2* and *Figure 3* show typical FCAW equipment.

The FCAW process is generally limited to welding low-carbon steels, some low-alloy steels, and some stainless steels. FCAW is an excellent process for groove welds on pipe.

Wind Resistance

FCAW-S is very useful in high-wind situations because the shielding gas is created inside the welding arc as the flux vaporizes. The molten puddle has a constant supply of shielding gas arising from the weld zone. The gas displaces the air from the arc outward, and the puddle is immediately covered by the residual slag. In contrast, an external shielding gas attempts to force gas down from the nozzle to displace the air, but the gas can be blown away from the weld puddle by wind gusts.

Figure 2 Gas-shielded FCAW (FCAW-G) equipment.

1.2.0 Safety Practices

Safety considerations are a critical part of preparing for a welding operation. The following is a summary of safety procedures and practices that must be observed while cutting or welding.

Figure 3 Self-shielded FCAW (FCAW-S) equipment.

Be advised that this is only a summary. Complete safety coverage is provided in the *Welding Level One* module, "Welding Safety." That module should be studied, if not already completed, before continuing. Above all, be sure to wear appropriate protective clothing and equipment at all times when welding or cutting.

1.2.1 Protective Clothing and Equipment

Protective clothing and equipment are a necessary part of every welding task. Follow the guidelines below as a general rule, and ensure that you are familiar with any unique requirements your employer may have regarding personal protective equipment (PPE).

- A welding hood equipped with a properly shaded lens is best for all forms of welding. The welding hood lens must be an appropriate shade for the application.
- Wear proper leather or flame-retardant clothing along with welding gloves for protection from flying sparks, molten metal, and heat.
- Wear high-top safety shoes or boots. Make sure that the tongue and lace area of the footwear is covered by a pant leg. If the tongue and lace area is exposed or if the footwear must be protected from burn marks, wear spats under the pant leg or leather chaps over the front and top of the footwear.
- Wear a 100-percent cotton cap without any mesh material. If a hard hat is required for the environment, use one that allows the attachment of rear deflector material and a face shield. A hard hat with a rear deflector is

generally preferred when working overhead, and may be required by some employers and job sites.
- Wear earmuffs, or at least earplugs, to prevent sparks from entering the ear canals.

1.2.2 Housekeeping and Fall Protection

Welding is often performed outside in the weather and in the construction environment. Production work done inside large, open shops exposes a welder to equipment moving across the floor and overhead. Welders often have to crawl under or reach over objects to complete their work.

At times, welders must work at height. Fall protection equipment is required for all work at height (*Figure 4*). Always keep fall protection devices untangled and properly attached. Combined with the other equipment a welder must carry, there is a great deal to manage, but fall protection equipment must always be the priority. Fall protection equipment saves lives. Never attach any kind of welding gear to the harness that could interfere with its ability to perform its intended function.

Regardless of where you are working, always keep the work area as neat as possible. This makes the area safer and helps to reduce distractions. Mindful actions, such as making sure that welding cables are not tripping hazards, contribute to a safer environment. Electrode stubs and unused electrodes must not be left on the floor. They can cause someone to lose their footing and fall. Parts waiting to be welded or left over from previous work are a tripping hazard and should be kept out of the walking area. If the parts have sharp edges, they could cut welding cable insulation or torch gas hoses.

Figure 4 Working at height.

1.2.3 Fire and Explosion Prevention

Flammable gases are commonly found around welding and cutting operations, as are sparks and other significant sources of ignition. For that reason, and because many fire-related incidents over the years have been traced back to welding operations, it is crucial that fire and explosion prevention be given consistent attention. Follow these guidelines to maintain a safe environment:

- Never carry matches or gas-filled lighters. Sparks can cause the matches to ignite or the lighter to explode in your pocket, causing serious injury.
- Never use oxygen to blow dust or dirt from clothing. Traces of the oxygen can remain trapped in the fabric for some time. If a spark hits the fabric during this time, the clothing can burn rapidly and violently out of control.
- Make sure that any flammable material in the work area is removed or shielded by a fire-resistant covering. Approved fire extinguishers must be available before attempting any heating, welding, or cutting operations. If a hot-work permit and a fire watch are required, be sure these are in place before beginning.
- Ensure that all site safety requirements are being observed.
- Never release a large amount of oxygen or use oxygen in place of compressed air. Its presence around flammable materials or sparks can cause rapid and uncontrolled combustion. Keep oxygen away from oil, grease, and other petroleum products.
- Never release a large amount of fuel gas, especially acetylene. Propane tends to concentrate in and along low areas and can ignite at a considerable distance from the release point. Acetylene is lighter than air, but is even more dangerous than propane. When mixed with air or oxygen, acetylene will ignite at much lower concentrations than any other fuel gas.
- Even though shielding gases are not toxic, any gas released in a sufficient quantity can displace the oxygen in a small room or enclosed work area, creating an atmosphere with little or no oxygen. Avoid large releases of any type of gas, regardless of its characteristics.
- To prevent fires, maintain a neat and clean work area, and make sure that any metal scrap or slag is cold before disposal.
- Before cutting containers such as tanks or barrels, check to see if they have contained any explosive, hazardous, or flammable materials, including petroleum products, citrus products, or chemicals that decompose into toxic fumes when heated. As a standard practice, always clean and then fill any tanks or barrels with

water, or purge them with a flow of inert gas such as nitrogen to displace any oxygen. Containers must be cleaned by steam cleaning, flushing with water, or washing with detergent until all traces of the material have been removed. Proper procedures for cutting or welding hazardous containers are described in the *American Welding Society (AWS) F4.1, Safe Practices for the Preparation of Containers and Piping for Welding and Cutting.*

WARNING!

Welding or cutting must never be performed on drums, barrels, tanks, vessels, or other containers until they have been emptied and cleaned thoroughly, eliminating all flammable materials and all substances (such as detergents, solvents, greases, tars, or acids) that might produce flammable, toxic, or explosive vapors when heated. Clean containers only in well-ventilated areas. Vapors can accumulate during cleaning, causing explosions and serious personal injury.

1.2.4 Work Area Ventilation

Many gases are introduced to the atmosphere through welding processes. Some of these gases are harmful, and all can displace or combine with atmospheric oxygen in a confined space. Follow these guidelines, as well as those of your employer, for work area ventilation:

- Make sure to follow all confined space procedures before conducting any welding or cutting in a confined space. Use powered extraction systems as needed.
- Never use oxygen to provide ventilation in confined spaces.
- Always perform cutting or welding operations in a well-ventilated area.

Cutting or welding operations involving zinc or cadmium materials or coatings result in toxic fumes. Another significant concern is Chromium 6, commonly referred to as *hexavalent chromium,* which is used in stainless steel. It releases toxic fumes when heated. Manganese is another hazardous metal used in stainless steel. It is also used in some welding rods, increasing the risk of exposure.

For long-term cutting or welding of materials containing these components, always wear an approved full-face, supplied-air respirator that uses breathing air supplied from outside the work area. For occasional short-term exposure, a HEPA-rated or metal-fume filter may be used on a standard respirator.

Powered fume-extraction equipment must be used when welding materials that contain zinc, cadmium, hexavalent chromium, manganese, and similar materials that produce hazardous fumes if the area is not naturally well-ventilated. *Figure 5* shows examples of both mobile and portable units.

1.3.0 Preparations for Welding

Before any welding procedure can take place, the work area must be made ready, the welding equipment must be properly set up, and the metal to be welded must be prepared.

1.3.1 Preparing the Welding Area

To practice welding, a welding table, bench, and/or appropriate pipe stands are needed. The welding surface must be made of steel. Provisions must be made for supporting the pipe welding coupons for out-of-position welding.

To set up the area for welding, follow these steps:

Step 1 Make sure that the area is properly ventilated. Make use of doors, windows, and fans.

Step 2 Check the area for fire hazards. Remove any flammable materials before proceeding.

Step 3 Know the location of the nearest fire extinguisher. Ensure the fire extinguisher is charged and that you and others in the immediate area know how to use it properly.

Classes of Fire Extinguishers

Fire extinguishers have different class ratings based on the types of fires they are used to extinguish. For example, Class A fire extinguishers are used for common materials like paper, wood, and plastic. Class B fire extinguishers are used for liquids such as gasoline and oil. Class C fire extinguishers are used for electrical equipment fires, and Class D extinguishers are used for combustible metals. Always make sure the fire extinguishers available on site are suitable for the types of fires that might occur.

(A) MOBILE UNIT

(B) PORTABLE UNIT

Figure 5 Mobile and portable extraction equipment.

Step 4 Set up flash shields and curtains around the welding area as necessary.

1.3.2 Preparing Pipe Weld Coupons

Pipe weld coupons should be cut from 3" to 12" (DN80 to 300) diameter Schedule 40 or Schedule 80 carbon steel pipe. Each welded joint requires two pipe coupons of the same diameter and schedule.

> **NOTE**
> Most codes allow you to substitute carbon steel pipe for stainless or low-alloy steel pipe as a base metal for welder qualification. However, you must use applicable stainless or low-alloy steel filler metals. Always refer to the WPS or site quality standards for the proper qualification information.

For V-groove joints, machined bevels and root faces provide the cleanest and most uniform way to prepare the bevel for FCAW welds. This is especially true for stainless steel, from which all contaminants from thermal cutting must be removed. *Figure 6* shows typical ASME, American Petroleum Institute (API), and AWS pipe bevel specifications for bevel angles, root faces, and root openings.

To prepare weld coupons for open-root V-groove weld joints, follow these steps:

Step 1 Clean all rust or mill scale from the inside and outside of the pipe with a grinder, wire brush, or other common method.

Step 2 Bevel the end of the pipe to 30 to 37½ degrees by any acceptable beveling method, including manual, mechanical, and thermal methods.

Step 3 Cut off the end of the beveled pipe to the length specified by your instructor. Two coupons will be required for each weld.

> **NOTE**
> For 1G-ROTATED welding, the coupons may need to be cut longer to reach the rollers being used.

The two coupons can now be further prepared for welding:

Step 1 Check the bevel. There should be no dross, and the bevel angle should be 30 to 37½ degrees.

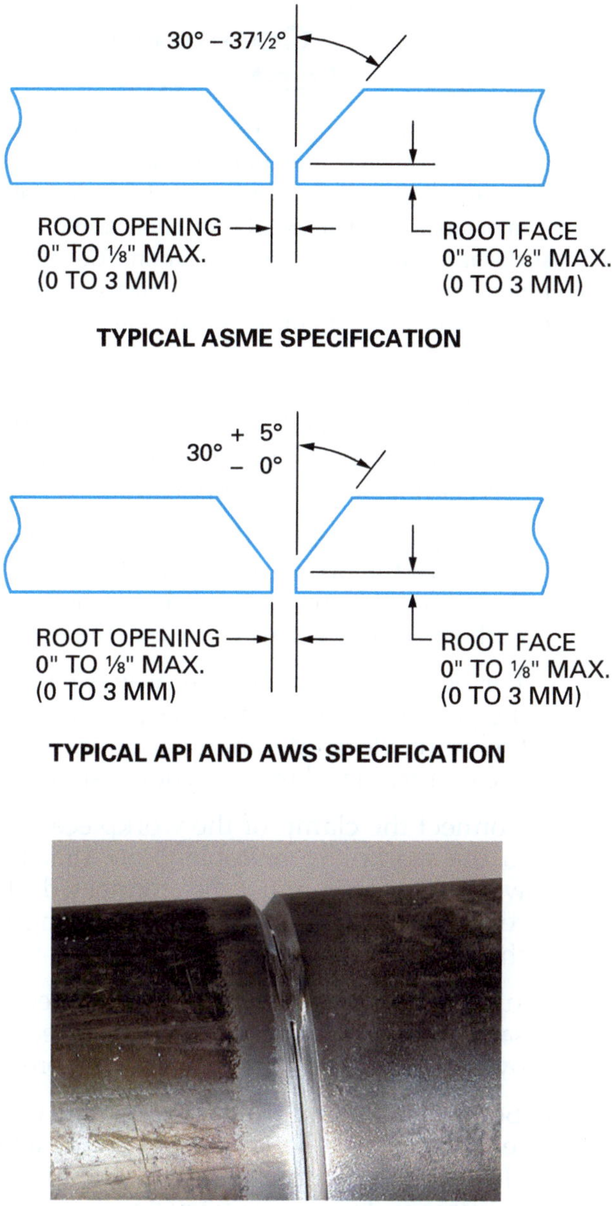

Figure 6 ASME, API, and AWS pipe bevel specifications.

Step 2 Grind or file a 0" to ⅛" (0 to 3 mm) root face on the bevel.

NOTE

Welding codes allow the root face for open-root welds to be from 0" to ⅛" (0 to 3 mm) wide and the root opening to be from 0" to ⅛" (0 to 3 mm). Select and adjust the root face and opening as directed by the instructor as welding practice begins.

Step 3 Align the two pipe sections. The inside diameter (ID) surfaces of the two coupons should be even all around and evenly separated by the required root opening. Pieces of filler wire, metal shims, or pieces of bare welding electrode can be used to achieve the desired root opening between the two pipe sections until they are tack welded. Small-diameter pipe coupons can be aligned by clamping both pieces to a piece of angle. Large-diameter pipe can be aligned using a pipe alignment jig or by holding a straightedge across the joint, parallel to the pipe axis. The straightedge must be used to check all around the pipe in case one or both sections are distorted.

Step 4 Once aligned, secure the two pipe sections with a clamp, or with tack welds if necessary.

Step 5 With the root opening established and the pipe ends aligned, make the first of four tack welds using the GMAW, GTAW, or SMAW process. The GMAW and SMAW processes are typically preferred. The root pass, and therefore the tack welds, should not be done using FCAW. Longer tacks may be required on heavy-wall or large-diameter pipe.

CAUTION

Due to a lack of weld control, FCAW is not typically allowed for the root pass of an open-root V-groove pipe weld. The necessary tack welds, as well as the root pass, should be done using GMAW, GTAW, or SMAW. GMAW and SMAW are generally preferred for root passes. FCAW for the root pass may be allowed in some specifications, generally for noncritical weld applications.

Step 6 After the first tack weld, check the root opening on the opposite side. Adjust the root opening space if necessary, and make the second tack weld on the side opposite the first tack.

CAUTION

Remove the spacer material before making the second tack, so that it will not get caught as the heated metal expands.

Step 7 Verify that the root opening remains correct before welding the third tack midway between the first and second tacks.

Step 8 Weld the fourth tack opposite the third tack and midway between the first and second tacks. There should now be four tack welds evenly spaced (90 degrees apart) around the pipe coupon (*Figure 7*).

Step 9 Clean and feather the tack welds with the edge of a grinding wheel as shown in *Figure 7*. Feathering the ends of the tack welds with a grinder helps to fuse the root pass into the tack weld.

1.4.0 Preparing FCAW Equipment

Select a proper welding machine for FCAW, and follow these steps to set it up for use:

Step 1 Verify that the welding machine can be used for FCAW, with or without internally controlled gas shielding.

Step 2 Verify the location of the primary power disconnect.

Step 3 Configure the welding machine for FCAW-S or FCAW-G parameters as directed by your instructor (*Figure 8*). Configure the polarity. The polarity of the gun and workpiece leads is dependent upon the application and the filler metal in use. Equip the gun with the correct liner material, contact tube for the diameter of the wire being used, and the correct nozzle for the application.

Step 4 Configure and load the wire feeder and gun with the proper type and diameter of flux-cored electrode wire in accordance with the manufacturer's instructions.

Step 5 For FCAW-G, connect the proper shielding gas for the application as specified by the wire electrode manufacturer, WPS, site quality standards, or your instructor.

Step 6 Connect the clamp of the workpiece lead to the workpiece or table. Note that the workpiece lead must be connected directly to the workpiece when welding in the 1G-ROTATED position.

Step 7 Turn on the welding machine. If necessary, purge the gun gas line as directed by the gun manufacturer's instructions.

Step 8 Set the initial welding voltage and wire feed speed for the type and size of wire electrode as recommended by the manufacturer and/or your instructor.

Figure 9 shows a typical gun configuration with dimensions.

1.4.1 Welding Voltage

Arc length is determined by voltage, which is set at the welding machine. Arc length is the distance from the wire electrode tip to the base metal or molten pool of the base metal. If the voltage is set too high, the arc will be too long. This could cause the wire to melt and fuse to the contact tube. A voltage that is too high can also cause porosity and excessive spatter. Voltage must be increased or decreased accordingly as the wire feed speed is increased or decreased. Set the voltage according to the manufacturer's recommendations.

GOING GREEN

Conserve Materials

Pipe for practice welding is expensive and hard to obtain. Make every effort to conserve the material available. Reuse weld coupons until all surfaces have been welded upon. Cut the coupon apart and reuse the pieces. Check with the instructor for the appropriate size of coupon.

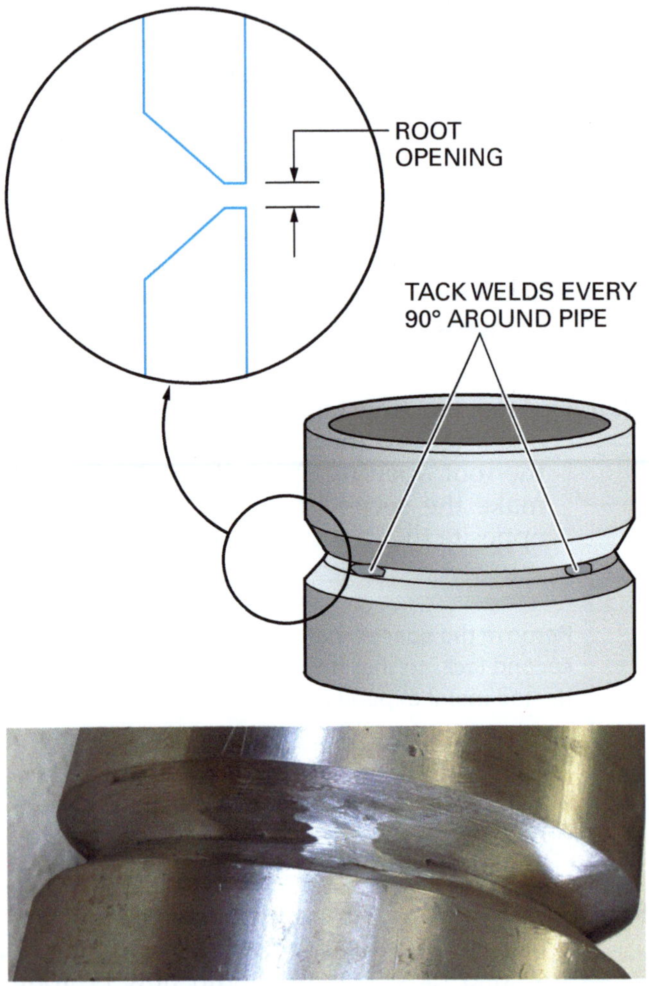

Figure 7 Tacked open-root V-groove weld coupon.

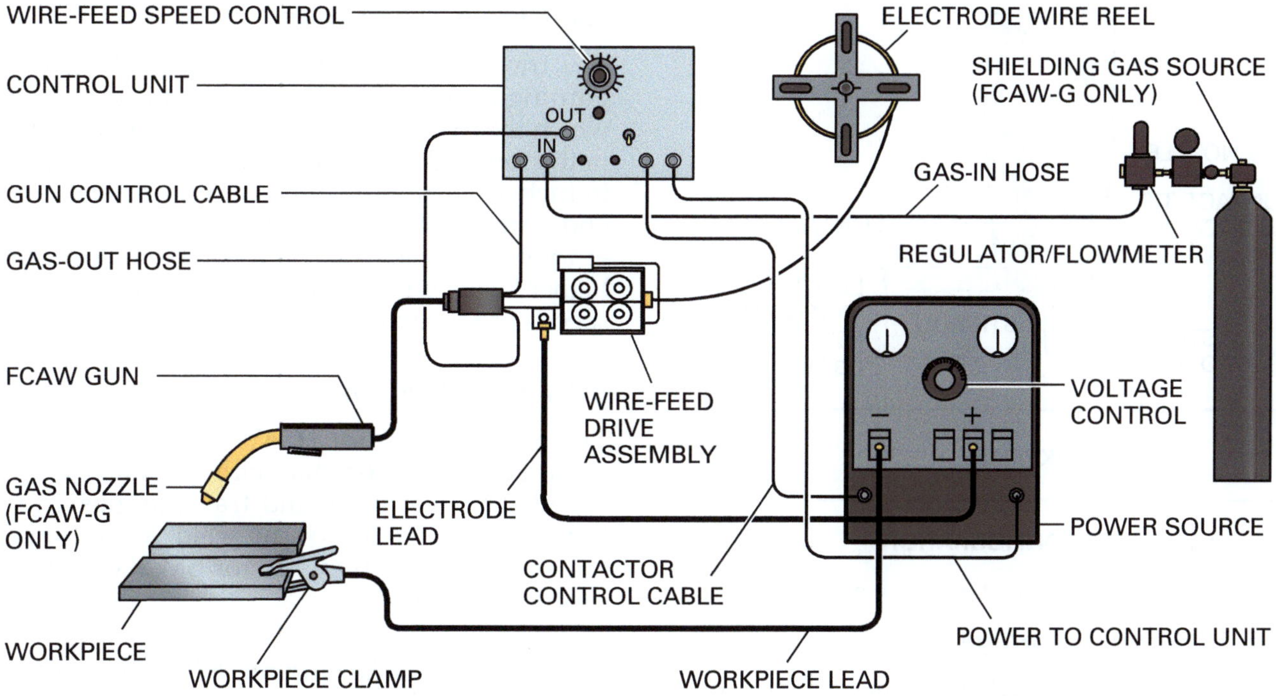

Figure 8 Typical FCAW equipment configuration diagram.

Underwater Welding

Combining exceptional welding skills with the skills of a commercial diver can produce an exciting and lucrative career. Wet welding, defined as welding at an elevated pressure when no barrier exists between the water and the welding arc, has many applications in the marine environment. Underwater welders are required to pass a Welder Performance Qualification (WPQ) as prescribed by *AWS D3.6:2010, Underwater Wet Welding.* Underwater burning can also be added to the skill set, further broadening the scope of opportunities available to individuals who can master the diverse skills of diving and welding.

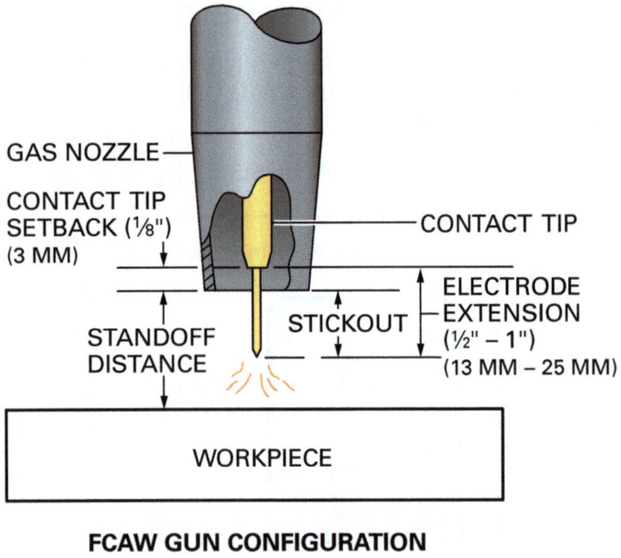

Figure 9 FCAW gun configuration and dimensions.

1.4.2 Welding Amperage

When using a standard constant-voltage welding machine, the wire feed speed controls the welding current (amperage) after the initial recommended amperage setting has been made by the welder. The welding machine provides the current to melt the wire electrode while maintaining the selected welding voltage. Within limits, when the wire feed speed is increased, the welding current and deposition rate are also increased. This results in higher welding heat, deeper penetration, and higher beads. When the wire feed speed is decreased, the welding current automatically decreases. With lower welding current and less heat, the deposition rate drops, and the weld beads are smaller with less penetration.

Note that some constant-voltage welding machines used for FCAW provide some degree of current modification. Standard constant-voltage FCAW units have a current slope that is fixed by the manufacturer for general welding applications and conditions.

1.4.3 Weld Travel Speed

The travel speed must be considered before beginning a weld. Weld travel speed is the speed at which the electrode tip passes across the base metal in the forward direction of the weld. It is measured in inches per minute (ipm) or millimeters per second (mm/s). Travel speed has a great effect on penetration and bead size. Slower travel speeds build higher beads with deeper penetration. Faster travel speeds produce smaller beads with less penetration.

1.4.4 Gun Position

The gun position for FCAW has a greater effect on weld penetration than either voltage or travel speed. The gun work and travel angles, in relation to the direction of the weld, are defined in the same way as they are for SMAW and other processes (*Figure 10*).

1.4.5 Electrode Extension, Stickout, and Standoff Distance

Electrode extension is the length of the wire that extends beyond the end of the welding gun contact tip. Changing the extension affects the welding current for low-conductivity metal wire electrodes. This is because the extension distance changes the preheating of the wire caused by its resistance to current flow. As the extension increases, wire preheating increases. Because the welding machine is self-regulating, it does not have to furnish as much welding current to melt the wire, so the current output automatically decreases. This results in less penetration.

Increasing extension is useful for bridging gaps and compensating for mismatch. However, it can cause overlap or a lack of fusion. When the extension is decreased, the power source is forced to increase its current output to burn off the wire. Too little extension can cause the wire to weld itself to the contact tip, and porosity can develop in the weld. For high-conductivity metal wires, the preheating effect of wire resistance is minimal. Wire speed and voltage settings have a more direct effect on penetration and deposition rate.

Wire Electrode Manufacturer's Recommendations

New welders are tempted to turn down the wire feed speed if they have difficulty controlling the weld puddle. Always follow the manufacturer's recommendations for shielding gas (if required) and the initial setting of the welding voltage and wire feed speed parameters. Voltage, wire feed speed, and travel speed are adjusted and balanced together to control the weld puddle. Adjust the technique using the proper settings, rather than attempting to adjust the machine to support an inferior or incorrect technique.

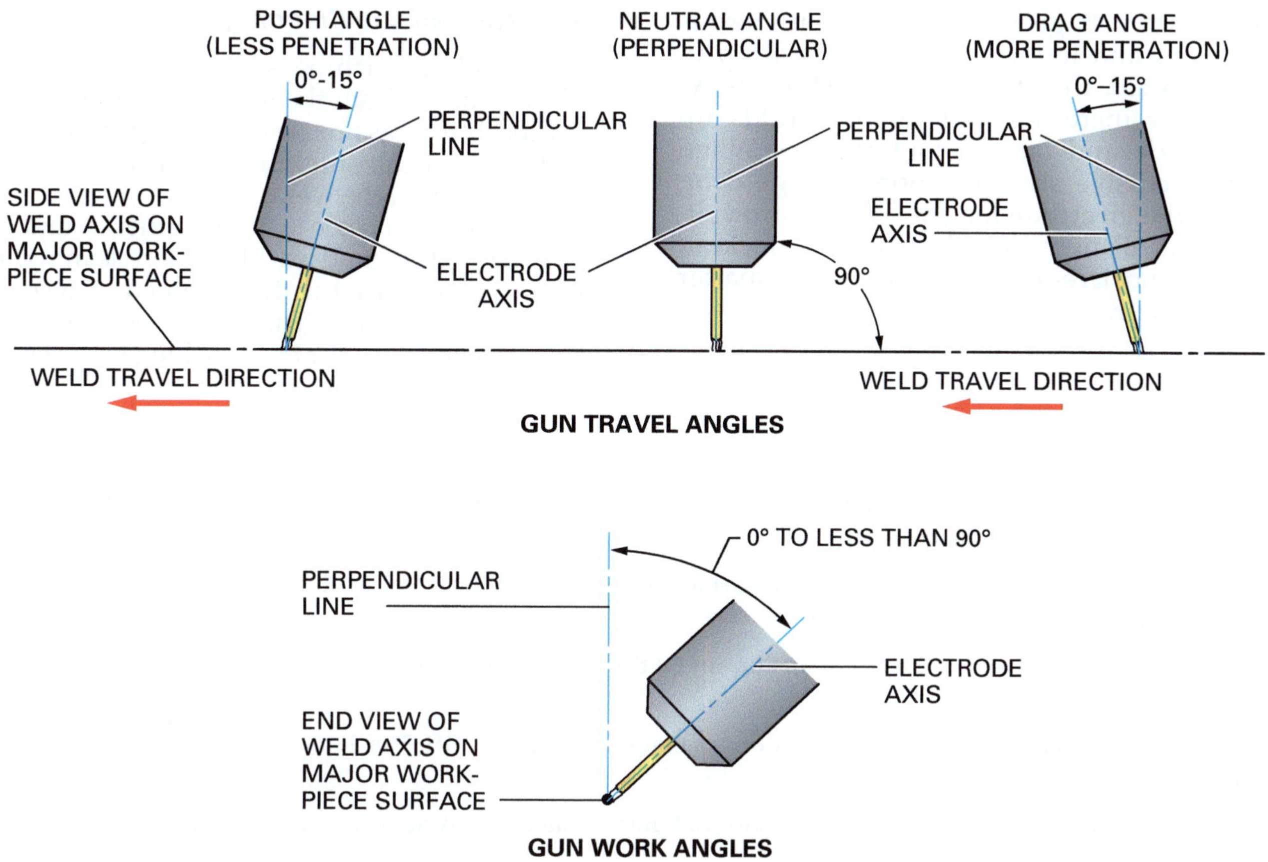

Figure 10 Gun travel and work angles.

The standard electrode extension for FCAW-S using a 0.045" (1.1 mm) wire electrode with an insulating nozzle is ¾" to 1¼" (19 to 32 mm), depending on the welding position (*Figure 11*). Check the manufacturer's specifications for specific extension length recommendations. Standoff distance is the distance from the gas nozzle or insulating nozzle to the workpiece. Stickout is the distance from the gas nozzle or insulating nozzle to the end of the electrode.

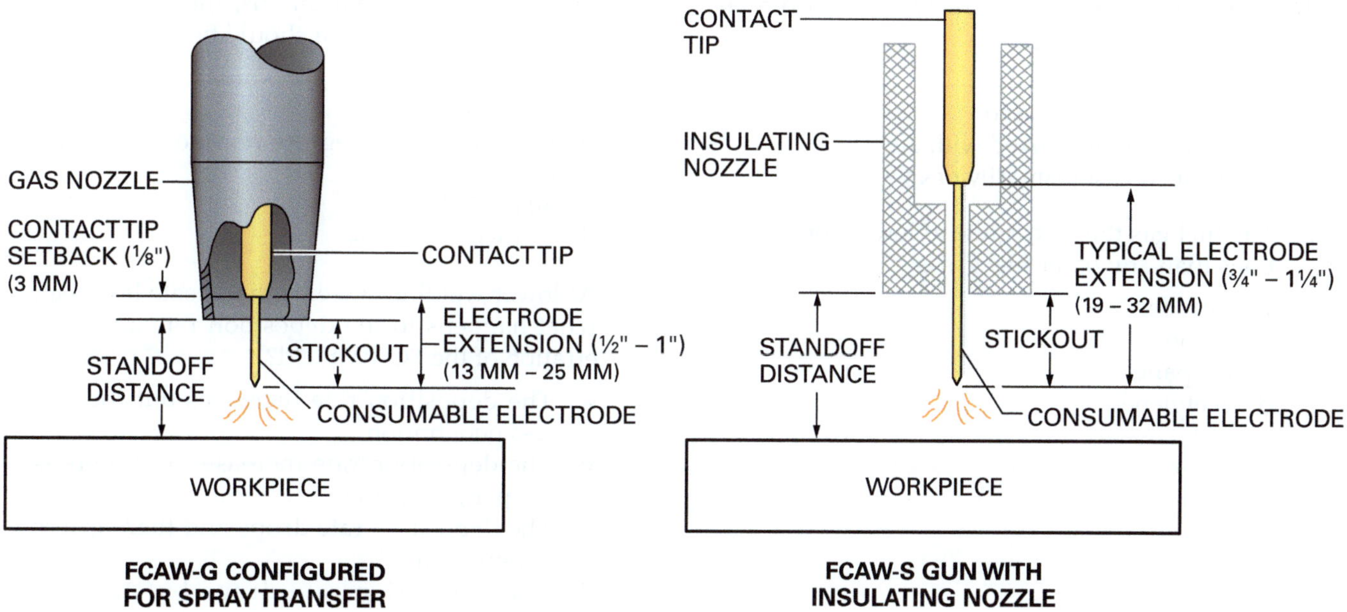

Figure 11 Electrode extensions and gun configurations.

Higher arc voltage is available when there is less stickout and less resulting voltage drop in the wire. When the stickout is too short, however, spatter increases and rapidly builds up in the nozzle and on the contact tube. When there is excessive stickout, more voltage drop occurs in the electrode length, and the arc voltage is lower. Penetration decreases as stickout increases, and irregular arc action and spatter also may occur.

1.4.6 Gas Nozzle Cleaning

As the welding machine is being used, weld spatter builds up on both the gas nozzle and the contact tube. If the gas nozzle is not properly cleaned, it will restrict the shielding gas flow. This will cause porosity in the weld.

Clean the gas nozzle with a reamer, round file, or the tang of a standard file. After cleaning, the nozzle can be sprayed with, or dipped in, a special anti-spatter compound. The anti-spatter compound helps prevent the spatter from sticking to the nozzle.

Additional Resources

AWS F4.1: Recommended Safe Practices for Preparation for Welding and Cutting of Containers and Piping. Miami, FL: American Welding Society.

Modern Welding Technology. Current Edition. Howard B. Cary and Scott Helzer. Englewood Cliffs, NJ: Prentice Hall.

OSHA 1926.351, Arc Welding and Cutting. Current Edition. Washington, DC: Occupational Safety & Health Administration.

The Procedure Handbook of Arc Welding. Current Edition. Cleveland, OH: The James F. Lincoln Arc Welding Foundation.

Welding Handbook, Volumes 1 through 4. Current Edition. Miami, FL: American Welding Society.

Welding Pressure Pipelines and Piping Systems (PDF). The Lincoln Electric Company, Cleveland, OH, USA. **www.lincolnelectric.com**. Last accessed May 29, 2015.

Innershield Wire – FCAW-S Welding Guide (PDF). The Lincoln Electric Company, Cleveland, OH, USA. **www.lincolnelectric.com**. Last accessed May 29, 2015.

1.0.0 Section Review

1. In the FCAW-S process, shielding gas from an external source _____.
   a. should always be used
   b. should never be used
   c. should be pre-mixed with air
   d. must consist primarily of CO_2

2. The fuel gas that can ignite at lower concentrations than any other fuel gas is _____.
   a. nitrogen
   b. argon
   c. propane
   d. acetylene

3. To meet ASME specifications, the bevel angle on a V-groove pipe joint should be _____.
   a. no less than 30 degrees nor more than 35 degrees
   b. no less than 30 degrees nor more than 37½ degrees
   c. 30 to 45 degrees
   d. 35 to 37½ degrees

4. At lower welding currents that create less heat, what happens to the deposition rate and penetration of the FCAW weld?
   a. The deposition rate increases but there is less penetration.
   b. The deposition rate increases and there is deeper penetration.
   c. The deposition rate drops and there is less penetration.
   d. The deposition rate drops but there is deeper penetration.

SECTION TWO

2.0.0 OPEN-ROOT V-GROOVE PIPE WELDS

Objective

Describe V-groove pipe welding positions and FCAW techniques.

a. Identify and describe the four pipe welding positions and acceptable weld profiles.
a. Describe the technique for FCAW of the root and fill passes.
a. Describe the techniques required to produce FCAW pipe welds in all positions.

Performance Tasks

1. Make FCAW open-root V-groove pipe welds in the 1G-ROTATED position.
2. Make FCAW open-root V-groove pipe welds in the 2G position.
3. Make FCAW open-root V-groove pipe welds in the 5G position.
4. Make FCAW open-root V-groove pipe welds in the 6G position.

The open-root V-groove weld is the most common groove weld made on carbon, low-alloy, and stainless steel pipe. These welds are typically used for joining medium- and thick-walled pipe used in various piping systems. Piping containing or carrying material that can cause injury to people or damage to the environment, or those that operate at elevated pressures, may be categorized as critical piping. Injury or damage occurs when the piping system or other system components fail to contain the material as designed. Welds in critical piping may be required to meet stringent requirements. The welds must be expertly completed, thoroughly inspected, and at least as strong as the piping being welded. Piping that is used for hydronic heating and air conditioning applications, simple water systems, and similar applications may be assessed using less stringent requirements.

2.1.0 Pipe Groove Weld Test Positions

Groove welds may be made in all positions on pipe. The weld position is determined by the axis of the pipe (*Figure 12*). Four standard weld test positions are used with pipe:

- *Flat (1G-ROTATED)* – The pipe axis is horizontal, and the pipe is slowly rotated while it is being welded on the top. The weld beads are flat.
- *Horizontal (2G)* – The pipe axis is vertical and the pipe rotation is fixed. The weld beads are horizontal.
- *Multiple (5G)* – The pipe axis is horizontal and the pipe rotation is fixed. The weld beads are flat, vertical, and overhead.
- *Multiple inclined (6G)* – The pipe axis is inclined at approximately 45 degrees ±5 degrees. Pipe rotation is fixed. Weld beads are horizontal, vertical, and overhead.

The pipe axis positions for 1G-ROTATED, 2G, and 5G can vary ±15 degrees from the basic position. The pipe axis position for 6G can vary ±5 degrees from the nominal 45-degree angle. *Table 1* provides a quick reference for the four positions.

To determine the quality of a pipe weld, specific sections of pipe are removed for destructive testing. For positions 1G-ROTATED and 2G, two specimens are removed. They must be taken from opposite sides of the pipe, as shown in *Figure 13*. Once removed, the specimens are put through different bend tests, which reveal the quality of the welds. Alternatively, specimens may be tested by radiography, a non-destructive testing (NDT) method.

When 5G or 6G welds will be destructively tested, the test specimens must be cut from the four regions shown in *Figure 13*. These regions are midway between the twelve o'clock and three o'clock; three o'clock and six o'clock; six o'clock and nine o'clock; and nine o'clock and twelve o'clock positions. The twelve o'clock position is the top of the coupon groove; the six o'clock position is at the bottom.

2.1.1 Acceptable and Unacceptable Pipe Weld Profiles

Pipe groove welds should be made with slight face reinforcement (not exceeding ⅛" or 3 mm) and a gradual transition to the base metal at each toe. Excessively large face reinforcement reduces the pipe's strength. It causes the stresses in the pipe to be concentrated along the sides of the weld. As a result, the pipe will not expand and contract uniformly along its length.

The root reinforcement inside of the pipe may range in thickness from being flush to a maximum of ⅛" (3 mm) thick. However, the root pass must have complete penetration. Pipe groove welds must not have any underfill at the face or root. They must be free from excessive undercut and any overlap. If a weld has any of the defects shown in *Figure 14*, it must be repaired.

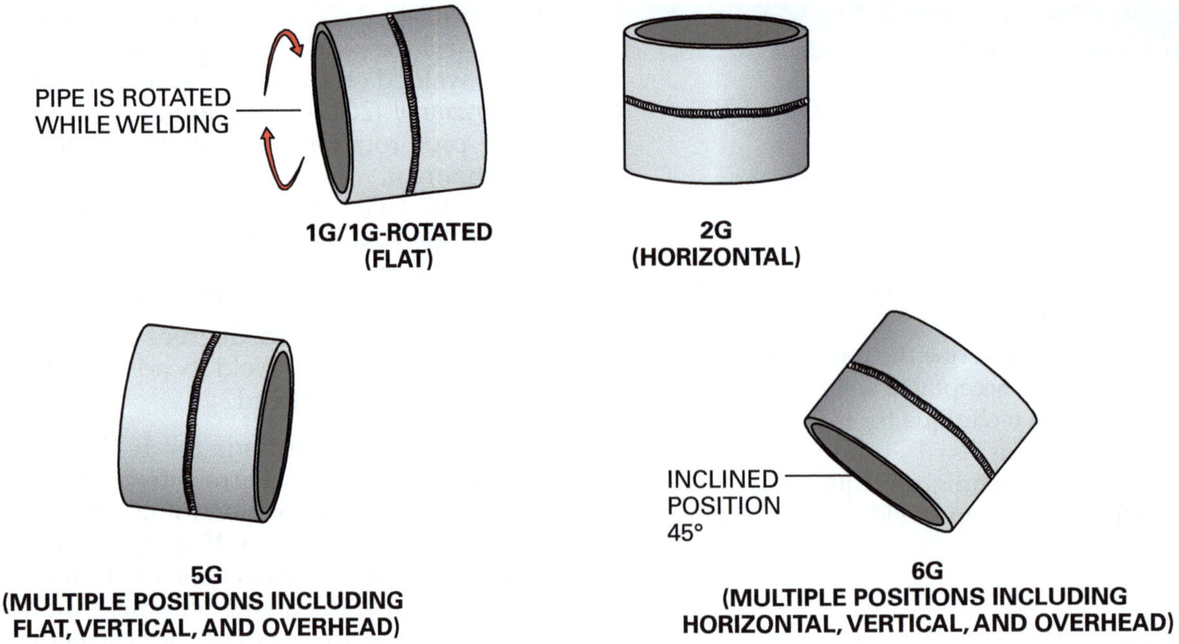

Figure 12 Four standard pipe groove-weld test positions.

Table 1 Pipe Groove-Weld Test Positions

Weld Test Position	Pipe Axis	Weld Beads	Pipe Position
1G-ROTATED	Horizontal	Flat	Slowly rotated
2G	Vertical	Horizontal	Fixed
5G	Horizontal	Flat, vertical, overhead	Fixed
6G	Inclined 45 degrees	Horizontal, vertical, overhead	Fixed

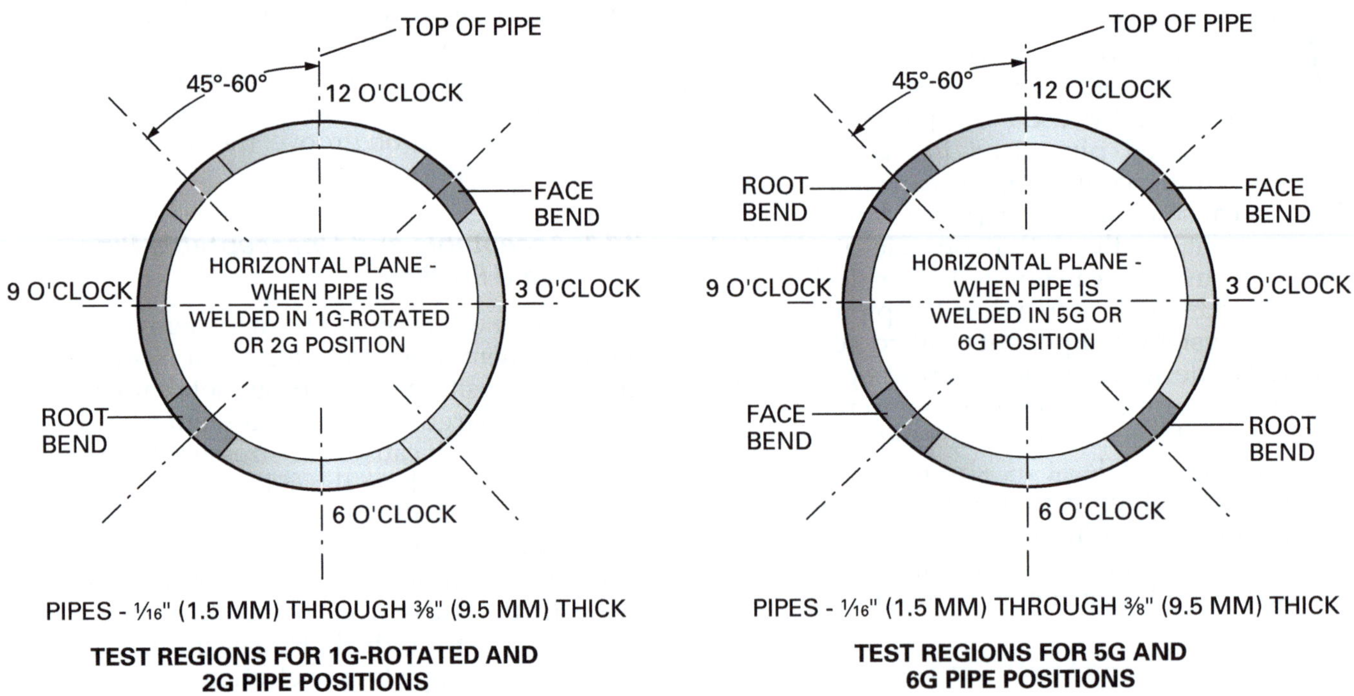

NOTE: Tack weld locations at 12, 3, 6, and 9 o'clock.

Figure 13 Test specimen regions for pipe positions 1G-ROTATED, 2G, 5G, and 6G.

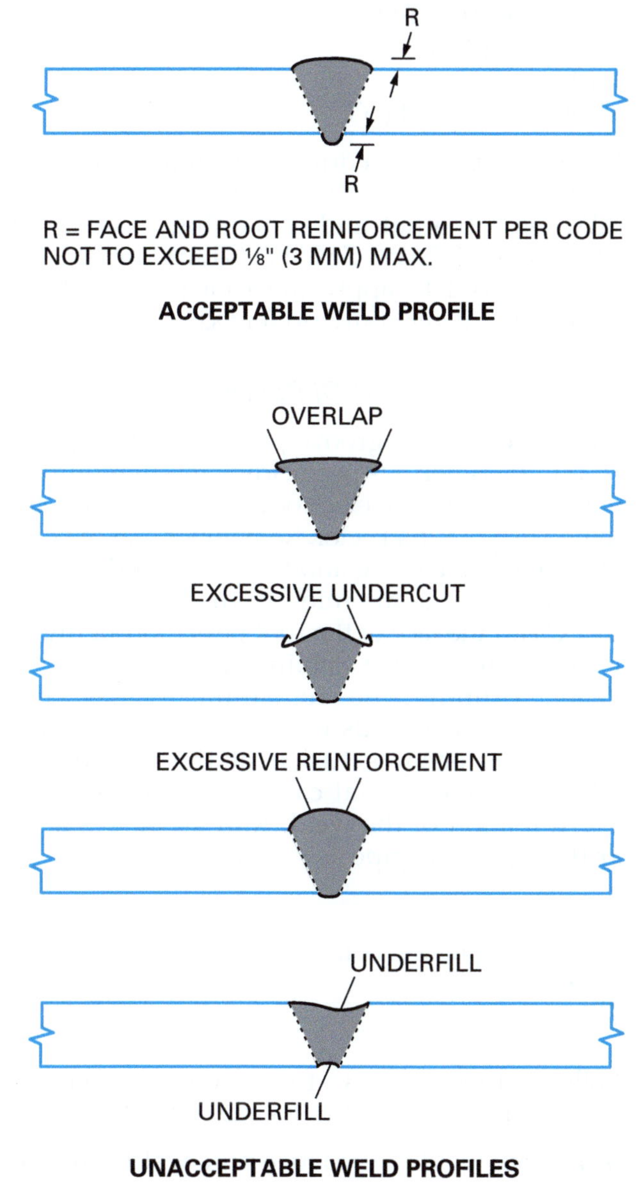

Figure 14 Acceptable and unacceptable pipe groove-weld profiles.

2.2.0 The Root Pass

The greatest challenge of an open-root V-groove weld is making the root pass. The root pass must have complete penetration and fusion, but not an excessive amount of root reinforcement. The extent of penetration and fusion is very important. If penetration is not sufficient, it will create a weak joint. If there is too much penetration along with excessive root reinforcement, the inside diameter of the pipe will be reduced and the flow restricted. Incomplete fusion will result in a defective weld.

Before proceeding further, it is important to remember that FCAW is not often used for the root pass on open-root V-groove pipe welds for critical applications. Root passes for these applications are most often done using GMAW or SMAW processes. There may be occasions however, when FCAW may be required for the root pass as part of a test or in noncritical applications.

One approach to running a keyhole that works well is to begin the root pass on top of the tack 1/8" to 3/16" (3 to 5 mm) behind the feathered portion of the tack. This helps preheat the joint for a better tie-in. When approaching the next tack, proceed past the feathered portion of the next tack to the same point. Ensure that the keyhole disappears before stopping the weld. After the root pass has been placed, it should be cleaned and inspected. Check the weld for incomplete fusion and penetration. The root reinforcement should be maintained at no more than 1/8" (3 mm).

Remove any excess buildup or undercut by grinding the face of the root pass with the edge of a grinding disk. Be careful not to grind through the root pass or widen the groove. A narrow grinding wheel (not to be confused with a cutoff wheel) is effective for this purpose.

6GR Test Position

Welding pipe in the 6G position is a common welding test. A ring may be added to test for restricted accessibility (6GR).

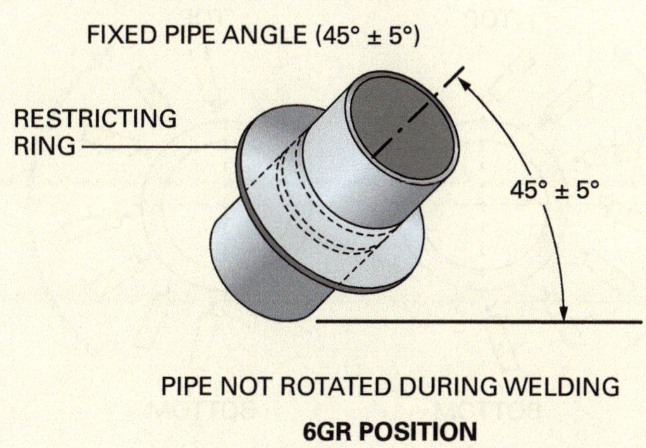

The progression of multiple-position (5G) welds is usually downhill on the sides for the root pass. The remaining FCAW fill passes progress uphill. Note that a backing gas must be used for stainless, low-alloy, and nonferrous alloys. *Figure 15* shows changes to the travel angle that may be needed for multiple position (5G) pipe weld fill passes when welding uphill or downhill.

> **CAUTION**
>
> When welding pipe, avoid making arc strikes outside of the weld groove on the surface of the pipe. An arc strike can cause a hardened spot that can crack as the pipe expands and contracts. An arc strike on the pipe surface is considered a defect, which will require repair or reworking.

2.3.0 Practicing Open-Root V-Groove Welds

The following sections explain how to perform FCAW welds in these open-root V-groove weld positions:

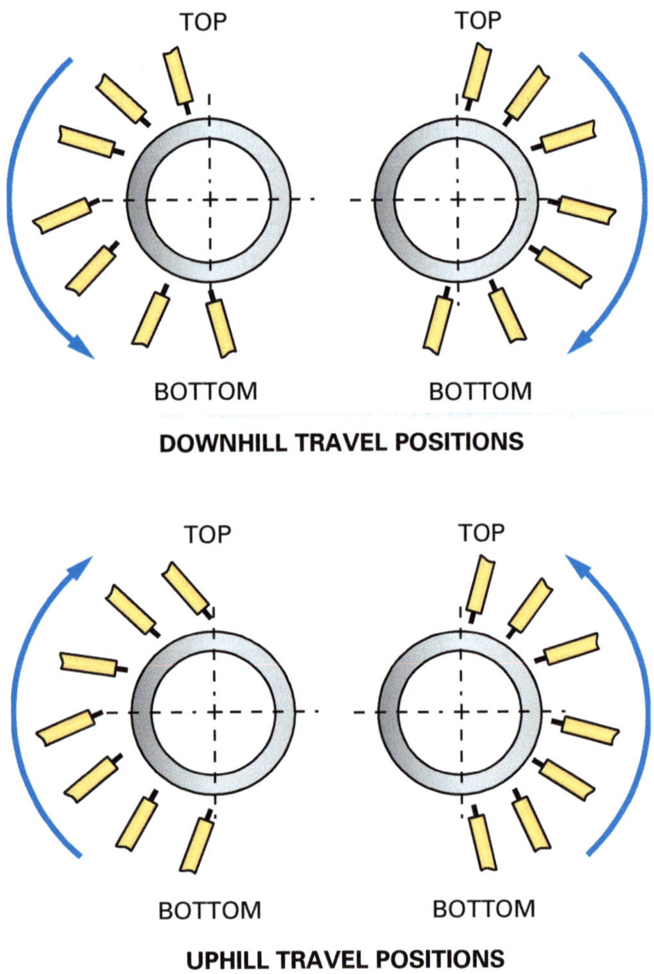

Figure 15 Approximate travel angle progressions.

- Flat (1G-ROTATED)
- Horizontal (2G)
- Multiple (5G)
- Multiple inclined (6G)

Note that the procedures that follow assume that the root pass will not be applied using FCAW. Only the hot and remaining passes will be done with FCAW; GMAW, GTAW, or SMAW should be used to apply the root pass; GMAW and SMAW are generally preferred.

2.3.1 Flat (1G-ROTATED) Position

Practice open-root V-groove pipe welds in the 1G-ROTATED position. Use a shielding gas (if required) and a flux-cored wire electrode of a wire diameter specified by your instructor. Remember that the root pass is to be done using another process other than FCAW. Pay special attention to fuse into the tack welds. Clean the face of the root pass thoroughly as required for the chosen welding process.

When running the remaining passes, use stringer or weave beads with a slight side-to-side motion. This ensures fusion at the toes of the weld bead. Take special care to fill the crater at the termination of the weld. Run all passes at or near the top of the pipe as it is rotated.

> **NOTE**
>
> Avoid using weave beads on stainless steel pipe.

Follow these steps to practice open-root V-groove pipe welds in the 1G-ROTATED position:

Step 1 Tack weld the practice pipe weld coupons together as explained earlier.

Step 2 Position the pipe weld coupon horizontally on two sets of rollers at a comfortable welding height. *Figure 16* shows roller supports commonly found in pipe welding shops. A welding positioner or pipe rotator is also required. The positioner shown in *Figure 17 (A)* is very popular. The pipe rotator, shown in *Figure 17 (B)*, is another option. This model can handle up to 16" (DN400) pipe and weights up to 1,200 pounds (544 kg). Adjustable roller stands can be used to help support longer stock.

Step 3 Make sure that the workpiece clamp is attached directly to the pipe coupon. This will prevent the welding current from passing through the roller bearings or arcing between the rollers and the pipe coupon.

Figure 16 Pipe roller supports.

(A) WELDING POSITIONER

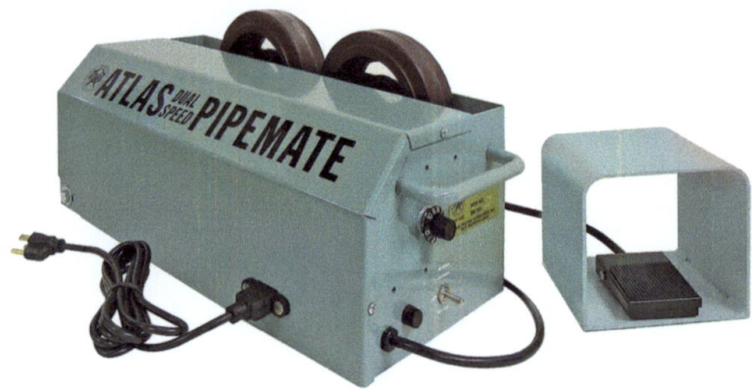

(B) PIPE ROTATOR

Figure 17 A welding positioner and a pipe rotator.

> **CAUTION**
>
> Failure to attach the workpiece clamp to the pipe coupon can result in variations in welding current, damage to the roller bearings, and arcing on the rollers and pipe coupon.

Step 4 Run the root pass using the GMAW, GTAW, or SMAW process. To begin, position the coupon with one of the tack welds at the eleven o'clock position. Start the root pass weld bead on this tack weld, and advance toward the twelve o'clock position.

Step 5 Rotate the pipe as needed to keep the weld in the flat position.

Step 6 Clean the crater and root pass as required to remove slag, silica deposits, and any potential discontinuities before restarting.

Step 7 Now using FCAW, use the same rotating procedure to make the remaining passes using stringer or weave beads. Pay special attention at the fusions to prevent slag inclusions or excess buildup. Overlap the passes, and clean the weld after each pass. *Figure 18* shows the related bead sequences and work angles.

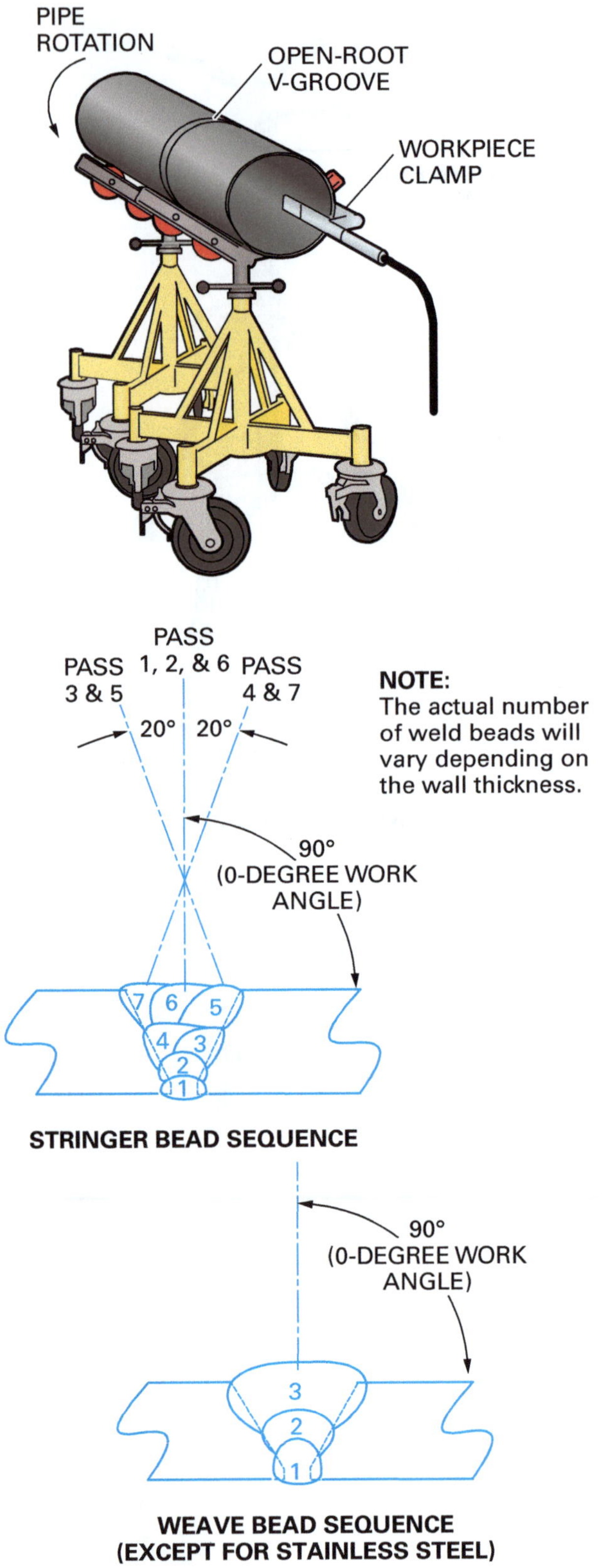

Figure 18 Multiple-pass 1G-ROTATED bead sequences and work angles.

2.3.2 Horizontal (2G) Position

Practice 2G position open-root V-groove pipe welds using a shielding gas (if required) and a flux-cored wire electrode of the wire diameter specified by your instructor. Remember that the root pass is to be done using another process other than FCAW. Pay special attention to fuse into the tack welds. Clean the face of the root pass thoroughly as required for the chosen welding process.

When running the remaining passes, use stringer beads and a slight side-to-side motion (*Figure 19*). This ensures fusion at the toes of the weld bead. Take special care to fill the crater at the termination of the weld.

Follow these steps to practice open-root V-groove pipe welds in the 2G position:

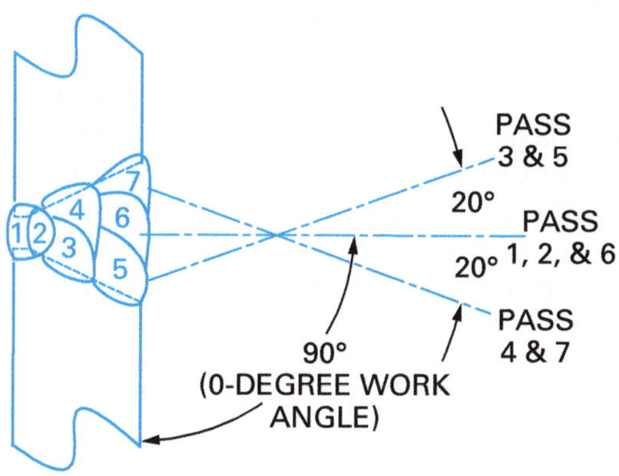

Figure 19 Multiple-pass 2G bead sequence and work angles.

Step 1 Tack weld the practice pipe weld coupons together as explained earlier.

Step 2 Clamp or tack weld the pipe coupon with the axis of the pipe vertical.

Step 3 Run the root pass using the GMAW, GTAW, or SMAW process. Start the root pass on a tack weld. Clean as required between restarts, if restarts are necessary. Continue this procedure until the root pass has been completed.

Step 4 Clean the crater and root pass as required to remove slag, silica deposits, and any potential discontinuities before restarting.

Step 5 Now using FCAW, run the remaining passes at the proper work and travel angles using stringer beads. Overlap the passes, and clean the weld after each pass.

2.3.3 Multiple (5G) Position

Practice 5G multiple-position open-root V-groove pipe welds. Use a shielding gas (if required) and a flux-cored wire electrode of the wire diameter specified by your instructor. Remember that the root pass is to be done using another process other than FCAW. Pay special attention to fuse into the tack welds. Clean the face of the root pass thoroughly as required for the chosen welding process. See *Figure 15* for the travel angle changes that may be needed as the passes are made on both sides of the pipe.

When running the remaining passes, use stringer beads and a slight side-to-side motion. This ensures fusion at the toes of the weld bead. The gun should be at approximately a 10- to 20-degree drag or push angle for the remaining passes. The type of angle depends on whether your instructor directs you to run downhill or uphill passes. In some cases, the root pass is made downhill, and the remaining passes are made uphill. Be sure to fill the crater at the termination of the weld. *Figure 20* shows the proper bead sequences and work angles for the 5G position.

Follow these steps to practice open-root V-groove pipe welds in the 5G position:

Step 1 Tack weld the practice pipe weld coupons together as explained earlier.

Step 2 Clamp or tack weld the pipe weld coupon into position with the pipe axis horizontal.

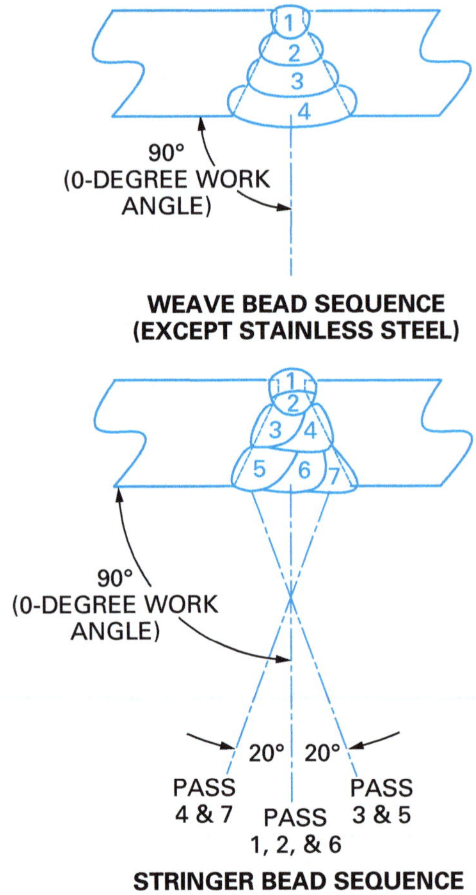

Figure 20 Multiple-pass 5G bead sequences and work angles.

Step 3 Run the root pass using the GMAW, GTAW, or SMAW process. Start the root pass on a tack weld. Clean as required between restarts, if restarts are necessary. Continue this procedure until the root pass has been completed, changing the travel angle as needed for an uphill or downhill pass, as directed by your instructor.

Step 4 Clean the crater and root pass as required to remove slag, silica deposits, and any potential discontinuities before restarting.

Step 5 Now using FCAW, run the remaining passes at the proper work and travel angles in either an uphill or downhill direction using stringer or weave beads. Overlap the passes, and clean the weld after each pass.

2.3.4 Multiple Inclined (6G) Position

Practice the 6G multiple inclined (45-degree) position open-root V-groove pipe welds. Use a shielding gas (if required) and a flux-cored wire electrode of the wire diameter specified by your instructor. Remember that the root pass is to be done using another process other than FCAW. Pay special attention to fuse into the tack welds. Clean the face of the root pass thoroughly as required for the chosen welding process. Refer to *Figure 15* for the travel angle changes that may be needed as the passes are made on both sides of the pipe.

When running the remaining passes, use stringer beads and a slight side-to-side motion. This ensures fusion at the toes of the weld bead. The gun should be at approximately a 10- to 20-degree drag or push angle for the remaining passes. The type of angle depends on whether your instructor directs you to run downhill or uphill passes. In some cases, the root pass is made downhill, and the remaining passes are made uphill. Be sure to fill the crater at the termination of the weld. *Figure 21* shows the proper bead sequences and work angles for the 6G position.

Follow these steps to practice open-root V-groove pipe welds in the 6G position:

> **NOTE**
> If required for training purposes, a restricting ring may be added to the 6G position coupon to form a 6GR position coupon.

Step 1 Tack weld the practice pipe weld coupons together as explained earlier.

Step 2 Clamp or tack weld the pipe weld coupon into position with the pipe axis inclined 45 degrees to the horizontal plane.

Step 3 Run the root pass using the GMAW, GTAW, or SMAW process. Start the root pass on a tack weld. Clean as required between restarts, if restarts are necessary. Continue this procedure until the root pass has been completed, changing the travel angle as needed for an uphill or downhill pass, as directed by your instructor.

Step 4 Clean the crater and root pass as required to remove slag, silica deposits, and any potential discontinuities before restarting.

Step 5 Now using FCAW, run the remaining passes at the proper work and travel angles in either an uphill or downhill direction using stringer beads. Overlap the passes, and clean the weld after each pass.

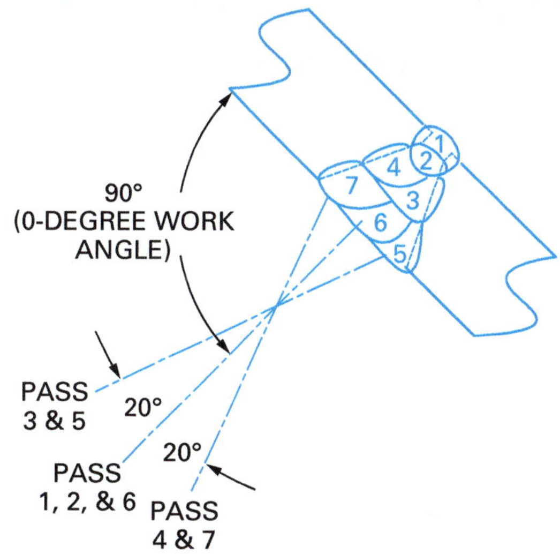

Figure 21 Multiple-pass 6G bead sequence and work angles.

Piping Purge Systems

Many pipe welding applications require a shielding-gas backing. The shielding gas is used to displace air and the oxygen it contains from the weld zone. Special systems and devices have been designed to simplify the process and ensure that oxygen is not present in the weld zone throughout the welding process.

The system shown below incorporates dams (A) that are positioned on each side of the weld. The shielding gas is applied, which first inflates the dams and then makes its way into the pipe area between the two dams, pushing out the air. Purge monitors (B) constantly monitor the oxygen content of the escaping gas to ensure it remains below the specified level throughout the welding process. A backing gas must be used for stainless, low-alloy, and nonferrous alloy pipe.

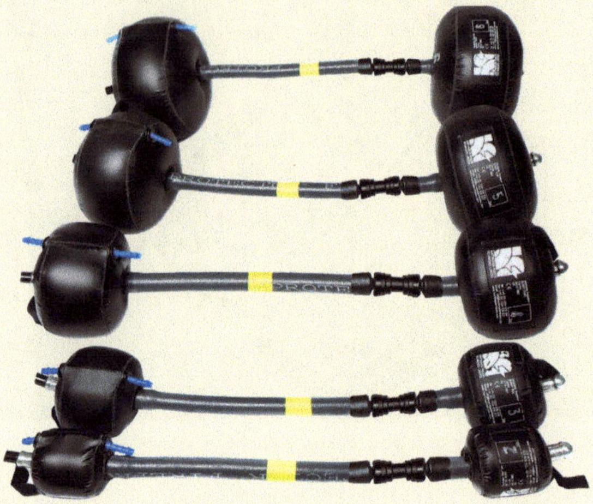

(A) INFLATABLE DAMS

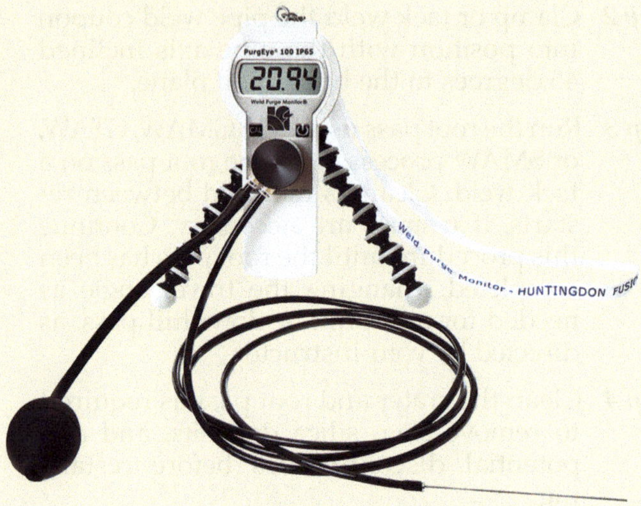

(B) PURGE MONITOR

Additional Resources

AWS F4.1: Recommended Safe Practices for Preparation for Welding and Cutting of Containers and Piping. Miami, FL: American Welding Society.

Modern Welding Technology. Current Edition. Howard B. Cary and Scott Helzer. Englewood Cliffs, NJ: Prentice Hall.

OSHA 1926.351, Arc Welding and Cutting. Current Edition. Washington, DC: Occupational Safety & Health Administration.

The Procedure Handbook of Arc Welding. Current Edition. Cleveland, OH: The James F. Lincoln Arc Welding Foundation.

Welding Handbook, Volumes 1 through 4. Current Edition. Miami, FL: American Welding Society.

Welding Pressure Pipelines and Piping Systems (PDF). The Lincoln Electric Company, Cleveland, OH, USA. **www.lincolnelectric.com**. Last accessed May 29, 2015.

Innershield Wire – FCAW-S Welding Guide (PDF). The Lincoln Electric Company, Cleveland, OH, USA. **www.lincolnelectric.com**. Last accessed May 29, 2015.

2.0.0 Section Review

1. In the 6G position, the weld beads required are _____.
   a. flat and horizontal
   b. horizontal and overhead
   c. flat, horizontal, and overhead
   d. horizontal, vertical, and overhead

2. The three welding processes commonly used for the root pass on open-root V-groove pipe welds are _____.
   a. GMAW, GTAW, and SMAW
   b. FCAW-G, GTAW, and GMAW
   c. FCAW-S, SMAW, and GTAW
   d. GTAW, FCAW-S, and FCAW-G

3. In the 1G-ROTATED position, all passes are run _____.
   a. all around the pipe, uphill and downhill
   b. all around the pipe, uphill on both sides
   c. at or near the three o'clock position
   d. at or near the top of the pipe

SUMMARY

This module describes flux-cored arc welding (FCAW) equipment and processes and their application in pipe welding. FCAW can be done using two different approaches. One approach requires the use of a shielding gas source (FCAW-G), while the other requires no shielding gas (FCAW-S). In this case, the flux-cored wire electrode contains the material needed to protect the weld zone. For this reason, FCAW-S is a popular welding method outdoors and in remote areas.

FCAW equipment and techniques are similar to those used in GMAW, although FCAW is rarely used to make root passes. FCAW processes are widely used in many types of applications. Practice FCAW techniques until you can consistently produce acceptable results on open-root V-groove pipe welds.

Review Questions

1. In FCAW-S, the shielding gas is generated _____.
   a. entirely from the flux in the wire electrode
   b. entirely from an external shielding gas
   c. from both the flux in the electrode and an external shielding gas
   d. from the molten weld metal

2. The FCAW process is limited to welding _____.
   a. low-carbon steel only
   b. high-carbon steels and aluminum
   c. stainless steel and aluminum
   d. low-carbon steels and some low-alloy and stainless steels

3. If the tongue and lace area of your footwear is exposed and require protection, you should _____.
   a. wear rubber boots
   b. treat them with a fire retardant
   c. wear shoes without laces
   d. wear spats or leather chaps

4. The gas that tends to concentrate in and along low areas and that can be ignited at a considerable distance from the release point is _____.
   a. argon
   b. oxygen
   c. propane
   d. acetylene

5. Chromium 6 is commonly referred to as _____.
   a. chrome-moly
   b. hexavalent chromium
   c. pentavalent chromium
   d. polychrome

6. The welding codes allow the root face on FCAW open-root welds on carbon steel pipe to be _____.
   a. 0" to ⅛" (0 to 3 mm) wide
   b. 0" to ¼" (0 to 6 mm) wide
   c. ⅛" to ¼" (3 to 6 mm) wide
   d. ⅛" to ½" (3 to 13 mm) wide

7. Feathering the ends of tack welds with a grinder helps _____.
   a. to build higher beads with deeper penetration
   b. improve fusion of the root pass into the tack welds
   c. to increase the deposition rate
   d. to minimize the arc length

8. When preparing an FCAW welding machine, verify the location of the _____.
   a. control unit
   b. tungsten electrode
   c. gun polarity switch
   d. primary power disconnect

9. If the FCAW arc length is too long and causes the wire to melt and fuse to the contact tube, it may be caused by a _____.
   a. poor workpiece lead connection
   b. defective wire electrode
   c. voltage setting that is too high
   d. voltage setting that is too high

10. Within the limits of an FCAW welding machine, when the wire electrode feed speed is increased, the welding current and deposition rate _____.
    a. are both decreased
    b. are both increased
    c. must remain at the original setpoint
    d. must be released from machine control

11. Pipe groove welds should be made with slight face reinforcement not to exceed _____.
    a. 1⁄16" (1.6 mm)
    b. ⅛" (3 mm)
    c. 3⁄16" (5 mm)
    d. ¼" (6 mm)

12. On FCAW multiple position 5G welds, the remaining passes generally progress _____.
    a. downhill on the sides
    b. uphill on the sides
    c. only across the top
    d. only across the bottom

13. For FCAW in the 1G-ROTATED position, avoid using _____.
    a. weave beads on stainless steel pipe
    b. stringer beads on stainless steel pipe
    c. weave beads on aluminum pipe
    d. stringer beads on low-carbon steel pipe

14. For FCAW in the 1G-ROTATED position, position the coupon with one of the tack welds at the _____.
    a. three o'clock position
    b. four o'clock position
    c. six o'clock position
    d. eleven o'clock position

15. Multiple inclined (6G) position open-root V-groove pipe welds are made with the pipe axis inclined _____.
    a. 10 degrees to the horizontal plane
    b. 20 degrees to the horizontal plane
    c. 45 degrees to the horizontal plane
    d. 75 degrees to the horizontal plane

Trade Terms Introduced in This Module

Flux-cored wire electrode: A wire electrode that is tube-like and filled with flux and/or degassing components with other elements to improve the chemical and mechanical characteristics of the weld.

Appendix

Performance Accreditation Tasks

PATs provide specific acceptable criteria for performance and help to ensure a true competency-based welding program for students.

The following tasks are designed to evaluate your ability to run open-root V-groove welds with FCAW-G or FCAW-S equipment in three standard test positions using carbon steel wire of the appropriate diameter. Perform each task when you are instructed to do so by your instructor. As you complete each task, show it to your instructor for evaluation. Do not proceed to the next task until told to do so.

For AWS 1G, 2G, and 5G certifications, refer to *AWS EG3.0:1996 Guide for the Training and Qualification of Welding Personnel; Level II – Advanced Welder* for bend test requirements. For AWS 6G certifications, refer to *AWS EG4.0:1996 Guide for the Training and Qualification of Welding Personnel; Level III – Expert Welder* for bend test requirements.

PATs 2 and 3 correspond to *AWS EG3.0-96, Section 3.3.6.3, Unit #3 FCAW, Learning Objective #7*. PAT 4 corresponds to *AWS EG4.0-96, Section 3.3.6.4, Unit #4 FCAW, Learning Objectives #7* and *#8*. PAT 1 has no AWS correlation.

OPEN-ROOT V-GROOVE PIPE WELD IN THE 1G-ROTATED POSITION

Using FCAW-G or FCAW-S carbon steel wire of the appropriate diameter and stringer beads, make an open-root V-groove weld on carbon steel pipe in the 1G-ROTATED position. For FCAW-G, use the appropriate shielding gas.

Note: Depending on site procedures or practices, the root pass for the following tasks may be run using another welding process such as GTAW, GMAW, SMAW, or backing material. Check with your instructor to determine the welding process to use for the root pass.

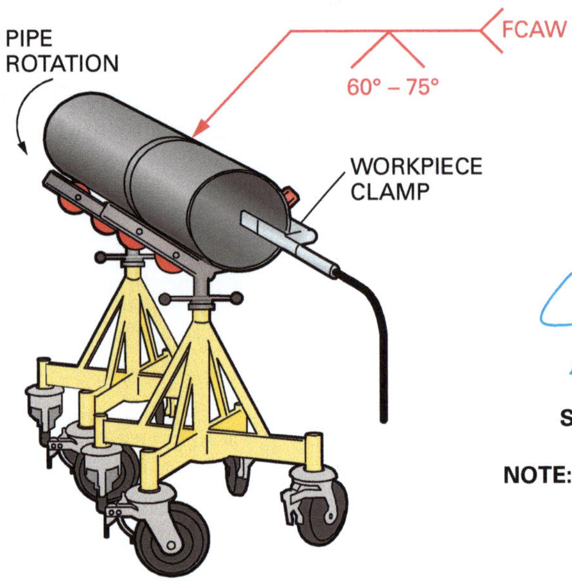

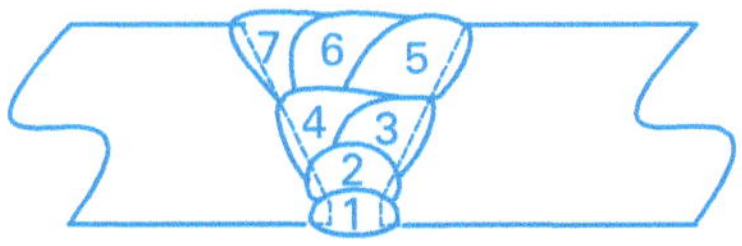

STRINGER BEAD SEQUENCE

NOTE: THE ACTUAL NUMBER OF WELD BEADS WILL VARY DEPENDING ON THE WALL THICKNESS.

Criteria for Acceptance:

- Uniform appearance on the bead face __________
- Craters and restarts filled to the full cross section of the weld __________
- Acceptable weld profile in accordance with the applicable codes and standards __________
- Smooth transition with complete fusion at the toes of the weld __________
- Complete joint penetration and face reinforcement with the root reinforcement at least flush with the inside of the pipe, to a maximum buildup of ⅛" (3 mm) __________
- No porosity __________
- No undercut __________
- No cracks __________
- No overlap __________
- No incomplete fusion __________
- No pinholes __________
- No inclusions __________

OPEN-ROOT V-GROOVE PIPE WELD IN THE 2G POSITION

Using FCAW-G or FCAW-S carbon steel wire of the appropriate diameter and stringer beads, make an open-root V-groove weld on carbon steel pipe in the 2G position. For FCAW-G, use the appropriate shielding gas.

Note: Depending on site procedures or practices, the root pass for the following tasks may be run using another welding process such as GTAW, GMAW, SMAW, or backing material. Check with your instructor to determine the welding process to use for the root pass.

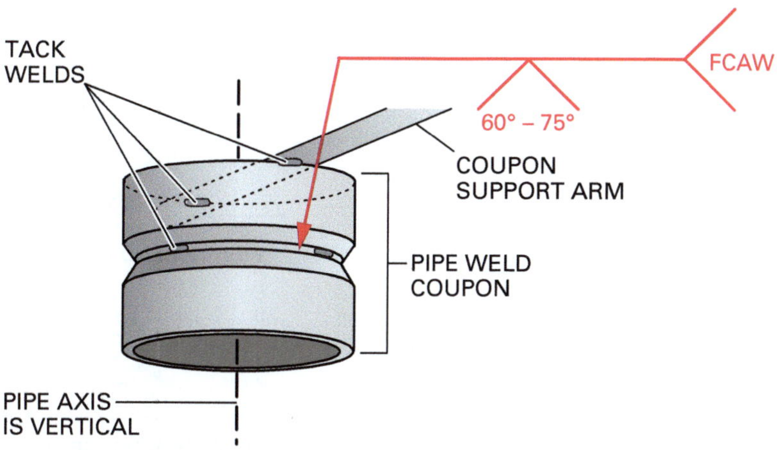

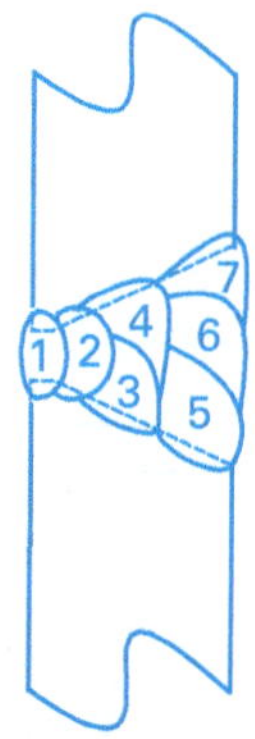

STRINGER BEAD SEQUENCE

Criteria for Acceptance:

- Uniform appearance on the bead face ____________
- Craters and restarts filled to the full cross section of the weld ____________
- Acceptable weld profile in accordance with the applicable codes and standards ____________
- Smooth transition with complete fusion at the toes of the weld
- Complete joint penetration and face reinforcement with the root reinforcement at least flush with the inside of the pipe, to a maximum buildup of ⅛" (3 mm) ____________
- No porosity ____________
- No undercut ____________
- No cracks ____________
- No overlap ____________
- No incomplete fusion ____________
- No pinholes ____________
- No inclusions ____________

OPEN-ROOT V-GROOVE PIPE WELD IN THE 5G POSITION

Using FCAW-G or FCAW-S carbon steel wire of the appropriate diameter and stringer beads, make an open-root V-groove weld on carbon steel pipe in the 5G position. For FCAW-G, use the appropriate shielding gas.

Note: Depending on site procedures or practices, the root pass for the following tasks may be run using another welding process such as GTAW, GMAW, SMAW, or backing material. Check with your instructor to determine the welding process to use for the root pass.

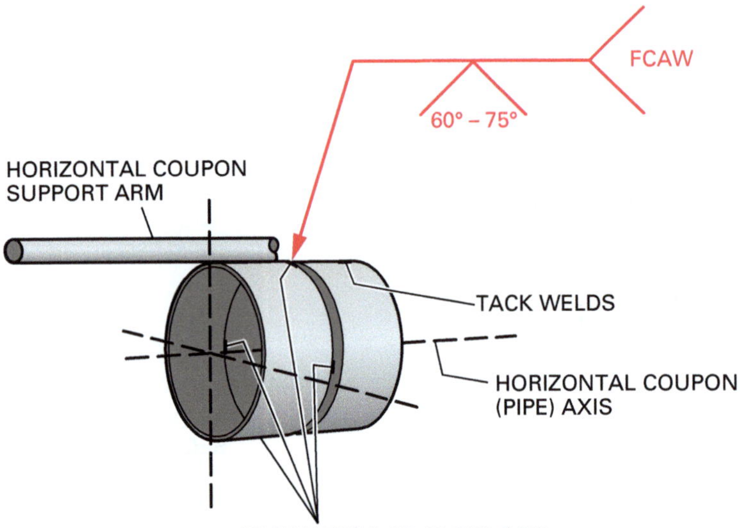

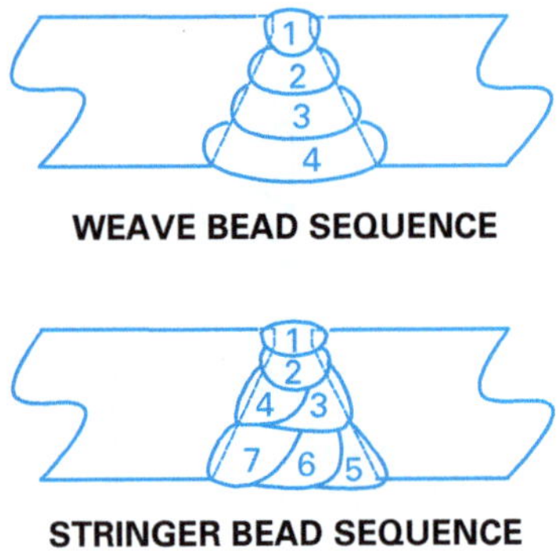

Criteria for Acceptance:

- Uniform appearance on the bead face ____________
- Craters and restarts filled to the full cross section of the weld ____________
- Acceptable weld profile in accordance with the applicable codes and standards ____________
- Smooth transition with complete fusion at the toes of the weld
- Complete joint penetration and face reinforcement with the root reinforcement at least flush with the inside of the pipe, to a maximum buildup of ⅛" (3 mm) ____________
- No porosity ____________
- No undercut ____________
- No cracks ____________
- No overlap ____________
- No incomplete fusion ____________
- No pinholes ____________
- No inclusions ____________

OPEN-ROOT V-GROOVE PIPE WELD IN THE 6G (OR 6GR) POSITION

Using FCAW-G or FCAW-S carbon steel wire of the appropriate diameter and stringer beads, make an open-root V-groove weld on carbon steel pipe in the 5G position. For FCAW-G, use the appropriate shielding gas.

Note: Depending on site procedures or practices, the root pass for the following tasks may be run using another welding process such as GTAW, GMAW, SMAW, or backing material. Check with your instructor to determine the welding process to use for the root pass.

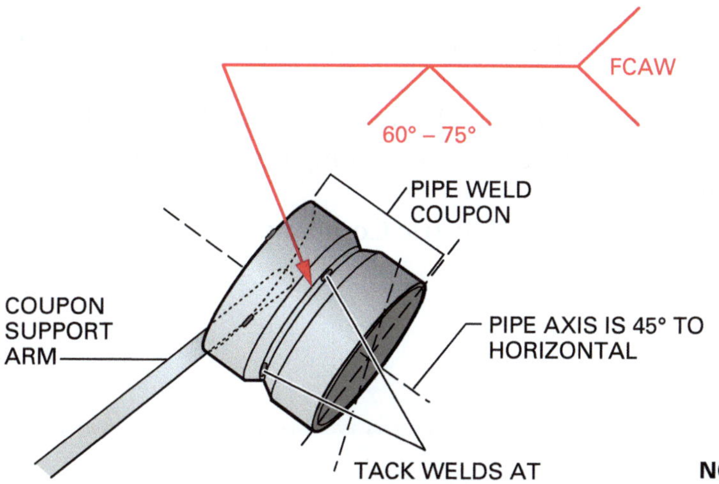

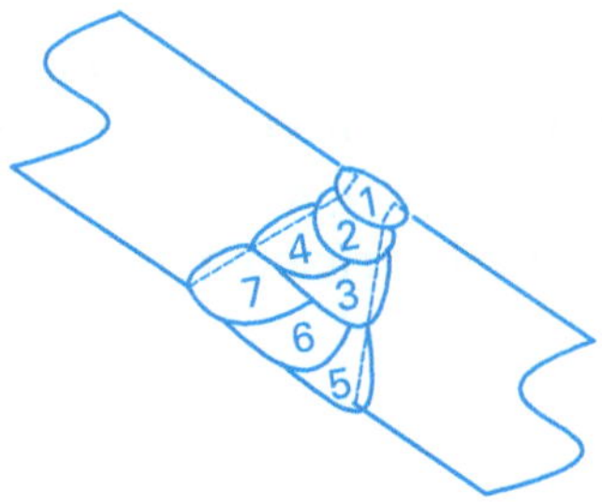

STRINGER BEAD SEQUENCE

NOTE:
IF REQUIRED FOR QUALIFICATION PURPOSES, A RESTRICTING RING MAY BE ADDED TO THE 6G POSITION COUPON TO FORM A 6GR POSITION COUPON.

Criteria for Acceptance:

- Uniform appearance on the bead face ____________
- Craters and restarts filled to the full cross section of the weld ____________
- Acceptable weld profile in accordance with the applicable codes and standards ____________
- Smooth transition with complete fusion at the toes of the weld
- Complete joint penetration and face reinforcement with the root reinforcement at least flush with the inside of the pipe, to a maximum buildup of ⅛" (3 mm) ____________
- No porosity ____________
- No undercut ____________
- No cracks ____________
- No overlap ____________
- No incomplete fusion ____________
- No pinholes ____________
- No inclusions ____________

Additional Resources

This module presents thorough resources for task training. The following reference material is recommended for further study.

AWS F4.1: Recommended Safe Practices for Preparation for Welding and Cutting of Containers and Piping. Miami, FL: American Welding Society.

Modern Welding Technology. Current Edition. Howard B. Cary and Scott Helzer. Englewood Cliffs, NJ: Prentice Hall.

OSHA 1926.351, Arc Welding and Cutting. Current Edition. Washington, DC: Occupational Safety & Health Administration.

The Procedure Handbook of Arc Welding. Current Edition. Cleveland, OH: The James F. Lincoln Arc Welding Foundation.

Welding Handbook, Volumes 1 through 4. Current Edition. Miami, FL: American Welding Society.

Welding Pressure Pipelines and Piping Systems (PDF). The Lincoln Electric Company, Cleveland, OH, USA. **www.lincolnelectric.com**. Last accessed May 29, 2015.

Innershield Wire – FCAW-S Welding Guide (PDF). The Lincoln Electric Company, Cleveland, OH, USA. **www.lincolnelectric.com**. Last accessed May 29, 2015.

Figure Credits

The Lincoln Electric Company, Cleveland, OH, USA, Module Opener, Figures 2–5

Topaz Publications, Inc., Figure 6 (photo)

Terry Lowe, Figures 7 (photo), 19-21 (photos)

International Diving Institute, LLC, SA01

Courtesy of Sumner Manufacturing Co., Figure 16

Atlas Welding Accessories, Inc., Figure 17

Huntingdon Fusion Techniques Limited, SA03

Section Review Answer Key

Answer	Section Reference	Objective
Section One		
1. b	1.1.0	1a
2. d	1.2.3	1b
3. b	1.3.2	1c
4. c	1.4.2	1d
Section Two		
1. d	2.1.0	2a
2. a	2.3.0	2b
3. d	2.3.1	2c

NCCER CURRICULA — USER UPDATE

NCCER makes every effort to keep its textbooks up-to-date and free of technical errors. We appreciate your help in this process. If you find an error, a typographical mistake, or an inaccuracy in NCCER's curricula, please fill out this form (or a photocopy), or complete the online form at **www.nccer.org/olf**. Be sure to include the exact module ID number, page number, a detailed description, and your recommended correction. Your input will be brought to the attention of the Authoring Team. Thank you for your assistance.

Instructors – If you have an idea for improving this textbook, or have found that additional materials were necessary to teach this module effectively, please let us know so that we may present your suggestions to the Authoring Team.

NCCER Product Development and Revision

13614 Progress Blvd., Alachua, FL 32615

Email: curriculum@nccer.org
Online: www.nccer.org/olf

❑ Trainee Guide ❑ Lesson Plans ❑ Exam ❑ PowerPoints Other ____________

Craft / Level: Copyright Date:

Module ID Number / Title:

Section Number(s):

Description:

Recommended Correction:

Your Name:

Address:

Email: Phone:

GTAW – Carbon Steel Pipe

Overview

Gas tungsten arc welding (GTAW) is a unique welding process that requires the welder to coordinate the use of both hands since the filler metal is fed manually. GTAW can be used on critical piping, and may sometimes be used for the root pass on pipe welds when other processes have been directed for the remaining passes. GTAW is excellent for welding thin materials, such as auto bodies. This module provides instruction and guided practice in the use of GTAW for carbon steel pipe welding.

Module 29304

Trainees with successful module completions may be eligible for credentialing through the NCCER Registry. To learn more, go to **www.nccer.org** or contact us at 1.888.622.3720. Our website has information on the latest product releases and training, as well as online versions of our *Cornerstone* magazine and Pearson's product catalog.

Your feedback is welcome. You may email your comments to **curriculum@nccer.org**, send general comments and inquiries to **info@nccer.org**, or fill in the User Update form at the back of this module.

This information is general in nature and intended for training purposes only. Actual performance of activities described in this manual requires compliance with all applicable operating, service, maintenance, and safety procedures under the direction of qualified personnel. References in this manual to patented or proprietary devices do not constitute a recommendation of their use.

29304 V5

29304
GTAW – Carbon Steel Pipe

Objectives

When you have completed this module, you will be able to do the following:

1. State the basic concepts of open-root GTAW of carbon-steel pipe and explain how to prepare for welding.
   a. State the basic concepts of GTAW.
   b. Identify common welding safety practices.
   c. Explain how to prepare the area, materials, and equipment for GTAW carbon-steel pipe welding.
2. Describe open-root V-groove pipe welding positions and GTAW pipe-welding techniques.
   a. Identify and describe the four pipe welding positions and acceptable weld profiles.
   b. Describe the techniques for GTAW and how to make the root and fill passes.
   c. Describe the techniques required to produce GTAW open-root V-groove carbon steel pipe welds in all positions.

Performance Tasks

Under the supervision of your instructor, you should be able to do the following:

1. Make GTAW open-root V-groove carbon steel pipe welds in the 1G-ROTATED position using carbon steel filler metal and argon gas.
2. Make GTAW open-root V-groove carbon steel pipe welds in the 2G position using carbon steel filler metal and argon gas.
3. Make GTAW open-root V-groove carbon steel pipe welds in the 5G position using carbon steel filler metal and argon gas.
4. Make GTAW open-root V-groove carbon steel pipe welds in the 6G position using carbon steel filler metal and argon gas.

Industry Recognized Credentials

If you are training through an NCCER-accredited sponsor, you may be eligible for credentials from NCCER's Registry. The ID number for this module is 29304. Note that this module may have been used in other NCCER curricula and may apply to other level completions. Contact NCCER's Registry at 888.622.3720 or go to **www.nccer.org** for more information.

Contents

Figures and Tables

Figures and Tables

Section One

1.0.0 GTAW Pipe-Welding Preparations

Objective

State the basic concepts of open-root GTAW of carbon-steel pipe and explain how to prepare for welding.

a. State the basic concepts of GTAW.
b. Identify common welding safety practices.
c. Explain how to prepare the area, materials, and equipment for GTAW carbon-steel pipe welding.

This section describes the basic gas tungsten arc welding (GTAW) process and reviews safety practices that apply to all forms of welding. Since the GTAW welder's second hand cannot generally be used for support and stabilization, the process requires welders to consider another aspect of their safety on the job as they coordinate the work of both hands.

1.1.0 Basic GTAW Concepts

The GTAW process uses an arc that is created between a non-consumable tungsten electrode and the base metal to melt the base and filler metals. The electrode, arc, and molten base metal are shielded from atmospheric contamination by a flow of shielding gas from the torch nozzle, as shown in *Figure 1*. The filler metal is a rod of similar composition to the base metal. It is usually handheld and manually fed into the leading edge of the weld puddle. GTAW produces high-quality welds without slag or oxidation. Because there is no flux, minimal post-weld cleaning is needed.

A major use of GTAW is to make manual, high-quality welds in ferrous and nonferrous piping. The GTAW process allows greater control of root penetration and weld fill than almost any other process. For this reason, GTAW is often used to make the root pass on pipe. This is true even when the remaining passes are made with SMAW, GMAW, or FCAW, all of which have higher deposition rates. GTAW is often used for critical welds and when working with thin, potentially delicate materials.

This module explains how to set up GTAW equipment and perform open-root V-groove welds on carbon steel pipe in the 1G-ROTATED, 2G, 5G, and 6G welding positions. The dimensions and specifications in this module are representative of codes in general and may not be specific to any particular code. Always follow the proper codes and standards for your site or situation.

1.2.0 Safety Practices

Safety considerations are a critical part of preparing for a welding operation. The following is a summary of safety procedures and practices that must be observed while cutting or welding. Be advised that this is only a summary. Complete safety coverage is provided in the *Welding Level One* module, "Welding Safety." That module should be studied, if not already completed, before continuing. Above all, be sure to wear appropriate protective clothing and equipment at all times when welding or cutting (*Figure 2*).

1.2.1 Protective Clothing and Equipment

Protective clothing and equipment are a necessary part of every welding task. Follow the guidelines below as a general rule, and ensure that you are familiar with any unique requirements your employer may have regarding personal protective equipment (PPE).

- A welding hood equipped with a properly shaded lens is best for all forms of welding. The welding hood lens must be an appropriate shade for the application.
- Wear proper leather or flame-retardant clothing along with welding gloves for protection from flying sparks, molten metal, and heat.

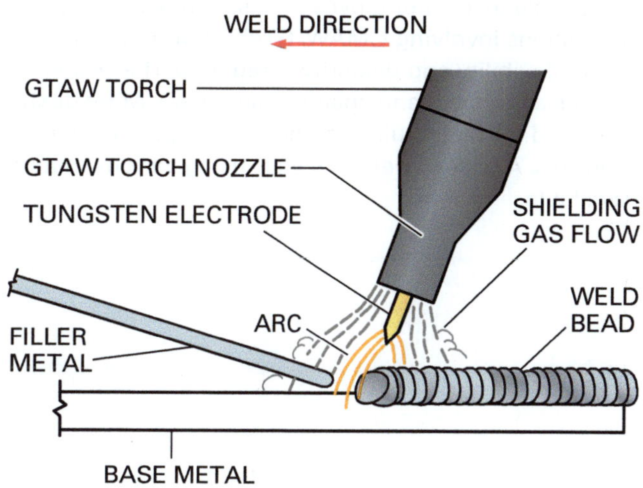

Figure 1 GTAW process.

Figure 2 Welder with the proper PPE using GTAW on an auto body.

- Wear high-top safety shoes or boots. Make sure that the tongue and lace area of the footwear is covered by a pant leg. If the tongue and lace area is exposed or if the footwear must be protected from burn marks, wear spats under the pant leg or leather chaps over the front and top of the footwear.
- Wear a 100-percent cotton cap without any mesh material. If a hard hat is required for the environment, use one that allows the attachment of rear deflector material and a face shield. A hard hat with a rear deflector is generally preferred when working overhead, and may be required by some employers and job sites.
- Wear earmuffs, or at least earplugs, to prevent sparks from entering the ear canals.

Lightweight Welding Gloves

Soft, flexible, leather gloves are used for light-duty work and for operations involving GMAW, GTAW, brazing, and soldering where more flexibility and dexterity is required. These welding processes produce less heat and spatter than others. More dexterity is required for the required motions and operating welding torch controls. A wide variety of styles, using various materials, is available.

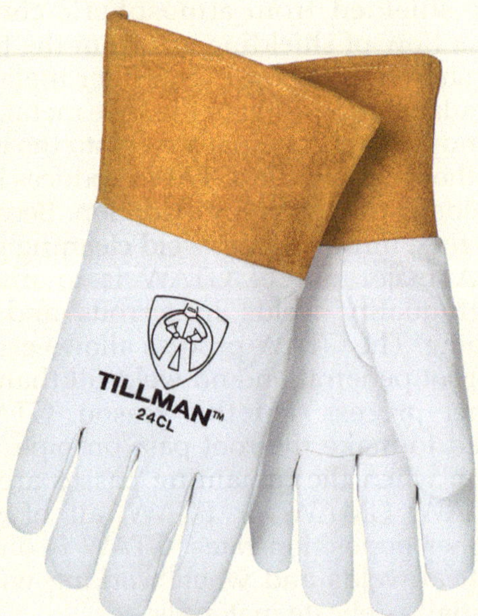

1.2.2 Housekeeping and Fall Protection

Welding is often performed outside in the weather and in the construction environment. Production work done inside large, open shops exposes a welder to equipment moving across the floor and overhead. Welders often have to crawl under or reach over objects to complete their work.

At times, welders must work at height. Fall protection equipment is required for all work at height (*Figure 3*). Always keep fall protection devices untangled and properly attached. Combined with the other equipment a welder must carry, there is a great deal to manage, but fall protection equipment must always be the priority. Fall protection equipment saves lives. Never attach any kind of welding gear to the harness that could interfere with its ability to perform its intended function.

Regardless of where you are working, always keep the work area as neat as possible. This makes the area safer and helps to reduce distractions. Mindful actions, such as making sure that welding cables are not tripping hazards, contribute to a safer environment. Electrode stubs and unused electrodes must not be left on the floor. They can cause someone to lose their footing and fall. Parts waiting to be welded or left over from previous work are a tripping hazard and should be kept out of the walking area. If the parts have sharp edges, they could cut welding cable insulation or torch gas hoses.

1.2.3 Fire and Explosion Prevention

Flammable gases are commonly found around welding and cutting operations, as are sparks and other significant sources of ignition. For that reason, and because many fire-related incidents over the years have been traced back to welding operations, it is crucial that fire and explosion prevention be given consistent attention. Follow these guidelines to maintain a safe environment:

Figure 3 Working at height.

- Never carry matches or gas-filled lighters. Sparks can cause the matches to ignite or the lighter to explode in your pocket, causing serious injury.
- Never use oxygen to blow dust or dirt from clothing. Traces of the oxygen can remain trapped in the fabric for some time. If a spark hits the fabric during this time, the clothing can burn rapidly and violently out of control.
- Make sure that any flammable material in the work area is removed or shielded by a fire-resistant covering. Approved fire extinguishers must be available before attempting any heating, welding, or cutting operations. If a hot-work permit and a fire watch are required, be sure these are in place before beginning.
- Ensure that all site safety requirements are being observed.
- Never release a large amount of oxygen or use oxygen in place of compressed air. Its presence around flammable materials or sparks can cause rapid and uncontrolled combustion. Keep oxygen away from oil, grease, and other petroleum products.
- Never release a large amount of fuel gas, especially acetylene. Propane tends to concentrate in and along low areas and can ignite at a considerable distance from the release point. Acetylene is lighter than air, but is even more dangerous than propane. When mixed with air or oxygen, acetylene will ignite at much lower concentrations than any other fuel gas.
- Even though shielding gases are not toxic, any gas released in a sufficient quantity can displace the oxygen in a small room or enclosed work area, creating an atmosphere with little or no oxygen. Avoid large releases of any type of gas, regardless of its characteristics.
- To prevent fires, maintain a neat and clean work area, and make sure that any metal scrap or slag is cold before disposal.
- Before cutting containers such as tanks or barrels, check to see if they have contained any explosive, hazardous, or flammable materials, including petroleum products, citrus products, or chemicals that decompose into toxic fumes when heated. As a standard practice, always clean and then fill any tanks or barrels with water, or purge them with a flow of inert gas such as nitrogen to displace any oxygen. Containers must be cleaned by steam cleaning, flushing with water, or washing with detergent until all traces of the material have

been removed. Proper procedures for cutting or welding hazardous containers are described in the *American Welding Society (AWS) F4.1, Safe Practices for the Preparation of Containers and Piping for Welding and Cutting.*

> **WARNING**
>
> Welding or cutting must never be performed on drums, barrels, tanks, vessels, or other containers until they have been emptied and cleaned thoroughly, eliminating all flammable materials and all substances (such as detergents, solvents, greases, tars, or acids) that might produce flammable, toxic, or explosive vapors when heated. Clean containers only in well-ventilated areas. Vapors can accumulate during cleaning, causing explosions and serious personal injury.

1.2.4 Work Area Ventilation

Many gases are introduced to the atmosphere through welding processes. Some of these gases are harmful, and all can displace or combine with atmospheric oxygen in a confined space. Follow these guidelines, as well as those of your employer, for work area ventilation:

- Make sure to follow all confined space procedures before conducting any welding or cutting in a confined space. Use powered extraction systems as needed.
- Never use oxygen to provide ventilation in confined spaces.
- Always perform cutting or welding operations in a well-ventilated area.

Cutting or welding operations involving zinc or cadmium materials or coatings result in toxic fumes. Another significant concern is Chromium 6, commonly referred to as *hexavalent chromium,* which is used in stainless steel. It releases toxic fumes when heated. Manganese is another hazardous metal used in stainless steel. It is also used in some welding rods, increasing the risk of exposure.

For long-term cutting or welding of materials containing these components, always wear an approved full-face, supplied-air respirator that uses breathing air supplied from outside the work area. For occasional short-term exposure, a HEPA-rated or metal-fume filter may be used on a standard respirator.

Powered fume-extraction equipment must be used when welding materials that contain zinc, cadmium, hexavalent chromium, manganese, and similar materials that produce hazardous fumes if the area is not naturally well-ventilated. *Figure 4* shows examples of both mobile and portable units.

1.3.0 Preparations for Welding

Before any welding procedure can take place, the work area must be made ready, the welding equipment must be properly set up, and the metal to be welded must be prepared.

1.3.1 Preparing the Welding Area

To practice welding, a welding table, bench, and/or appropriate pipe stands are needed. The welding surface must be made of steel. Provisions must be made for supporting the pipe welding coupons for out-of-position welding.

To set up the area for welding, follow these steps:

Step 1 Make sure that the area is properly ventilated. Make use of doors, windows, and fans.

Step 2 Check the area for fire hazards. Remove any flammable materials before proceeding.

Step 3 Know the location of the nearest fire extinguisher. Ensure the fire extinguisher is charged and that you and others in the immediate area know how to use it properly.

Step 4 Set up flash shields and curtains around the welding area as necessary.

1.3.2 The Welding Machine

Identify a proper constant-current welding machine for GTAW, and follow these steps to prepare it for use.

Step 1 Verify that the welding machine can be used for GTAW, with or without internally controlled gas shielding.

Step 2 Identify an air-cooled or water-cooled GTAW torch. Make sure that it is compatible with the welding machine and any cooling unit that may be needed.

Step 3 Verify the location of the primary power disconnect for the welding machine. This may be a circuit breaker or even the cord for a portable machine.

Step 4 Configure the welding machine for GTAW parameters as directed by your instructor (*Figure 5*). Configure the torch polarity; DCEN is generally preferred. Equip the torch with the correct diameter

(A) MOBILE UNIT **(B) PORTABLE UNIT**

Figure 4 Mobile and portable extraction equipment.

and type of tungsten as directed by your instructor or site WPS.

Step 5 Connect the proper shielding gas for the application and as specified by the filler metal manufacturer, WPS, site quality standards, or your instructor.

Step 6 Connect the clamp of the workpiece lead to the workpiece.

Step 7 Turn on the welding machine, adjust the flowmeter, and purge the torch as directed by the manufacturer's instructions.

1.3.3 Filler Metals

Filler metals are selected to be compatible with the base metal to be welded. The WPS or site standards will specify the proper filler metal type and size to use.

> **CAUTION**
>
> When the WPS or site quality standards specify a filler metal, it must be used. Always cut filler metal rods so that their ID numbers are visible on each section. Be sure to use the end of the filler metal that has no ID numbers on it. Keep the filler metal stubs with their numbers visible for quality control checks.

For the welding exercises in this module, use 3/32" (2.4 mm) and/or 1/8" (3.2 mm) carbon steel filler metals of classification ER70S-2, or the equivalent, to weld the carbon steel pipe coupons. Remove only a small number of filler metal rods at a time from the packaging. Keep the remainder in the package to keep them clean. Before using the filler metal, check it for burned ends or contamination, such as corrosion, dirt, oil, or grease, all of which can cause weld discontinuities. Clean the filler metal with a clean, oil-free rag and a chemical cleaner. Snip any burned ends. If the filler metal cannot be cleaned, do not use it.

> **CAUTION**
>
> Do not use RG-45 or RG-60 filler metals. They will cause porosity because they are not deoxidized rods.

1.3.4 Preparing Pipe Weld Coupons

Pipe weld coupons should be cut from 2" to 6" (DN50 to 150) diameter Schedule 40 or Schedule 80 carbon steel pipe. Each welded joint will require two coupons of the same diameter and schedule. Typical ASME, API, and AWS pipe

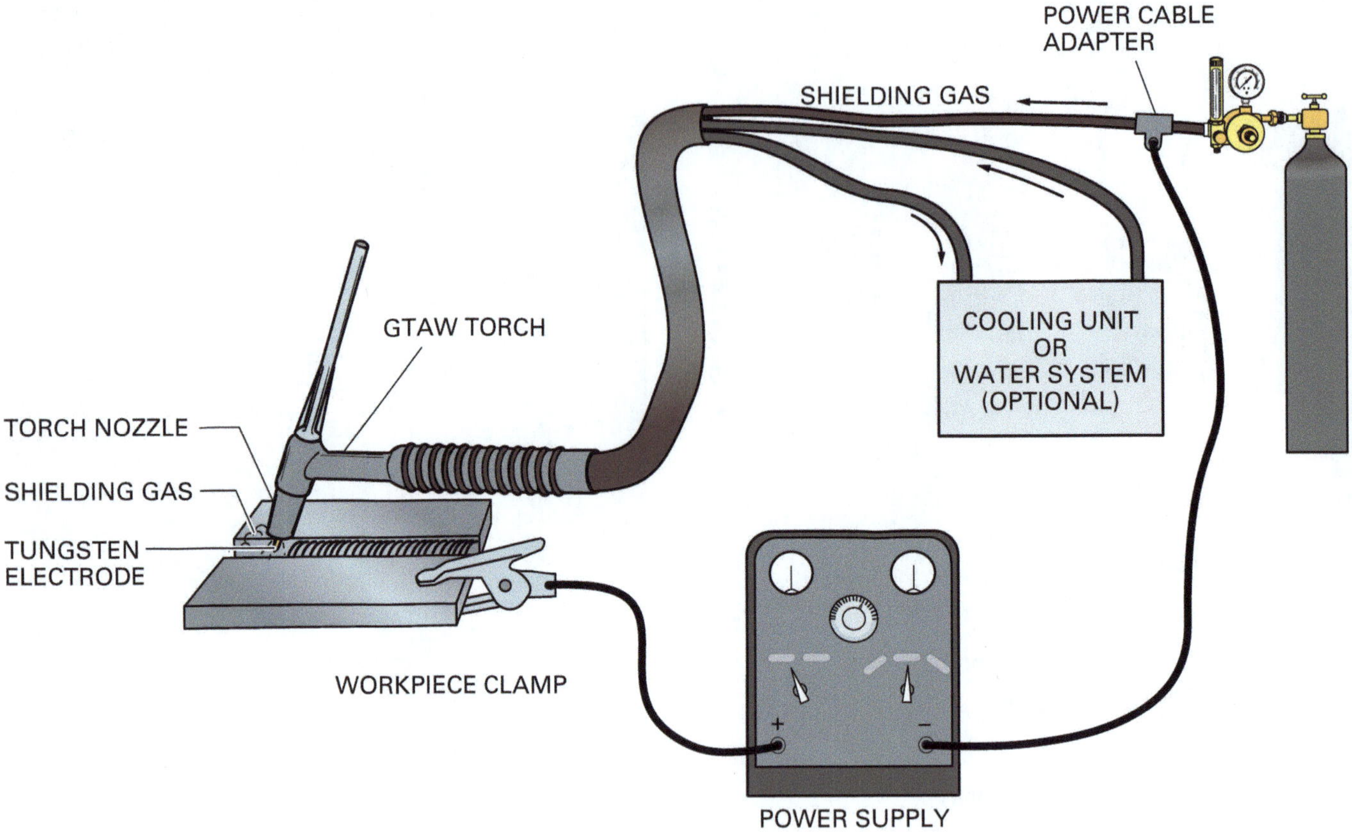

Figure 5 Typical GTAW configuration diagram.

bevel specifications for bevel angles, root faces, and root openings are shown in *Figure 6*.

To prepare weld coupons for open-root V-groove weld joints, follow these steps:

Step 1 Clean all rust or mill scale from the inside and outside of the pipe with a grinder, wire brush, or other common method.

CAUTION

There is little or no surface cleaning action resulting from the GTAW process. As a result, pipe surfaces must be thoroughly cleaned to prevent weld discontinuities.

Step 2 Bevel the end of the pipe to 30 to 37½ degrees by any acceptable beveling method, including manual, mechanical, and thermal methods.

Step 3 Cut off the end of the beveled pipe to the length specified by your instructor. Two coupons will be required for each weld.

NOTE

For 1G-ROTATED welding, the coupons may need to be cut longer to reach the rollers being used.

The two coupons can now be further prepared for welding:

Step 1 Check the bevel. There should be no dross, and the bevel angle should be 30 to 37½ degrees.

Step 2 Grind or file a 0" to ⅛" (0 to 3 mm) root face on the bevel.

Welding Large-Diameter Pipe

The practice coupons for large-diameter pipe must be greater than 1½" in length. More metal is needed to help absorb the increased heat used to make these welds.

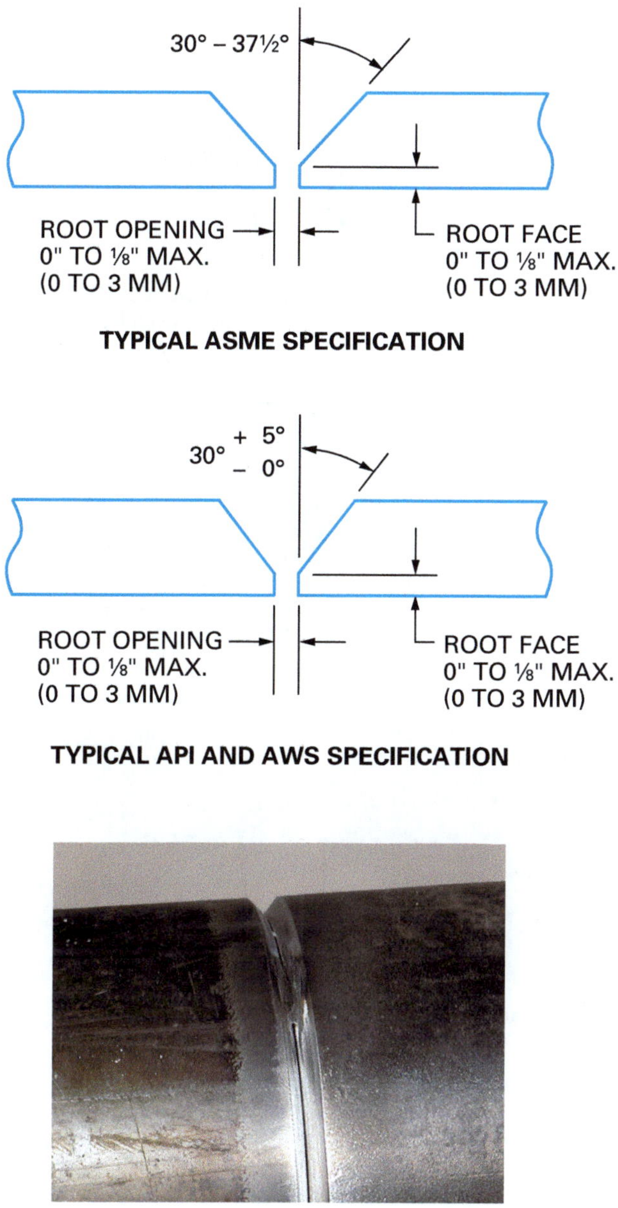

Figure 6 ASME, API, and AWS pipe bevel specifications.

> **NOTE**
>
> Welding codes allow the root face for open-root welds to be from 0" to ⅛" (0 to 3 mm) wide and the root opening to be from 0" to ⅛" (0 to 3 mm). Select and adjust the root face and opening as directed by the instructor as welding practice begins.

Step 3 Align the two pipe sections. The inside surfaces of the two coupons should be even all around and evenly separated by the required root opening. Pieces of filler rod, metal shims, or pieces of bare welding electrode can be used to achieve the desired root opening between the two pipe sections until they are tack welded. Small-diameter pipe coupons can be aligned by clamping both pieces to a piece of angle. Large-diameter pipe coupons can be aligned using a pipe alignment jig or by holding a straightedge across the joint, parallel to the pipe axis. The straightedge must be used to check all around the pipe in case one or both sections are distorted.

Step 4 Once aligned, secure the two pipe sections with a clamp, or with tack welds if necessary.

> **NOTE**
>
> If a consumable insert is used, make sure that it is secured to both pipe sections.

Step 5 When the root opening is correct and the pipe ends are aligned, make the first of four tack welds. Longer tacks may be required on heavy-wall or large-diameter pipe.

GOING GREEN

Advanced Inverter Power Sources

An advanced inverter power source can make welding pipe with GTAW easier while saving energy. Features such as HF start only, current pulsing, and weld current sequencing (stop/start ramping) provide these machines with much more control over some of the variables that affect the quality of a weld.

Step 6 After the first tack weld, check the root opening on the opposite side. Adjust the root opening space if necessary, and make the second tack weld.

> **CAUTION**
>
> Before making the second tack, remove any spacer that was used. This will prevent the spacer from getting caught as the heated metal expands.

Step 7 Verify that the root opening remains correct before welding the third tack midway between the first and second tacks.

Step 8 Weld the fourth tack opposite the third tack and midway between the first and second tacks. There should now be four tack welds evenly spaced (90 degrees) around the pipe coupon (*Figure 7*).

Step 9 Feather the tack welds with the edge of a grinding wheel. Feathering the ends of the tack welds with a grinder helps fuse the root pass into the tack welds.

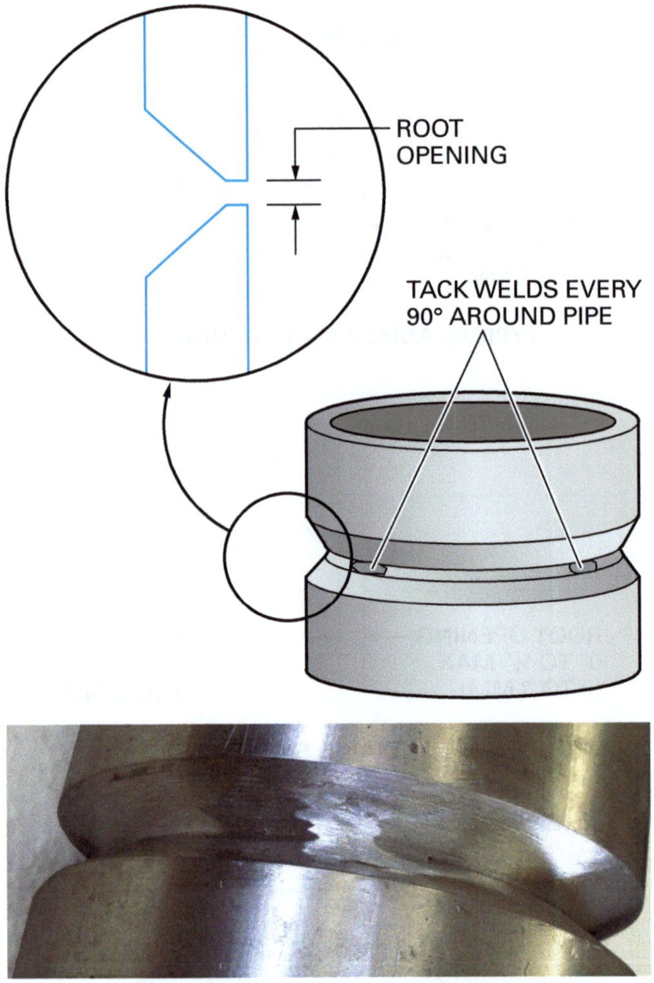

Figure 7 Tacked open-root V-groove weld coupon.

Additional Resources

AWS F4.1: Recommended Safe Practices for Preparation for Welding and Cutting of Containers and Piping. Miami, FL: American Welding Society.

Modern Welding Technology. Current Edition. Howard B. Cary and Scott Helzer. Englewood Cliffs, NJ: Prentice Hall.

OSHA 1926.351, Arc Welding and Cutting. Current Edition. Washington, DC: Occupational Safety & Health Administration.

The Procedure Handbook of Arc Welding. Current Edition. Cleveland, OH: The James F. Lincoln Arc Welding Foundation.

Welding Handbook, Volumes 1 through 4. Current Edition. Miami, FL: American Welding Society.

Welding Pressure Pipelines and Piping Systems (PDF). The Lincoln Electric Company, Cleveland, OH, USA. **www.lincolnelectric.com**. Last accessed May 29, 2015.

1.0.0 Section Review

1. One of the unique characteristics of GTAW is that _____.
   a. there is no flux involved
   b. the filler metal can only be wire-fed
   c. preheating is always required
   d. no shielding gas is required

2. The fuel gas that is heavier than air and tends to collect in low areas is _____.
   a. oxygen
   b. nitrogen
   c. acetylene
   d. propane

3. The four tack welds on pipe joints should be _____.
   a. 30 to 37-1/2 degrees apart
   b. 45 to 60 degrees apart
   c. 75 degrees apart
   d. 90 degrees apart

SECTION TWO

2.0.0 GTAW TECHNIQUES AND POSITIONS

Objective

Describe open-root V-groove pipe-welding positions and GTAW pipe-welding techniques.

a. Identify and describe the four pipe welding positions and acceptable weld profiles.
b. Describe the techniques for GTAW and how to make the root and fill passes.
c. Describe the techniques required to produce GTAW open-root V-groove carbon-steel pipe welds in all positions.

Performance Tasks

1. Make GTAW open-root V-groove carbon-steel pipe welds in the 1G-ROTATED position using carbon-steel filler metal and argon gas.
2. Make GTAW open-root V-groove carbon-steel pipe welds in the 2G position using carbon-steel filler metal and argon gas.
3. Make GTAW open-root V-groove carbon-steel pipe welds in the 5G position using carbon-steel filler metal and argon gas.
4. Make GTAW open-root V-groove carbon-steel pipe welds in the 6G position using carbon-steel filler metal and argon gas.

Like all welding processes, GTAW requires a significant amount of practice to master. This section reviews the various pipe welding positions and outlines the procedures for GTAW practice in the various positions.

2.1.0 Pipe Groove Weld Test Positions

Groove welds may be made in all positions on pipe. The weld position is determined by the axis of the pipe (*Figure 8*). Four standard weld test positions are used with pipe:

- *Flat (1G-ROTATED)* – The pipe axis is horizontal, and the pipe is slowly rotated while it is being welded on the top. The weld beads are flat.
- *Horizontal (2G)* – The pipe axis is vertical and the pipe rotation is fixed. The weld beads are horizontal.
- *Multiple (5G)* – The pipe axis is horizontal and the pipe rotation is fixed. The weld beads are flat, vertical, and overhead.
- *Multiple inclined (6G)* – The pipe axis is inclined at approximately 45 degrees ±5 degrees. Pipe rotation is fixed. Weld beads are horizontal, vertical, and overhead. If required for training purposes, a restricting ring may be added to the 6G position coupon to form a 6GR position coupon.

The pipe axis positions for 1G-ROTATED, 2G, and 5G can vary ±15 degrees from the basic position. The pipe axis position for 6G can vary ±5 degrees from the nominal 45-degree angle. *Table 1* provides a quick reference for the four positions.

To determine the quality of a pipe weld, specific sections of pipe are removed for destructive testing. For positions 1G-ROTATED and 2G, two specimens are removed. They must be taken from opposite sides of the pipe, as shown in *Figure 9*. Once removed, the specimens are put through different bend tests, which reveal the quality of the welds. Alternatively, specimens may be tested by radiography, a non-destructive testing (NDT) method.

When 5G or 6G welds will be destructively tested, the test specimens must be cut from the four regions shown in *Figure 9*. These regions are midway between the twelve o'clock and three o'clock; three o'clock and six o'clock; six o'clock and nine o'clock; and nine o'clock and twelve o'clock positions. The twelve o'clock position is the top of the coupon groove; the six o'clock position is at the bottom.

2.1.1 Acceptable and Unacceptable Pipe Weld Profiles

Pipe groove welds should be made with slight face reinforcement (not exceeding ⅛" or 3 mm) and a gradual transition to the base metal at each toe. The root pass should have complete penetration along the entire circumference of the root. Any detectable incomplete penetration is typically rejected and should be repaired. The root reinforcement inside of the pipe may range in thickness from being flush to a maximum of ⅛" (3 mm) thick.

Pipe groove welds must not have excess reinforcement or any underfill at the face or root. They must be free from excessive undercut or any overlap. Excessively large face reinforcement reduces the pipe's strength. It causes the stresses in the pipe to be concentrated along the toes of the weld. As a result, the pipe will not expand and contract uniformly along its length. If a weld has any of the defects shown in *Figure 10*, it must be repaired.

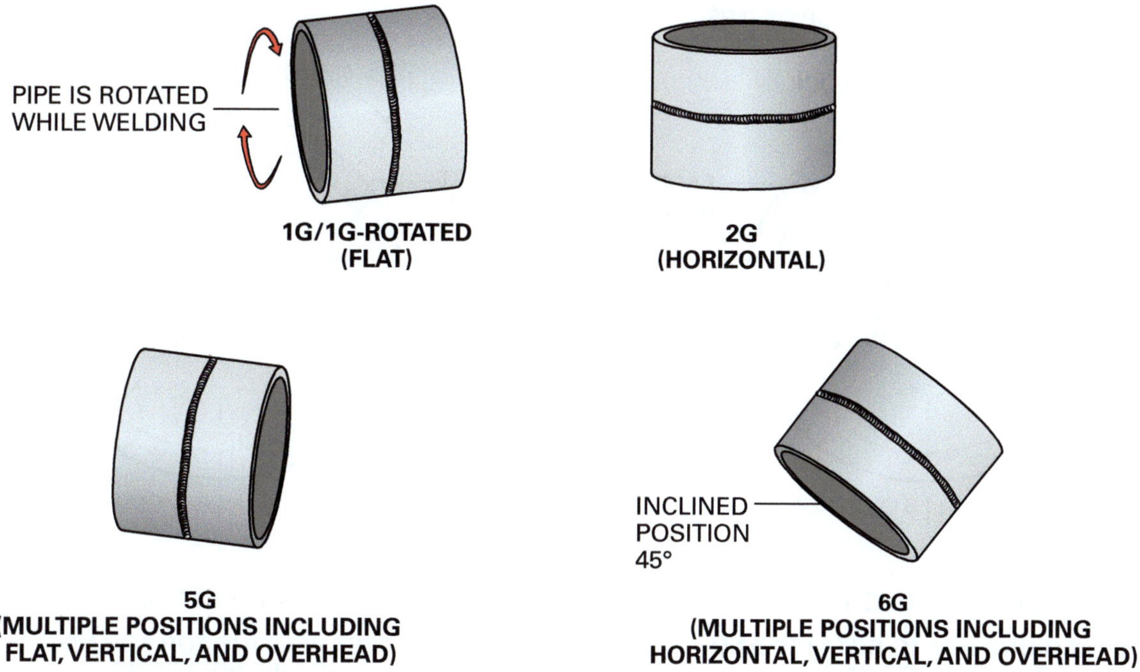

Figure 8 Four standard pipe groove-weld test positions.

2.2.0 GTAW Techniques

GTAW weld bead characteristics and quality are affected by several factors, each of which is influenced by the way you handle the torch. These factors include the following:

- Torch travel speed and arc length
- Torch angles
- Torch and filler metal handling techniques

2.2.1 Torch Travel Speed and Arc Length

Torch travel speed and arc length affect the GTAW weld puddle and weld penetration. Slow travel speeds allow more heat to concentrate and form a larger, more deeply penetrating puddle. Faster travel speeds prevent heat buildup and form smaller, shallower puddles.

Arc length is the major control for bead width. As the torch is raised, arc length, voltage, and bead width increase.

2.2.2 Torch Angles

The two basic torch angles that must be controlled when performing GTAW are the work angle and the travel angle. The definition of these angles is the same as for all other types of arc welding.

The torch work angle (*Figure 11*) is an angle of 90 degrees or less between a line perpendicular to the major workpiece surface at the point of electrode contact and a plane determined by the electrode axis and the weld axis. For pipe, the plane is determined by the electrode axis and a line tangent to the pipe surface at the same point. Remember that, in the case of GTAW, the electrode is the tungsten and not the filler metal as in other forms of welding. Note how the work angle in *Figure 11* is adjusted depending on the position of the bead being placed in the groove.

The torch travel angle is an angle of 90 degrees or less between the electrode axis and a line perpendicular to the weld axis at the point

Table 1 Pipe Groove-Weld Test Positions

Weld Test Position	Pipe Axis	Weld Beads	Pipe Position
1G-ROTATED	Horizontal	Flat	Slowly rotated
2G	Vertical	Horizontal	Fixed
5G	Horizontal	Flat, vertical, overhead	Fixed
6G	Inclined 45 degrees	Horizontal, vertical, overhead	Fixed

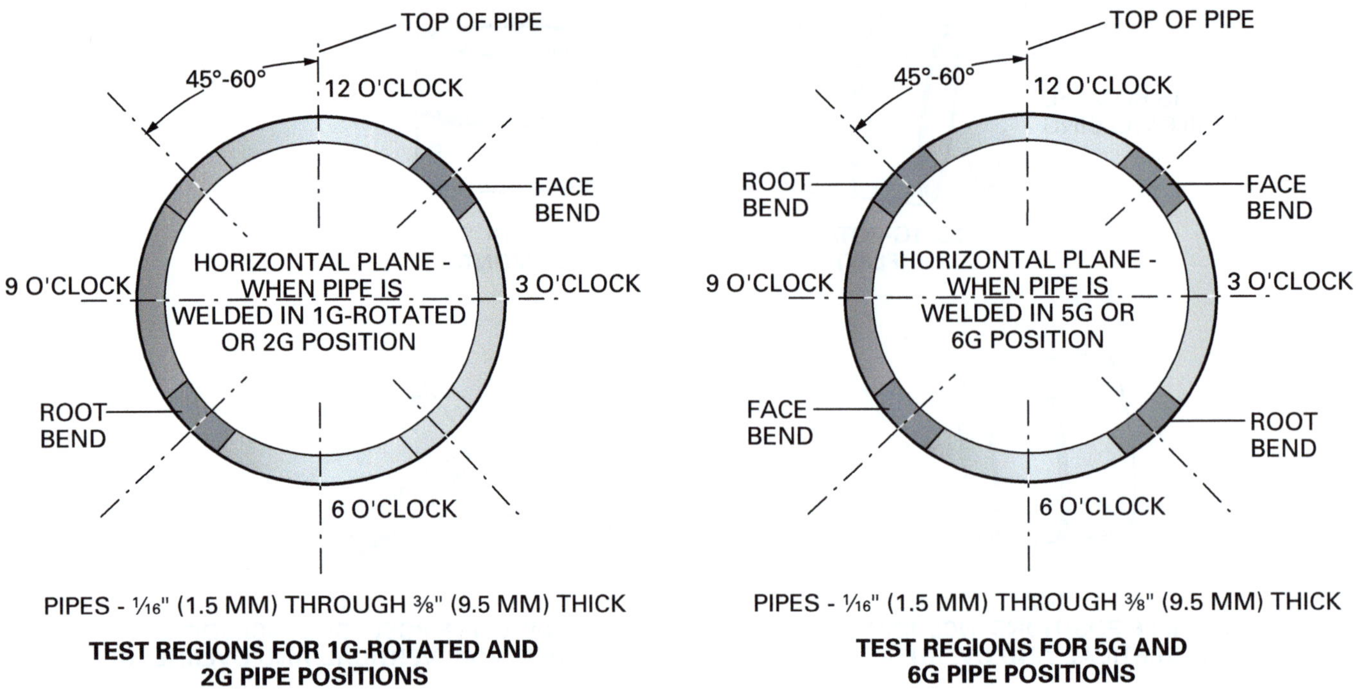

NOTE: TACK WELD LOCATIONS AT 12, 3, 6, AND 9 O'CLOCK.

Figure 9 Test specimen regions for pipe positions 1G-ROTATED, 2G, 5G, and 6G.

of electrode contact in a plane determined by the electrode axis and the weld axis. For pipe, the plane is determined by the electrode axis and a line tangent to the pipe surface at the same point.

The travel angle used for GTAW is a push angle (*Figure 12*). The push angle is created when the torch is tilted back, away from the direction of travel, so that the electrode tip precedes the torch in the direction of the weld. In this position, the electrode tip and shielding gas are directed just ahead of the weld bead. Push angles of 15 to 20 degrees are normally used for GTAW.

A push angle that is too great (more than 20 degrees) will tend to draw air from under the back edge of the torch nozzle where it will mix with the shielding gas stream and contaminate the weld. In *Figure 12*, note that the angle of the filler rod is 90 degrees to the axis of the torch. As the torch position changes while progressing around the pipe, the angle of the filler rod changes with it.

2.2.3 Torch and Filler Metal Handling Techniques

The manual handling of the filler rod for GTAW makes the process unique and requires that the correct technique be learned and practiced. The two basic torch and filler metal handling methods used to perform GTAW are the freehand and

6GR Test Position

Welding pipe in the 6G position is a common welding test. A ring may be added to test for restricted accessibility (6GR).

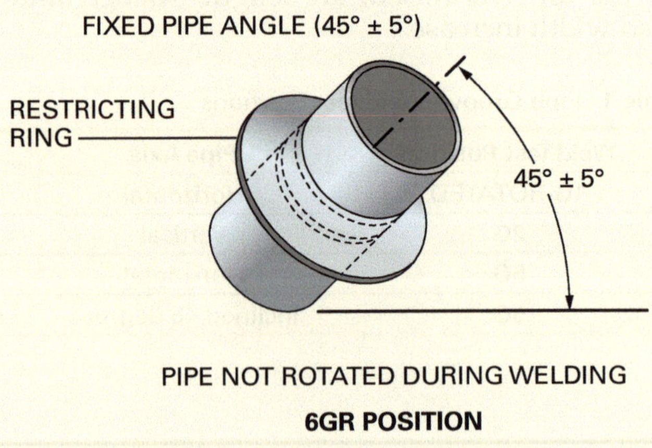

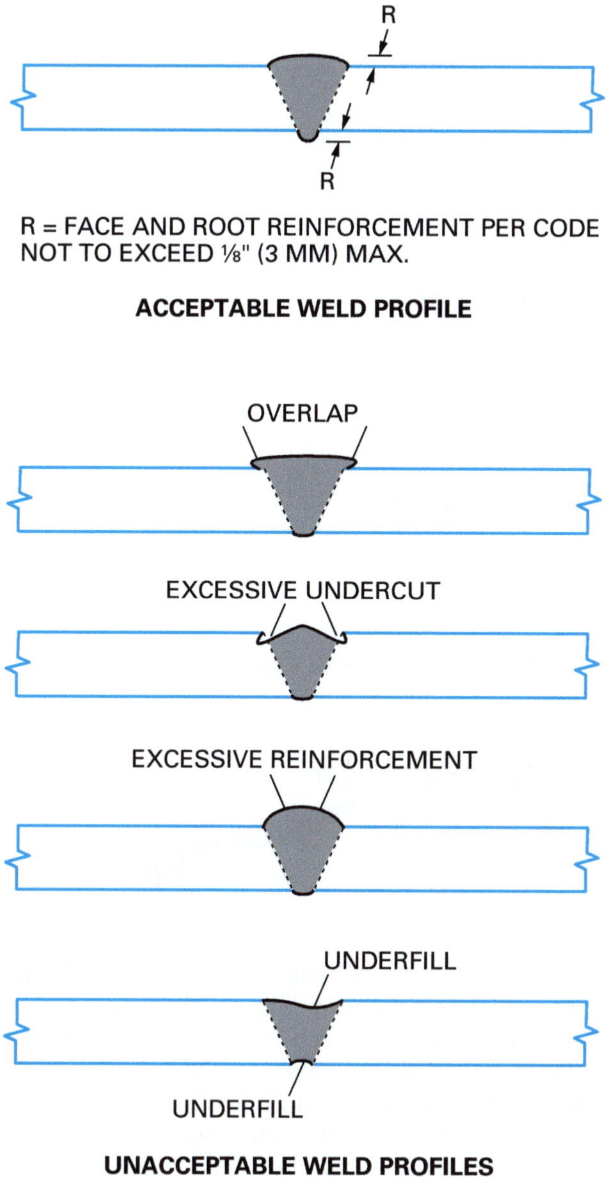

Figure 10 Acceptable and unacceptable pipe groove-weld profiles.

walking-the-cup techniques. Try both, and use the one that gives the best results.

In the freehand technique (*Figure 13*), the torch electrode tip is held just above the weld puddle or base metal. The torch is supported by the welder's hand, which is usually steadied by resting some part of it on or against the base metal to maintain the proper arc length. If required, the welder can move the torch tip in a small circular motion within the molten puddle to maintain its size and advance the puddle. Filler metal is added as needed.

The GTAW filler rod is held in the opposite hand. The tip of the rod is always kept within the shielding gas envelope, even when it is not being fed into the weld puddle, to protect the rod from

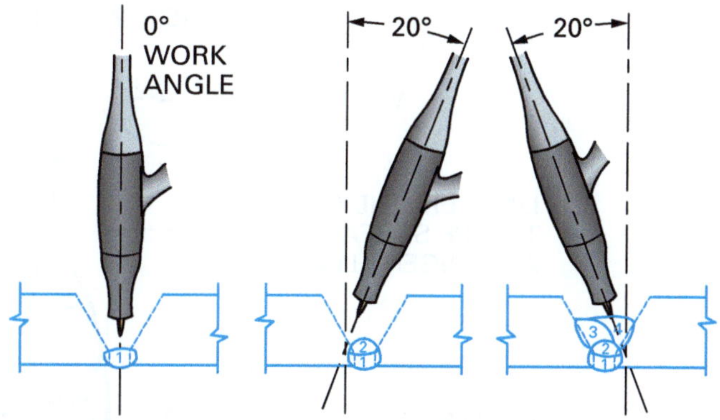

Figure 11 Typical GTAW torch work angles.

atmospheric contamination and to keep it preheated. Note that this is important regardless of the welding technique used.

Insert the filler metal into the leading edge of the weld puddle, using extreme care not to touch the tungsten electrode with the end of the filler metal. If the tungsten electrode touches the filler metal or weld puddle, it will become contaminated with filler metal. The electrode must then be removed and cleaned by grinding or chemical cleaning before proceeding. Also, do not insert the end of the filler metal rod into the molten puddle under the electrode and then attempt to melt it off. Even if it does not touch the tungsten, this technique can cause hard spots and weld defects. A technique for preventing contamination of the electrode is to coordinate the movement of the torch toward the back edge of the weld puddle as the filler metal is inserted into the leading edge of the weld puddle.

CAUTION

Insert the end of the filler metal rod only into the leading edge of the weld puddle. Do not insert the filler rod into any other portion of the weld puddle beyond the leading edge. Doing so can cause weld defects.

In the walking-the-cup method, the welder rests the edge of the torch nozzle (cup) against the base metal or groove edges to steady the torch and maintain a constant arc length. The torch is rocked from side to side on the edge of the cup as it is advanced to maintain the puddle size and heat both sides of the groove. The filler metal is added during the fill passes in the same manner as in the freehand technique, using care not to contaminate the electrode.

Feeding the filler rod consistently as the weld progresses is important to ensure the weld puddle develops evenly and produces an even bead. Like all welding techniques, the filler rod technique described next requires practice to master but is well worth the effort.

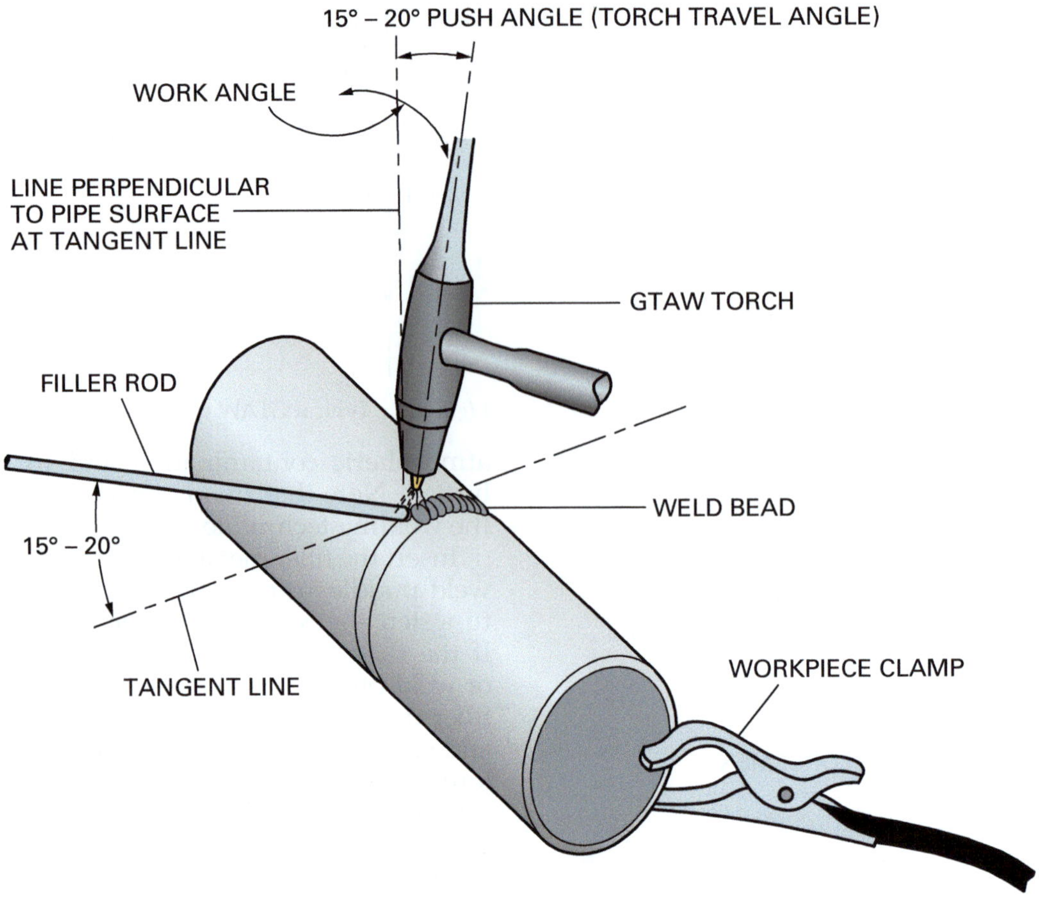

Figure 12 GTAW torch travel angle and filler rod position.

Figure 13 Freehand technique.

Cradle the filler rod between the pointer finger and the middle finger, similar to the way you might hold a pencil. Then use the thumb to consistently push the rod forward in small motions. Once the thumb is extended pushing the rod forward, quickly bend and reposition the thumb to begin a new forward push. Consistent motion is important, so the thumb must be repositioned quickly to maintain the pace. The thumb can also be used to slowly roll the electrode as well during the pushing motion. This technique is easily practiced without actively welding; simply practice with a rod to become comfortable with the technique before trying it on a weld.

Another technique to consider is referred to as the *finger-pull technique*. The rod is secured between the pointer and middle fingers, which are used to pull the filler rod forward. At the end of the forward stroke, the fingers are slipped backwards along the rod to reposition, without pulling it backwards and out of the weld puddle in the process. Once the two fingers are repositioned, the pulling motion begins again.

2.2.4 Root and Hot Passes

The greatest challenge of an open-root V-groove weld is making the root pass. The root pass is made from the V-groove side of the joint. The root pass must have complete penetration but not an excessive amount of root reinforcement. The penetration is controlled by using one of two different techniques. The first is called running a keyhole. A keyhole (*Figure 14*) is a hole that is made when the root faces of two plates are melted away by the welding arc. The molten metal flows to a weld

Figure 14 A root-pass keyhole.

puddle at the back of the keyhole, forming the weld.

To run a keyhole, rest the filler rod on the joint root just ahead of the leading edge of the arc. When you require more filler metal, move the arc back to the rear of the weld puddle, and insert the rod into the leading edge of the weld puddle to add more metal. This is sometimes referred to as the *on-the-wire method*.

Another technique, slightly different than the on-the-wire method, allows continuous feeding of the rod into the molten weld puddle. The filler metal rod used is smaller than the root opening, and it is passed between the root faces. The filler metal is melted as the arc is passed over it. This keeps the puddle width to a minimum and achieves complete penetration. This technique may be used when access to the joint from the outside of the pipe or tube is limited.

When welding with a consumable insert, hold the torch perpendicular to the pipe surface and point it radially toward the center of the pipe. Use an arc length of about 1⁄8" (3 mm) to melt the insert. Forward movement is controlled by the melting rate of the insert and the characteristics of the weld puddle. Increased fluidity and a rising weld puddle indicate that there is sufficient melting. When this occurs, advance the arc in a step-by-step manner.

When an insert is not used, clean and inspect the root pass after it has been run. Remove any excess buildup or undercut with the edge of a narrow grinding wheel (not a cutoff wheel). This will make it easier to apply the hot pass. Be careful not to widen the groove or grind through the root pass.

For most carbon steel pipe, the hot pass is usually performed with GTAW. The remaining passes are often made with SMAW, GMAW, or FCAW. For the purposes of this module, all passes, including the root and hot passes, will be done with GTAW.

It is essential that the hot pass be well-fused to the root pass and pipe bevels. The hot pass is made by running a bead that fuses to both sides of the root pass, using higher current or higher voltage, depending on the process employed. After completing and thoroughly cleaning the hot pass bead, apply the remaining passes.

2.3.0 Practicing Open-Root V-Groove Welds

The following sections explain how to perform these open-root V-groove weld positions:

- Flat (1G-ROTATED)
- Horizontal (2G)
- Multiple (5G)
- Multiple inclined (6G)

> **NOTE**
> The following procedures assume that a consumable insert is not being used for the root pass.

2.3.1 Flat (1G-ROTATED) Position

Practice open-root V-groove pipe welds in the 1G-ROTATED position. Use a shielding gas and a solid filler rod that has a diameter specified by your instructor. Keep the torch angle at 90 degrees to the pipe axis (a 0-degree work angle) with a 15- to 20-degree push angle. Use a slight oscillating motion to form the weld puddle for the root pass. Pay special attention to fuse into the tack welds. Clean the face of the root pass with a brush.

Run the remaining passes using stringer or weave beads and either the walking-the-cup or the freehand torch-handling technique. Take special care to fill the crater at the termination of the weld. Stagger the location of the starts and stops for each pass. Run all passes at or near the top of the pipe as it rotates.

Follow these steps to practice open-root V-groove pipe welds in the 1G-ROTATED position:

Step 1 Tack weld the practice pipe weld coupons together as explained earlier.

Step 2 Position the pipe coupon horizontally on two sets of rollers at a comfortable welding height. A welding positioner or pipe rotator is also required. The positioner shown in *Figure 15 (A)* is very popular. The pipe rotator, shown in *Figure 15 (B)*,

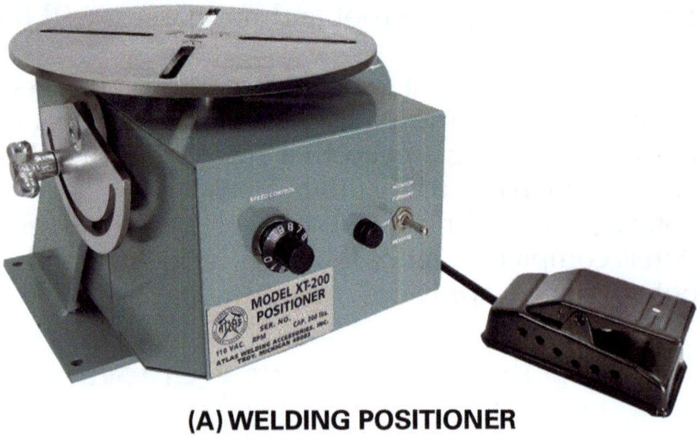

(A) WELDING POSITIONER

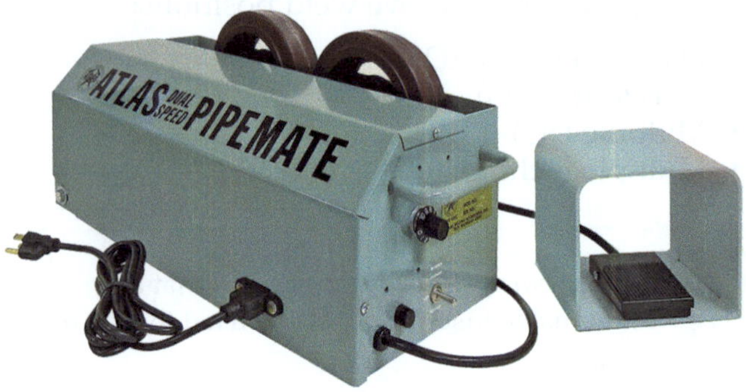

(B) PIPE ROTATOR

Figure 15 A welding positioner and a pipe rotator.

is another option. This model can handle up to 16" (DN400) pipe and weights up to 1,200 pounds (544 kg). Adjustable roller stands can be used to help support longer stock.

Step 3 Make sure that the workpiece clamp is attached directly to the pipe coupon. This will prevent the welding current from passing through the roller bearings or arcing between the rollers and the pipe coupon.

CAUTION

Failure to attach the workpiece clamp to the pipe coupon in the 1G-ROTATED position can result in variations in welding current, damage to the roller bearings, and arcing on the rollers and pipe coupon.

Step 4 Run the root pass using an oscillating motion. Position a tack weld at the eleven o'clock position. Start the weld bead on the tack weld, and advance toward the twelve o'clock position.

Step 5 Rotate the pipe as needed to keep the weld in the flat position.

Step 6 Brush and, as needed, grind the root pass to clean and shape the weld.

Step 7 Use the same procedure to make the hot pass, shown as Bead 2 in *Figure 16 (A)*, and remaining fill passes. Use stringer beads. Pay special attention to complete fusion at the toes of the weld. Overlap the passes and clean the weld after each pass. *Figure 16* shows the related bead sequence and work angles.

2.3.2 Horizontal (2G) Position

Practice open-root V-groove pipe welds in the 2G position. Use a shielding gas and a filler rod of a diameter specified by your instructor. The torch should be at a 15- to 20-degree push angle for the weld passes (*Figure 17*). To prevent the weld puddle from sagging, the torch work angle can be dropped slightly, but not more than 10 degrees. For the root pass, use an oscillating motion. Be sure to achieve complete fusion into the tack welds.

When running the remaining passes, use stringer beads and either the walking-the-cup or the freehand torch-handling technique. Ensure proper fusion at the toes of the weld. Take special care to fill the crater at the termination of the weld, and stagger the location of the starts and stops for each pass.

Follow these steps to practice open-root V-groove pipe welds in the 2G position:

Step 1 Tack weld the practice pipe weld coupons together as explained earlier.

Step 2 Clamp or tack weld the pipe coupon with the axis of the pipe vertical.

Step 3 Run the root pass, shown as Bead 1 in *Figure 17*. Start the root pass on a tack weld. Continue this procedure until the root pass has been completed.

Welding Heavy-Wall Pipe

When welding a V-groove joint in heavy-wall pipe, it is sometimes difficult to maintain a proper arc length. This may be because of the torch size or the depth/angle of the groove. In these cases, use a narrow nozzle. It will allow the torch to be positioned closer to the root opening. Another method is to use a gas lens that will allow a longer electrode stickout. In the latter case, you may have to use a larger electrode to prevent overheating.

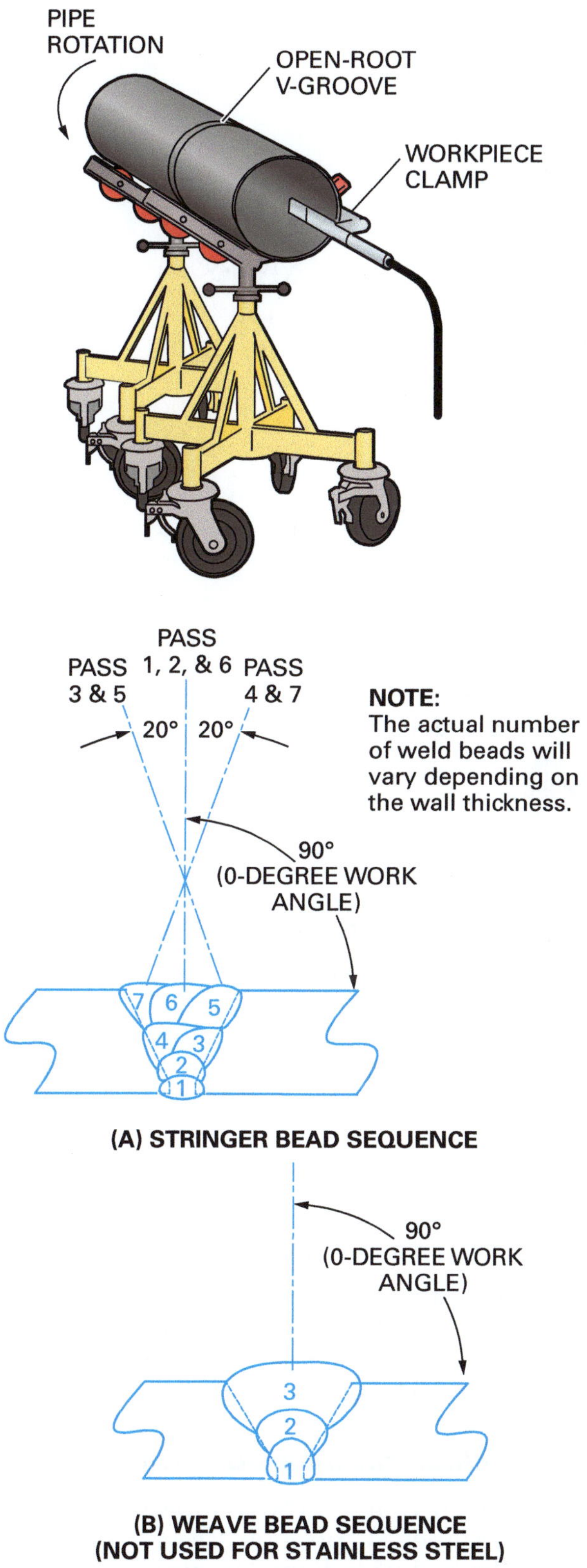

Figure 16 Multiple-pass 1G-ROTATED bead sequence and work angles.

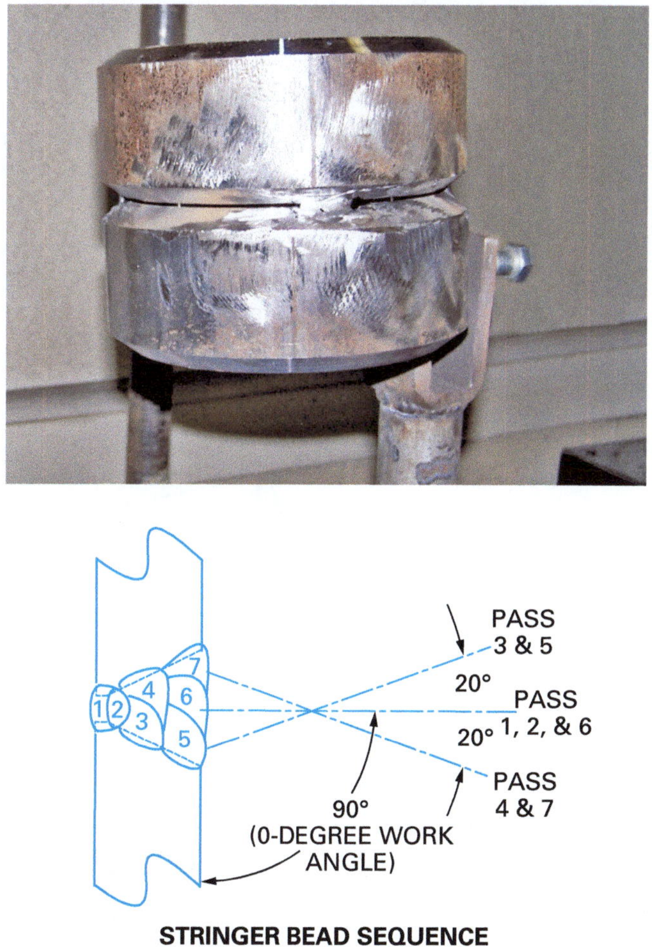

Figure 17 Multiple-pass 2G bead sequence and work angles.

Step 4 Brush and, as needed, grind the root pass to clean and shape the weld.

Step 5 Run the hot pass, Bead 2, and the remaining passes with stringer beads at the proper work angle. Overlap the passes and clean the weld after each pass.

2.3.3 Multiple (5G) Position

Practice open-root V-groove pipe welds in the multiple (5G) position. Use a shielding gas and a filler rod of a diameter specified by your instructor. Start on a tack weld at the bottom of the pipe, and weld uphill toward the top. Use a torch push angle of 15 to 20 degrees to the pipe surface as shown in *Figure 18*.

The torch should be at approximately a 15- to 20-degree push angle for all passes as shown in *Figure 19*. Use stringer beads or weave beads for the remaining passes, and either the walking-the-cup

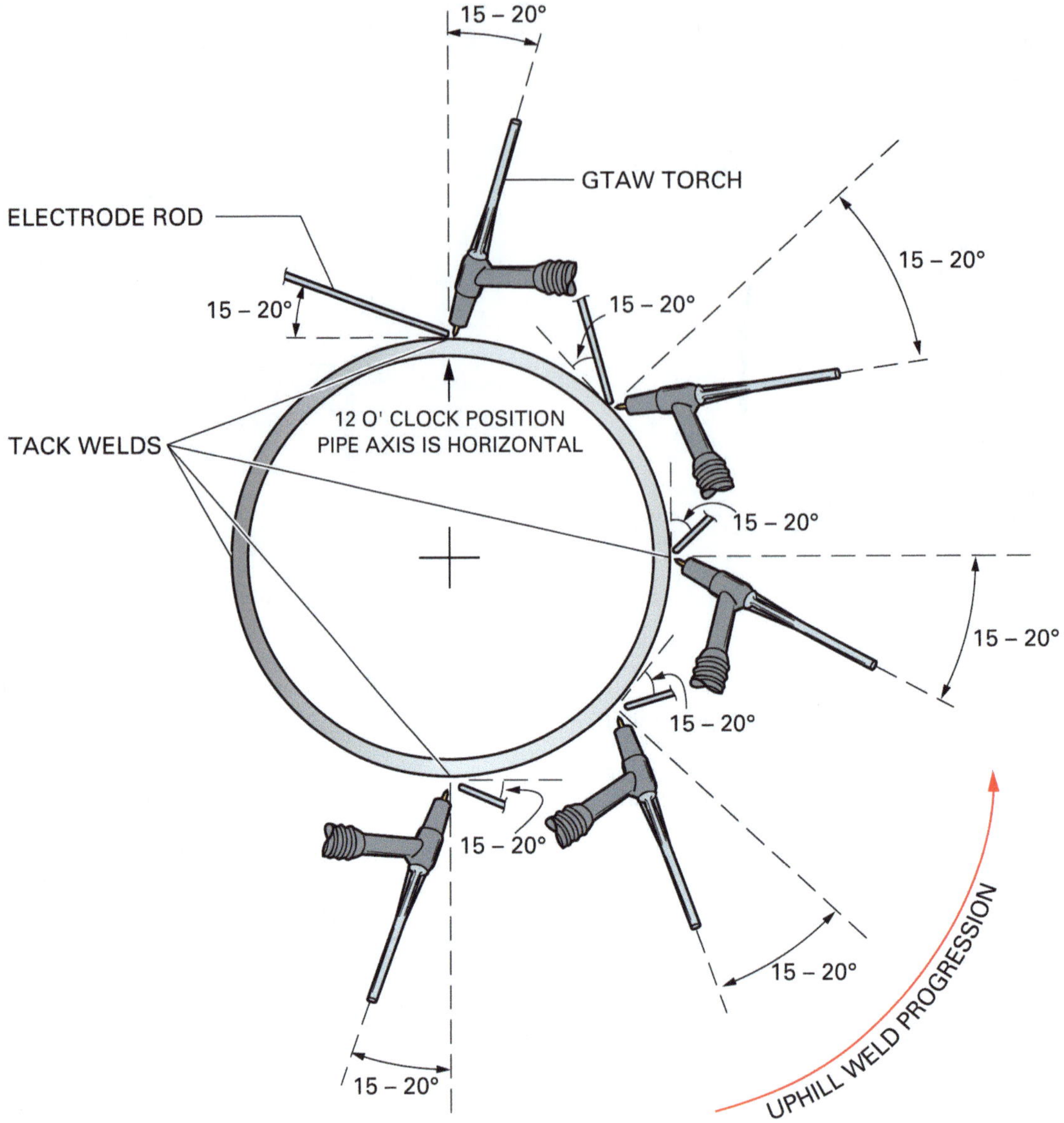

Figure 18 GTAW torch and filler rod angles for the 5G position.

or the freehand torch-handling technique. Ensure proper fusion is achieved at the toes of the weld bead. Pay special attention to filling the crater at the termination of the weld, and stagger the location of the starts and stops for each pass.

Follow these steps to practice open-root V-groove pipe welds in the 5G position:

Step 1 Tack weld the practice pipe weld coupons together as explained earlier.

Step 2 Clamp or tack weld the pipe weld coupon into position with the pipe axis horizontal.

Step 3 Run the root pass uphill using a stringer bead, as shown in *Figure 19* (*B*). Repeat the procedure on the opposite side of the pipe; the pipe coupon cannot be rotated.

Step 4 Brush and, as needed, grind the root pass to clean and shape the weld.

Step 5 Run the hot pass and the remaining passes uphill at the proper work angles. Overlap the passes and clean the weld after each pass.

2.3.4 Multiple Inclined (6G) Position

Practice open-root V-groove pipe welds in the multiple inclined 6G position. Use a shielding gas and a filler rod of a diameter specified by your instructor. Start at the bottom of the pipe, and weld uphill toward the top. For the root pass, the torch should be at approximately a 15- to 20-degree push angle and at 90 degrees (0-degree work angle) to the pipe (*Figure 20*).

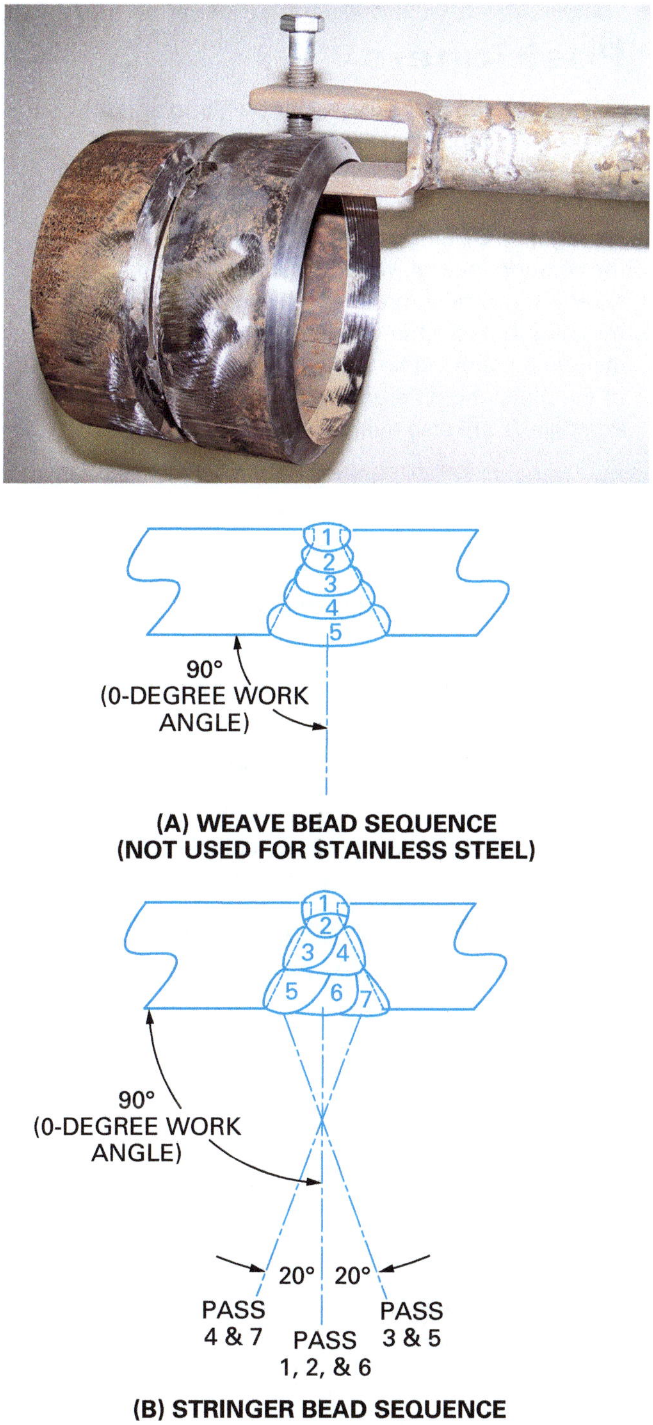

Figure 19 Multiple-pass 5G bead sequences and work angles.

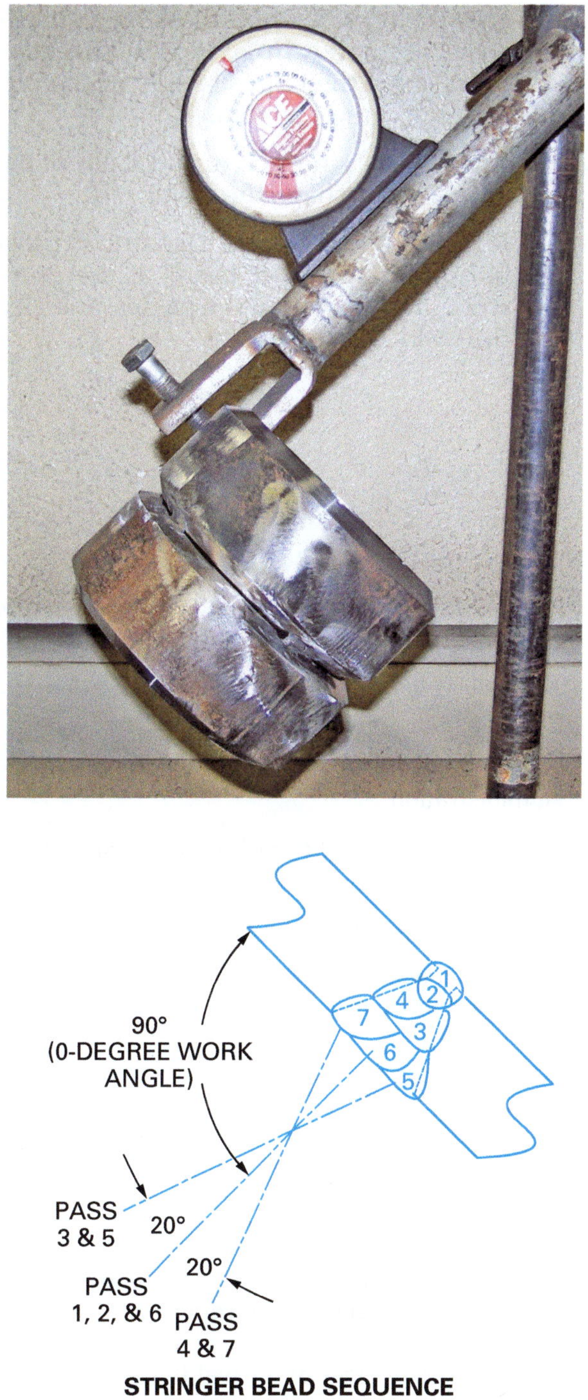

Figure 20 Multiple-pass 6G bead sequence and work angles.

When running the remaining passes, use stringer beads and either the walking-the-cup or the free-hand torch-handling technique. Ensure proper fusion at the toes of the weld. Pay special attention to fill the crater at the termination of the weld, and stagger the location of the starts and stops for each pass.

Follow these steps to practice open-root V-groove pipe welds in the 6G position:

Step 1 Tack weld the practice pipe weld coupons together as explained earlier.

Step 2 Clamp or tack weld the pipe weld coupon into position with the pipe axis inclined 45 degrees to the horizontal plane.

Step 3 Run the root pass uphill using a stringer bead, as shown in *Figure 20*. Repeat the process for the opposite side of the pipe; the pipe cannot be rotated.

Step 4 Brush and, as needed, grind the root pass to clean and shape the weld.

Step 5 Run the hot pass and remaining passes uphill with stringer beads at the proper work angle. Overlap the passes, and clean the weld after each pass.

Positioners

Positioners can make life easier by tilting and/or rotating the table as needed for various welding positions. The Atlas XT-200 model shown in *Figure 15* can tilt forward up to 130 degrees and backwards 15 degrees. By tilting the table 90 degrees and using the rotation feature, welders can easily work in the 1G-ROTATED position. The unit can also be quickly reconfigured to other welding positions. Options include a hollow table shaft to support gas purging of the pipe, and different rotational speed ranges. Positioners are also available in very large sizes.

Additional Resources

AWS F4.1: Recommended Safe Practices for Preparation for Welding and Cutting of Containers and Piping. Miami, FL: American Welding Society.

Modern Welding Technology. Current Edition. Howard B. Cary and Scott Helzer. Englewood Cliffs, NJ: Prentice Hall.

OSHA 1926.351, Arc Welding and Cutting. Current Edition. Washington, DC: Occupational Safety & Health Administration.

The Procedure Handbook of Arc Welding. Current Edition. Cleveland, OH: The James F. Lincoln Arc Welding Foundation.

Welding Handbook, Volumes 1 through 4. Current Edition. Miami, FL: American Welding Society.

Welding Pressure Pipelines and Piping Systems (PDF). The Lincoln Electric Company, Cleveland, OH, USA. **www.lincolnelectric.com**. Last accessed May 29, 2015.

2.0.0 Section Review

1. Excessively large face reinforcement reduces the pipe's strength, causing the stresses in the pipe to be concentrated _____.
   a. in the hot pass
   b. along the toes of the weld
   c. at the root face
   d. across the face of the weld

2. A slow travel speed with GTAW allows the formation of _____.
   a. significant discontinuities
   b. an excessive amount of slag
   c. a smaller, shallower puddle
   d. a larger, more deeply penetrating puddle

3. Welding with the axis of the pipe coupons positioned vertically describes the _____.
   a. 1G-ROTATED position
   b. 2G position
   c. 5G position
   d. 6G position

SUMMARY

Compared to other welding processes, GTAW is rather unique in that the welder must manually feed the filler metal into the weld. All welding requires good coordination, but GTAW requires coordinated movement from both hands. GTAW is excellent for use with thin materials such as sheet metal. Learning and mastering the GTAW process significantly increases a welder's value on the job and broadens the scope of opportunities. Continue practicing these welds and the handling of the filler rod until the needed skills have been acquired.

Review Questions

1. A flow of shielding gas from the nozzle of a GTAW torch shields the electrode, arc, and the molten base metal from _____.
   a. excessive heat
   b. blowing dust
   c. tungsten contamination
   d. atmospheric contamination

2. Acetylene is lighter than air and is not as likely to collect in low areas. It is more dangerous than propane however, because it _____.
   a. will ignite at much lower concentrations
   b. is toxic even at low concentrations
   c. is highly corrosive to several metals
   d. must be stored at very high pressures

3. Welding operations on materials containing zinc or cadmium create _____.
   a. fires
   b. explosions
   c. toxic fumes
   d. weak welds

4. If the filler metal cannot be cleaned of contamination or corrosion, _____.
   a. do not use it
   b. preheat it with a torch first
   c. increase the flow of shielding gas
   d. add a paste flux to the weld zone

5. The number of standard pipe groove weld test positions is _____.
   a. three
   b. four
   c. five
   d. seven

6. Regardless of the technique used for GTAW, always keep the tip of the filler rod _____.
   a. within the shielding gas envelope
   b. outside of the shielding gas envelope
   c. in contact with the tungsten electrode
   d. touching the molten puddle under the electrode

7. If the tungsten electrode touches the filler rod or the weld puddle, it will become contaminated with _____.
   a. shielding gas
   b. flux and/or slag
   c. the filler metal
   d. hexavalent chromium

8. The edge of the torch nozzle is intentionally rested against the base metal or groove edges in the GTAW technique known as the _____.
   a. stringer bead technique
   b. on-the-wire technique
   c. freehand technique
   d. walking-the-cup technique

9. Which two fingers are used to pull the filler rod forward when using the finger-pull technique?
   a. The pointer and the thumb
   b. The pointer and middle finger
   c. The middle and ring fingers
   d. The ring and little fingers

10. When making open-root V-groove pipe welds in the 1G-ROTATED position, keep the torch work angle at 0 degrees and use a _____.
    a. 10- to 15-degree drag angle
    b. 10- to 15-degree push angle
    c. 15- to 20-degree drag angle
    d. 15- to 20-degree push angle

Appendix

Performance Accreditation Tasks

Performance Accreditation Tasks (PATs) provide specific acceptable criteria for performance and help to ensure a true competency-based welding program for students.

The following tasks are designed to evaluate your ability to run open-root V-groove welds with GTAW equipment in four standard test positions using carbon steel filler rod of the appropriate diameter and argon shielding gas. Perform each task when you are instructed to do so by your instructor. As you complete each task, show it to your instructor for evaluation. Do not proceed to the next task until told to do so. For AWS 2G and 5G certifications, refer to *AWS EG3.0-96 Guide for the Training and Qualification of Welding Personnel; Level II – Advanced Welder* for bend test requirements. For AWS 6G certifications, refer to *AWS EG4.0-96 Guide for the Training and Qualification of Welding Personnel; Level III – Expert Welder* for bend test requirements.

PATs 2 and 3 correspond to *AWS EG3.0-96, Section 3.3.6.4, Unit #4, GTAW, Learning Objective #14*. PAT 4 corresponds to *AWS EG4.0-96, Section 3.3.6.5, Unit #5, GTAW, Learning Objective #10*. PAT 1 has no AWS correlation.

OPEN-ROOT V-GROOVE PIPE WELD IN THE 1G-ROTATED POSITION

Using GTAW equipment and carbon steel filler rod of the appropriate diameter, argon shielding gas, and stringer beads, make open-root V-groove welds on carbon steel pipe in the 1G-ROTATED position.

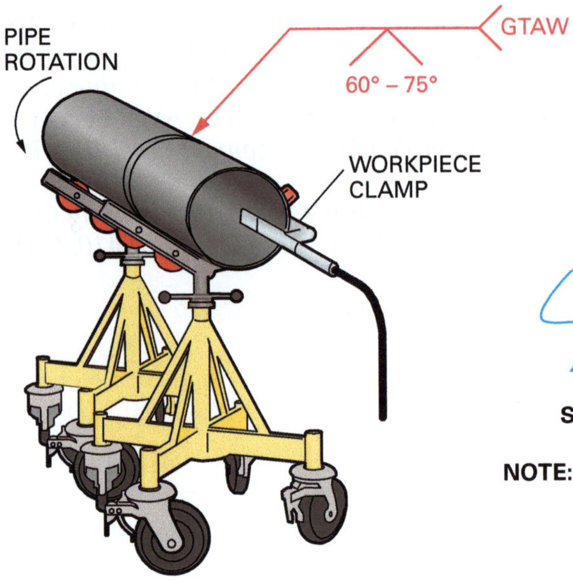

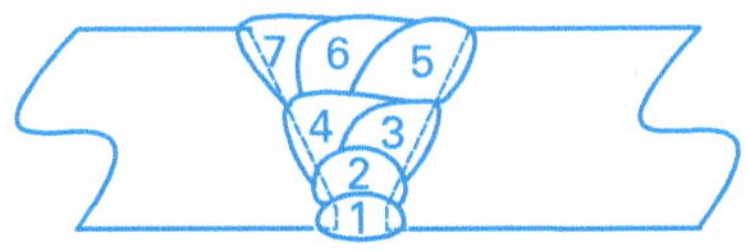

STRINGER BEAD SEQUENCE

NOTE: THE ACTUAL NUMBER OF WELD BEADS WILL VARY DEPENDING ON THE WALL THICKNESS.

Criteria for Acceptance:

- Uniform appearance on the bead face ________
- Craters and restarts filled to the full cross section of the weld ________
- Acceptable weld profile in accordance with the applicable codes and standards ________
- Smooth transition with complete fusion at the toes of the weld ________
- Complete joint penetration and face reinforcement with the root reinforcement at least flush with the inside of the pipe, to a maximum buildup of ⅛" (3 mm) ________
- No porosity ________
- No undercut ________
- No cracks ________
- No overlap ________
- No incomplete fusion ________

GTAW OPEN-ROOT V-GROOVE PIPE WELD IN THE 2G POSITION

Using GTAW equipment and carbon steel filler rod of the appropriate diameter, argon shielding gas, and stringer beads, make an open-root V-groove weld on carbon steel pipe in the 2G position.

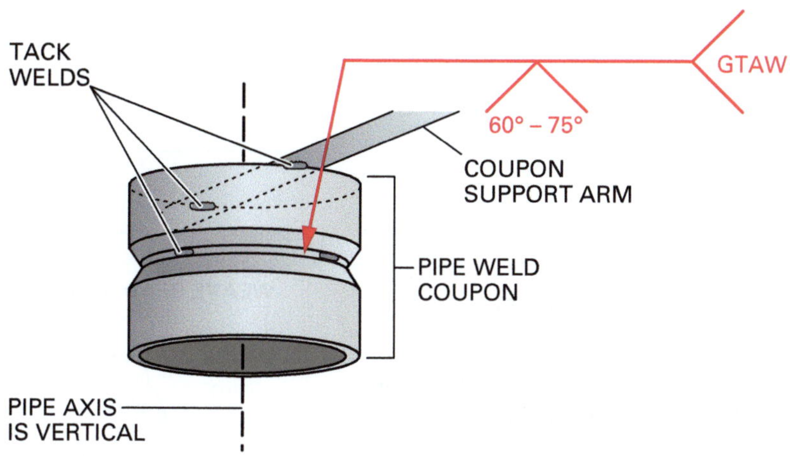

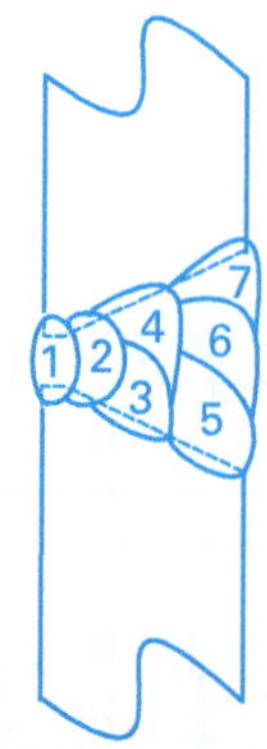

STRINGER BEAD SEQUENCE

Criteria for Acceptance:

- Uniform appearance on the bead face ____________
- Craters and restarts filled to the full cross section of the weld ____________
- Acceptable weld profile in accordance with the applicable codes and standards ____________
- Smooth transition with complete fusion at the toes of the weld ____________
- Complete joint penetration and face reinforcement with the root reinforcement at least flush with the inside of the pipe, to a maximum buildup of ⅛" (3 mm) ____________
- No porosity ____________
- No undercut ____________
- No cracks ____________
- No overlap ____________
- No incomplete fusion ____________

GTAW OPEN-ROOT V-GROOVE PIPE WELD IN THE 5G POSITION

Using GTAW equipment and carbon steel filler rod of the appropriate diameter, argon shielding gas, and stringer or weave beads, make an open-root V-groove weld on carbon steel pipe in the 5G position.

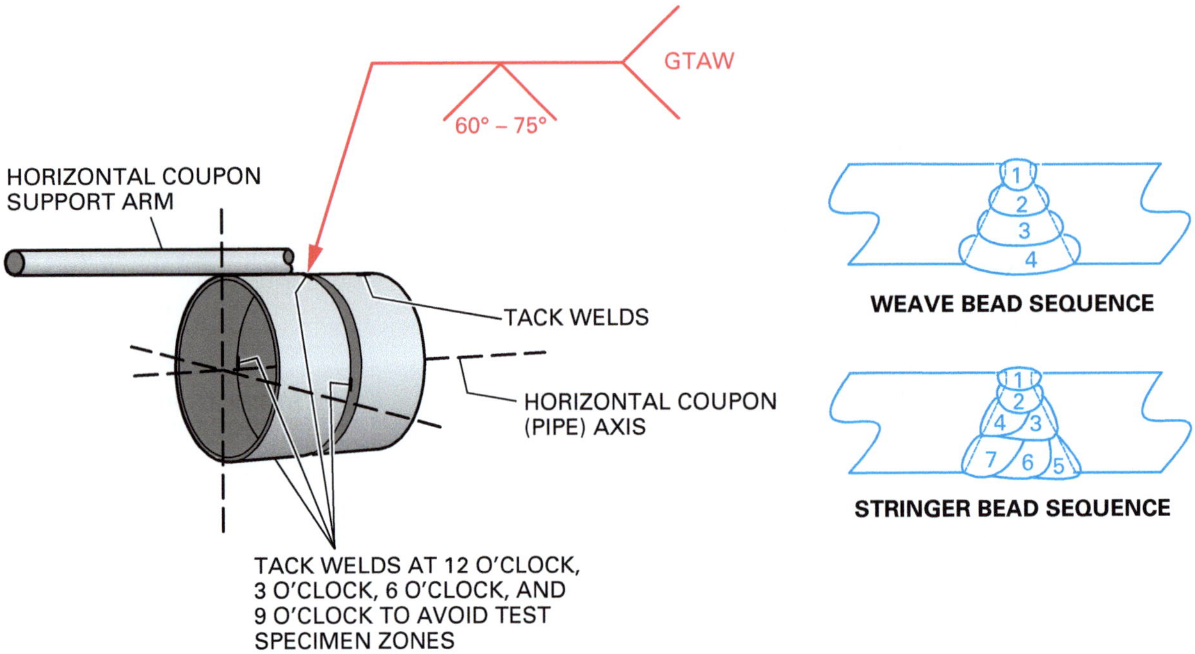

Criteria for Acceptance:

- Uniform appearance on the bead face ______
- Craters and restarts filled to the full cross section of the weld ______
- Acceptable weld profile in accordance with the applicable codes and standards ______
- Smooth transition with complete fusion at the toes of the weld ______
- Complete joint penetration and face reinforcement with the root reinforcement at least flush with the inside of the pipe, to a maximum of ⅛" (3 mm) ______
- No porosity ______
- No undercut ______
- No cracks ______
- No overlap ______
- No incomplete fusion ______

GTAW OPEN-ROOT V-GROOVE PIPE WELD IN THE 6G (OR 6GR) POSITION

Using GTAW equipment and carbon steel filler rod of the appropriate diameter, argon shielding gas, and stringer beads, make an open-root V-groove weld on carbon steel pipe in the 6G (or 6GR) position.

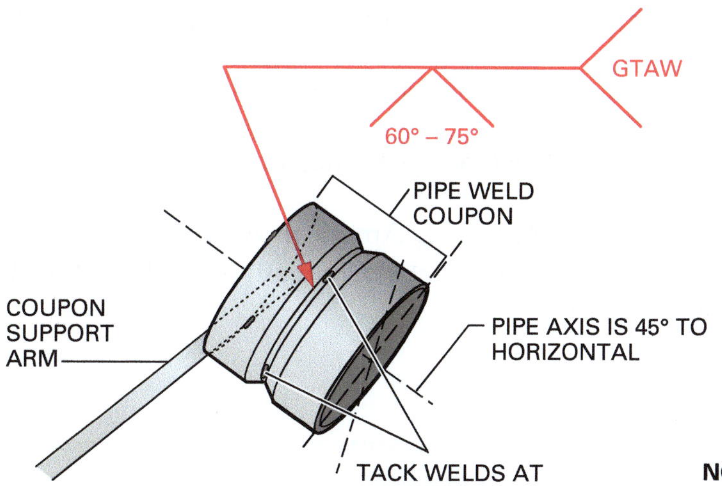

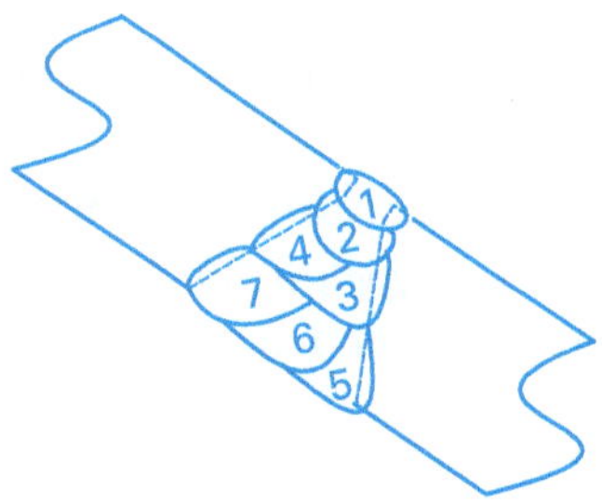

STRINGER BEAD SEQUENCE

NOTE:
IF REQUIRED FOR QUALIFICATION PURPOSES, A RESTRICTING RING MAY BE ADDED TO THE 6G POSITION COUPON TO FORM A 6GR POSITION COUPON.

Criteria for Acceptance:

- Uniform appearance on the bead face ________
- Craters and restarts filled to the full cross section of the weld ________
- Acceptable weld profile in accordance with the applicable codes and standards ________
- Smooth transition with complete fusion at the toes of the weld ________
- Complete joint penetration and face reinforcement with the root reinforcement at least flush with the inside of the pipe, to a maximum of ⅛" (3 mm) ________
- No porosity ________
- No undercut ________
- No cracks ________
- No overlap ________
- No incomplete fusion ________

Additional Resources

This module presents thorough resources for task training. The following reference material is recommended for further study.

AWS F4.1: Recommended Safe Practices for Preparation for Welding and Cutting of Containers and Piping. Miami, FL: American Welding Society.

Modern Welding Technology. Current Edition. Howard B. Cary and Scott Helzer. Englewood Cliffs, NJ: Prentice Hall.

OSHA 1926.351, Arc Welding and Cutting. Current Edition. Washington, DC: Occupational Safety & Health Administration.

The Procedure Handbook of Arc Welding. Current Edition. Cleveland, OH: The James F. Lincoln Arc Welding Foundation.

Welding Handbook, Volumes 1 through 4. Current Edition. Miami, FL: American Welding Society.

Welding Pressure Pipelines and Piping Systems (PDF). The Lincoln Electric Company, Cleveland, OH, USA. **www.lincolnelectric.com**. Last accessed May 29, 2015.

Figure Credits

The Lincoln Electric Company, Cleveland, OH, USA, Module Opener, Figures 2–4, 13, SA02

John Tillman Company, SA01

Topaz Publications, Inc., Figure 6 (photo)

Terry Lowe, Figures 7 (photo), 14, 17 (photo), 19 (photo), 20 (photo)

Atlas Welding Accessories, Inc., Figure 15

Section Review Answer Key

Answer	Section Reference	Objective
Section One		
1. a	1.1.0	1a
2. d	1.2.3	1b
3. d	1.3.4	1c
Section Two		
1. b	2.1.1	2a
2. d	2.2.1	2b
3. b	2.3.2	2c

NCCER CURRICULA — USER UPDATE

NCCER makes every effort to keep its textbooks up-to-date and free of technical errors. We appreciate your help in this process. If you find an error, a typographical mistake, or an inaccuracy in NCCER's curricula, please fill out this form (or a photocopy), or complete the online form at **www.nccer.org/olf**. Be sure to include the exact module ID number, page number, a detailed description, and your recommended correction. Your input will be brought to the attention of the Authoring Team. Thank you for your assistance.

Instructors – If you have an idea for improving this textbook, or have found that additional materials were necessary to teach this module effectively, please let us know so that we may present your suggestions to the Authoring Team.

NCCER Product Development and Revision

13614 Progress Blvd., Alachua, FL 32615

Email: curriculum@nccer.org
Online: www.nccer.org/olf

❑ Trainee Guide ❑ Lesson Plans ❑ Exam ❑ PowerPoints Other ______________________

Craft / Level: Copyright Date:

Module ID Number / Title:

Section Number(s):

Description:

Recommended Correction:

Your Name:

Address:

Email: Phone:

GTAW – Low Alloy and Stainless Steel Pipe

Overview

Working with low alloy and stainless steel can be challenging for welders, because the weld puddle tends to be fluid and hard to manage. With practice and the proper techniques, however, sound and reliable welds can be created. This module provides information regarding these materials and their unique characteristics. It describes GTAW techniques and processes for welding low alloy and stainless steel pipe in all positions. General welding safety requirements are emphasized.

Module 29305

Trainees with successful module completions may be eligible for credentialing through the NCCER Registry. To learn more, go to **www.nccer.org** or contact us at 1.888.622.3720. Our website has information on the latest product releases and training, as well as online versions of our *Cornerstone* magazine and Pearson's product catalog.

Your feedback is welcome. You may email your comments to **curriculum@nccer.org**, send general comments and inquiries to **info@nccer.org**, or fill in the User Update form at the back of this module.

This information is general in nature and intended for training purposes only. Actual performance of activities described in this manual requires compliance with all applicable operating, service, maintenance, and safety procedures under the direction of qualified personnel. References in this manual to patented or proprietary devices do not constitute a recommendation of their use.

29305 V5

29305
GTAW – Low Alloy and Stainless Steel Pipe

Objectives

When you have completed this module, you will be able to do the following:

1. State the basic concepts of open-root GTAW of low alloy and stainless steel pipe and how to prepare for welding.
   a. State the basic concepts of low alloy and stainless steel pipe welding using the GTAW process.
   b. Identify common welding safety practices.
   c. Explain how to prepare the area, materials, and equipment for low alloy and stainless steel pipe welding using the GTAW process.
2. Describe open-root V-groove pipe welding positions and GTAW pipe-welding techniques.
   a. Describe the techniques used to apply GTAW to low alloy and stainless steel pipe.
   b. Describe how to make the root pass with a gas backing.
   c. Describe the techniques required to produce open-root GTAW low alloy and stainless steel pipe welds in various positions.

Performance Tasks

Under the supervision of your instructor, you should be able to do the following:

1. Make GTAW open-root V-groove low alloy or stainless steel pipe welds in the 2G position using the appropriate filler metal and a gas backing.
2. Make GTAW open-root V-groove low alloy or stainless steel pipe welds in the 5G position using the appropriate filler metal and a gas backing.
3. Make GTAW open-root V-groove low alloy or stainless steel pipe welds in the 6G position using the appropriate filler metal and a gas backing.

Trade Terms

Austenitic steel
Carbide precipitation
Carburized
Ferritic steel
Intergranular corrosion
Low alloy steel
Martensitic steel

Industry Recognized Credentials

If you are training through an NCCER-accredited sponsor, you may be eligible for credentials from NCCER's Registry. The ID number for this module is 29305. Note that this module may have been used in other NCCER curricula and may apply to other level completions. Contact NCCER's Registry at 888.622.3720 or go to **www.nccer.org** for more information.

Contents

Figures and Tables

SECTION ONE

1.0.0 GTAW LOW ALLOY AND STAINLESS PIPE WELDING PREPARATIONS

Objective

State the basic concepts of open-root GTAW of low alloy and stainless steel pipe and how to prepare for welding.

a. State the basic concepts of low alloy and stainless steel pipe welding using the GTAW process.
b. Identify common welding safety practices.
c. Explain how to prepare the area, materials, and equipment for low alloy and stainless steel pipe welding using the GTAW process.

Trade Terms

Austenitic steel: These stainless steels are nonmagnetic in the annealed condition and nonhardenable by heat treatment; they can be hardened significantly by cold working. These steels combine excellent corrosion- and heat-resistance with good mechanical properties over a broad temperature range. Austenitic stainless steels make up the largest of the three common stainless steel groups and are the most likely to be encountered by welders.

Carbide precipitation: The process of carbides combining with chromium to create chromium carbide. The result is a reduction of chromium and the corrosion-resistance it provides. Intergranular corrosion results from the loss of chromium.

Carburized: The process of iron or steel combining with carbon liberated from a metal when the metal is heated.

Ferritic steel: These stainless steels are always magnetic and can be hardened to some extent by cold working. They combine corrosion- and heat-resistance with fair mechanical properties over a narrower temperature range than austenitic steels. Ferritic stainless steels are straight chromium stainless steels containing 11.5 to 27 percent chromium, as well as other metals.

Intergranular corrosion: A selective corrosion attack of the crystalline grain boundaries of a stainless steel. It results from chromium depletion which occurs during carbide precipitation.

Low alloy steel: Steel that is alloyed with other elements, amounting to 1 to 8 percent of the steel by weight, to improve its mechanical properties.

Martensitic steel: These stainless steels comprise two groups: chromium-martensitic and martensitic. The chromium-martensitic stainless steels contain from 11.5 to 18 percent chromium, along with other metals. Martensitic stainless steels contain only 4 to 6 percent chromium; for this reason, they are not considered true stainless steels, although their corrosion-resistance is much greater than that of mild carbon steels. Martensitic stainless steels are magnetic and can be hardened by quenching and tempering.

GTAW open-root V-groove welds provide excellent service in low alloy steel pipe and stainless steel pipe applications. It is especially well-suited for the root pass due to the way it controls penetration. This section describes the basic concepts of GTAW pipe welding and presents guidelines for the preparation of the equipment and work area.

1.1.0 Basic GTAW Low Alloy and Stainless Pipe Welding Concepts

Gas tungsten arc welding (GTAW) is a welding process that uses an arc created between a nonconsumable tungsten electrode and the base metal to melt both the base metal and the filler metal. The electrode, arc, and molten base metal are shielded from atmospheric contamination by a flow of shielding gas from the torch nozzle, as shown in *Figure 1*. The filler metal is a rod of similar composition to the base metal. The bare, solid-metal filler rod is usually handheld and manually fed into the leading edge of the weld puddle. Because there is no flux, minimal post-weld cleaning is needed.

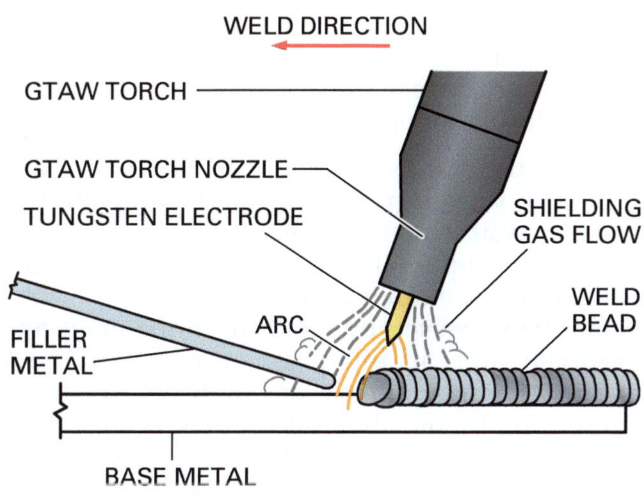

Figure 1 GTAW process.

A primary use of GTAW is to make high-quality welds in stainless steel piping and tubing. These materials are often used in processes requiring high levels of cleanliness. Stainless steel is also used for automotive and industrial exhaust systems and similar applications. It is very resistant to fire, heat, and corrosion.

The GTAW process provides greater control of root penetration and the added fill than most other processes. For this reason, GTAW is often used to make the root pass on pipe. This is true even when the remaining passes are made with SMAW, GMAW, or FCAW, all of which have naturally higher deposition rates.

This module provides guidance on the setup of GTAW equipment and the application of open-root V-groove welds on stainless steel and low alloy steel pipe in the 2G, 5G, and 6G welding positions. The dimensions and specifications in this module are representative of codes in general and may not be specific to any particular code. Always follow the proper codes, standards, and/or specifications for your site.

1.1.1 Welding Low Alloy Steel

Alloy steel is more difficult to weld than carbon steel. Because the molten weld puddle tends to be more fluid, it is slightly more difficult to control. Problems also occur if the wrong filler metal is used, or if the proper temperatures for preheat, interpass, or post-weld heat treatment are not closely followed. Alloy steel has the following characteristics in comparison to carbon steel:

- More likely to develop hydrogen-induced underbead cracking
- Requires a closer match of filler metal and base metal
- Has higher hardenability, which can cause cracks during cooling
- Requires the heating cycles to be more closely managed
- Needs greater care during joint fit-up to ensure a uniform root opening

These traits are similar for all alloy steels, but they can vary significantly depending on the metallurgical structure of the steel. Alloy steels in the same group (HSLA, Q&T, HTLA, and Cr-Mo) have similar welding characteristics.

Hardness caused by a failure to maintain appropriate preheat, interpass, or post-weld heat treatment temperatures is one of the major causes of cracking in low alloy steel welds. As the alloy or carbon content of the alloy steel increases, so does the steel's hardenability. It is very important to carefully control the heating cycle with every increase in the alloy and/or carbon content.

Because controlling the heating cycle is so important when welding alloy steel, heat treatment specialists are often brought in to work directly with the welder. Instruments are often connected to the weldment to provide a continuous record of time and temperature readings during the entire heating and welding cycle.

NOTE

The interpass temperature should typically not exceed 350°F (177°C). Be sure to refer to the WPS or site specifications for exact temperature requirements.

Low alloy steel has higher alloy content than carbon steel, so back-side protection is often required for both tacking and welding. To determine if a purge gas is needed for back-side protection, check the WPS or site quality standards.

Low alloy steel piping is prone to weld discontinuities, so the welding requires a refined technique. Greater control must be exercised over arc length and electrode handling. When running a pass, the oscillation of the tungsten should not be wider than the filler rod. The filler rod is typically of a much higher quality (consistently more pure) than the base metal. "Staying on the rod" minimizes the melting of the base metal, which has more impurities than the filler rod. If the weld puddle becomes too hot and falls to the inside during welding, stop and file or grind the area to remove any **carburized** metal impurities.

1.1.2 Welding Stainless Steel

Welding stainless steel is slightly more difficult than welding carbon steel. An example of stainless steel welding is shown in *Figure 2* (notice that the workpiece is mounted on a positioner). Stainless steel has the following characteristics in comparison to carbon steel:

Weld Puddle Size

When using GTAW in the horizontal, vertical, and overhead positions, you must keep the weld puddle smaller than in the flat position. The weld puddle is normally held in the joint by capillary action. When the puddle is too large, gravitational force can overcome the capillary action and the liquid metal may run out of the joint.

- Conducts heat more slowly
- Expands more when heated
- Has a higher resistance to electrical current flow

These traits are similar for all stainless steels, but they can vary greatly depending on the structure of the metal. Stainless steels in the same metallurgical group (**austenitic steel**, **martensitic steel**, and **ferritic steel**) have similar welding characteristics. Some specific characteristics of these three types of stainless steel are examined further in this section.

> **NOTE**
> The interpass temperature should typically not exceed 350°F (177°C). Be sure to refer to the WPS or site specifications for exact temperature requirements.

Austenitic stainless steels are generally easier to weld than other types of stainless steels. They normally do not require preheating or postheating, but the interpass temperature must be controlled to prevent **carbide precipitation**.

Martensitic stainless steels air-harden in proportion to their chromium and carbon content. This makes the weld and weld zone very hard and brittle unless the piece is heat-treated. The heating requirements of martensitic stainless steels are based on their carbon content. Approximate heating requirements are as follows:

- *0 to 0.1 percent carbon* – No preheat or postheat required.
- *0.1 to 0.2 percent carbon* – Preheat to 500°F (260°C), and cool slowly.
- *0.2 to 0.5 percent carbon* – Preheat to 500°F (260°C), and then anneal after welding; anneal to 1,500°F to 1,600°F (816°C to 871°C), furnace-cool to 1,100°F (593°C), and then air-cool.

Figure 2 An example of stainless steel welding.

If preheating and postheating cannot be used, weld martensitic stainless steels with a ductile filler metal such as ER309 or ER310, if this is allowed by the WPS or other specifications. The heat-affected zone will still be hard and brittle, but the filler metal will be ductile and able to absorb some stress. This will help to protect the heat-affected zone.

Ferritic stainless steels can be more problematic than other stainless steel types. Those that have a higher chromium and carbon content are prone to carbide precipitation and **intergranular corrosion**. Welding ferritic stainless steels causes embrittlement resulting from severe grain growth. As a result, they may require annealing after welding to re-dissolve the carbides and restore corrosion-resistance.

When heat treatment is not an option, weld ferritic stainless steels with ER308, ER309, or ER310 electrodes. These ductile filler metals resist cracking.

GOING GREEN

Advanced Inverter Power Sources

An advanced inverter power source, like the one shown here, can make welding pipe with GTAW easier while using primary power and saving energy. With features such as HF start only, current pulsing, and weld current sequencing (stop/start ramping), these machines provide much more control over some of the variables that affect the quality of a weld.

1.2.0 Safety Practices

Safety considerations are a critical part of preparing for a welding operation. The following is a summary of safety procedures and practices that must be observed while cutting or welding. Be advised that this is only a summary. Complete safety coverage is provided in the *Welding Level One* module, "Welding Safety." That module should be studied, if not already completed, before continuing. Above all, be sure to wear appropriate protective clothing and equipment at all times when welding or cutting.

1.2.1 Protective Clothing and Equipment

Protective clothing and equipment are a necessary part of every welding task. Follow the guidelines below as a general rule, and ensure that you are familiar with any unique requirements your employer may have regarding personal protective equipment (PPE).

- A welding hood equipped with a properly shaded lens is best for all forms of welding. The welding hood lens must be an appropriate shade for the application.
- Wear proper leather or flame-retardant clothing along with welding gloves for protection from flying sparks, molten metal, and heat.
- Wear high-top safety shoes or boots. Make sure that the tongue and lace area of the footwear is covered by a pant leg. If the tongue and lace area is exposed or if the footwear must be protected from burn marks, wear spats under the pant leg or leather chaps over the front and top of the footwear.
- Wear a 100-percent cotton cap without any mesh material. If a hard hat is required for the environment, use one that allows the attachment of rear deflector material and a face shield. A hard hat with a rear deflector is generally preferred when working overhead, and may be required by some employers and job sites.
- Wear earmuffs, or at least earplugs, to prevent sparks from entering the ear canals.

1.2.2 Housekeeping and Fall Protection

Welding is often performed outside in the weather and in the construction environment. Production work done inside large, open shops exposes a welder to equipment moving across the floor and overhead. Welders often have to crawl under or reach over objects to complete their work.

At times, welders must work at height. Fall protection equipment is required for all work at height (*Figure 3*). Always keep fall protection devices untangled and properly attached. Combined with the other equipment a welder must carry, there is a great deal to manage, but fall protection equipment must always be the priority. Fall protection equipment saves lives. Never attach any kind of welding gear to the harness that could interfere with its ability to perform its intended function.

Regardless of where you are working, always keep the work area as neat as possible. This makes the area safer and helps to reduce distractions. Mindful actions, such as making sure that welding cables are not tripping hazards, contribute to a safer environment. Electrode stubs and unused electrodes must not be left on the floor. They can cause someone to lose their footing and fall. Parts waiting to be welded or left over from previous work are a tripping hazard and should be kept out of the walking area. If the parts have sharp edges, they could cut welding cable insulation or torch gas hoses.

1.2.3 Fire and Explosion Prevention

Flammable gases are commonly found around welding and cutting operations, as are sparks and other significant sources of ignition. For that reason, and because many fire-related incidents over the years have been traced back to welding operations, it is crucial that fire and explosion prevention be given consistent attention. Follow these guidelines to maintain a safe environment:

- Never carry matches or gas-filled lighters. Sparks can cause the matches to ignite or the lighter to explode in your pocket, causing serious injury.

Figure 3 Working at height.

FFA Design Competition

Lincoln Electric sponsored a national welding helmet design contest for the Future Farmers of America (FFA) organization. Shown here is the 2014 winner of the competition, a design from the Highland FFA Chapter located in Gilbert, AZ. The design will be applied to a limited-edition Viking helmet.

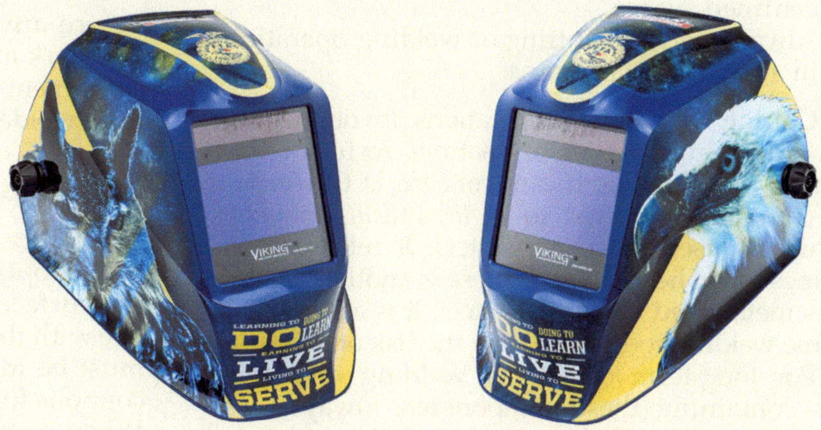

- Never use oxygen to blow dust or dirt from clothing. Traces of the oxygen can remain trapped in the fabric for some time. If a spark hits the fabric during this time, the clothing can burn rapidly and violently out of control.
- Make sure that any flammable material in the work area is removed or shielded by a fire-resistant covering. Approved fire extinguishers must be available before attempting any heating, welding, or cutting operations. If a hot-work permit and a fire watch are required, be sure these are in place before beginning.
- Ensure that all site safety requirements are being observed.
- Never release a large amount of oxygen or use oxygen in place of compressed air. Its presence around flammable materials or sparks can cause rapid and uncontrolled combustion. Keep oxygen away from oil, grease, and other petroleum products.
- Never release a large amount of fuel gas, especially acetylene. Propane tends to concentrate in and along low areas and can ignite at a considerable distance from the release point. Acetylene is lighter than air, but is even more dangerous than propane. When mixed with air or oxygen, acetylene will ignite at much lower concentrations than any other fuel gas.
- Even though shielding gases are not toxic, any gas released in a sufficient quantity can displace the oxygen in a small room or enclosed work area, creating an atmosphere with little or no oxygen. Avoid large releases of any type of gas, regardless of its characteristics.
- To prevent fires, maintain a neat and clean work area, and make sure that any metal scrap or slag is cold before disposal.
- Before cutting containers such as tanks or barrels, check to see if they have contained any explosive, hazardous, or flammable materials, including petroleum products, citrus products, or chemicals that decompose into toxic fumes when heated. As a standard practice, always clean and then fill any tanks or barrels with water, or purge them with a flow of inert gas such as nitrogen to displace any oxygen. Containers must be cleaned by steam cleaning, flushing with water, or washing with detergent until all traces of the material have been removed. Proper procedures for cutting or welding hazardous containers are described in the *American Welding Society (AWS) F4.1, Safe Practices for the Preparation of Containers and Piping for Welding and Cutting.*

WARNING

Welding or cutting must never be performed on drums, barrels, tanks, vessels, or other containers until they have been emptied and cleaned thoroughly, eliminating all flammable materials and all substances (such as detergents, solvents, greases, tars, or acids) that might produce flammable, toxic, or explosive vapors when heated. Clean containers only in well-ventilated areas. Vapors can accumulate during cleaning, causing explosions and serious personal injury.

1.2.4 Work Area Ventilation

Many gases are introduced to the atmosphere through welding processes. Some of these gases are harmful, and all can displace or combine with atmospheric oxygen in a confined space. Follow these guidelines, as well as those of your employer, for work area ventilation:

- Make sure to follow all confined space procedures before conducting any welding or cutting in a confined space. Use powered extraction systems as needed.

- Never use oxygen to provide ventilation in confined spaces.
- Always perform cutting or welding operations in a well-ventilated area.

Cutting or welding operations involving zinc or cadmium materials or coatings result in toxic fumes. Another significant concern is Chromium 6, commonly referred to as *hexavalent chromium,* which is used in stainless steel. It releases toxic fumes when heated. Manganese is another hazardous metal used in stainless steel. It is also used in some welding rods, increasing the risk of exposure.

For long-term cutting or welding of materials containing these components, always wear an approved full-face, supplied-air respirator that uses breathing air supplied from outside the work area. For occasional short-term exposure, a HEPA-rated or metal-fume filter may be used on a standard respirator.

Powered fume-extraction equipment must be used when welding materials that contain zinc, cadmium, hexavalent chromium, manganese, and similar materials that produce hazardous fumes if the area is not naturally well-ventilated. *Figure 4* shows examples of both mobile and portable units.

1.3.0 Preparations for Welding

Before any welding procedure can take place, the work area must be made ready, the welding equipment must be properly set up, and the metal to be welded must be prepared.

1.3.1 Preparing the Welding Area

To practice welding, a welding table, bench, and/or appropriate pipe stands are needed. The welding surface must be made of a material compatible with the material of the workpiece. Provisions must be made for supporting the pipe welding coupons for out-of-position welding.

When working with stainless steel and low alloy materials, it is essential that any tools such as grinding wheels, files, and wire brushes have not been used on carbon steel or other metals. Bits of carbon steel will remain on the tools and contaminate the weld area. Any tool that was previously used to remove hydrocarbon contamination can also cause discontinuities. Consider whether there is any potential for cross-contamination before using these tools. *Figure 5* shows a stainless steel pipe weld that was contaminated with carbon steel during the cleaning process.

(A) MOBILE UNIT

(B) PORTABLE UNIT

Figure 4 Mobile and portable extraction equipment.

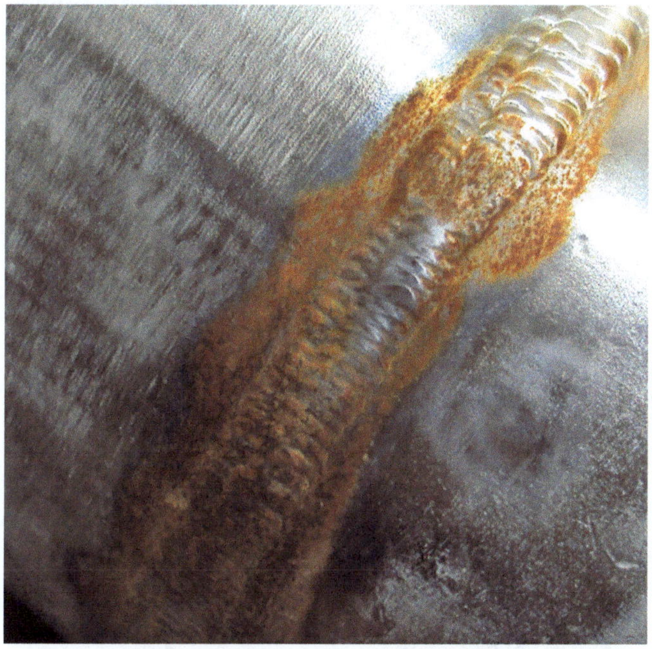

Figure 5 Contaminated stainless steel pipe weld.

To set up the area for welding, follow these steps:

Step 1 Make sure that the area is properly ventilated. Make use of doors, windows, and fans.

Step 2 Check the area for fire hazards. Remove any flammable materials before proceeding.

Step 3 Know the location of the nearest fire extinguisher. Ensure the fire extinguisher is charged and that you and others in the immediate area know how to use it properly.

Step 4 Set up flash shields and curtains around the welding area as necessary.

1.3.2 The Welding Machine

Identify a proper constant-current welding machine for GTAW, and follow these steps to set it up for use:

Step 1 Verify that the welding machine can be used for GTAW, with or without internally controlled gas shielding.

Step 2 Identify an air-cooled or water-cooled GTAW torch. Make sure that it is compatible with the welding machine and any cooling unit.

Step 3 Verify the location of the primary power disconnect. This can be a circuit breaker, a disconnect switch, or even the power cord for corded models.

Step 4 Configure the welding machine for GTAW parameters as directed by your instructor (see *Figure 6*). Configure the torch polarity. Equip the torch with the correct diameter and type of tungsten as directed by your instructor or site WPS.

Step 5 Connect the proper shielding gas for the application as described earlier and as specified by the filler rod manufacturer, WPS, site quality standards, or your instructor.

Step 6 Connect the clamp of the workpiece lead to the workpiece.

Step 7 Turn on the welding machine, adjust the flowmeter, and purge the torch as directed by the manufacturer's instructions.

Note that a backing gas is required for welding stainless steel pipe to protect the back side of the weld during the root pass. The application of the backing gas is discussed later in this module.

1.3.3 Filler Metals

Filler metals are selected to be compatible with the base metal to be welded. The WPS or site standards, which may include specific work packages, will specify the proper filler metal type and size to use.

CAUTION

When the WPS or site quality standards specify a filler metal, it must be used. Always cut the filler rods so that their ID numbers are visible on each section. Be sure to use the end of the rod that has no ID numbers on it. Keep the filler rod stubs (with their numbers) throughout the welding process for quality control checks.

For the welding exercises in this module, use 3⁄32" or 1⁄8" (2.4 or 3.2 mm) stainless and/or low alloy steel filler rods to weld the pipe coupons, as specified by your instructor. Remove only a small number of filler rods at a time. Store the remainder in the package to keep them clean. Before using the filler rod, check it for burned ends or contamination, such as corrosion, dirt, oil, or grease, which can all cause weld defects. Clean the rod with a clean, oil-free rag and a chemical cleaner. Snip any burned ends. If a rod cannot be cleaned, do not use it.

1.3.4 Preparing Pipe Weld Coupons

Pipe weld coupons should be cut from 2" to 6" (DN50 to 150) diameter Schedule 40 or Schedule

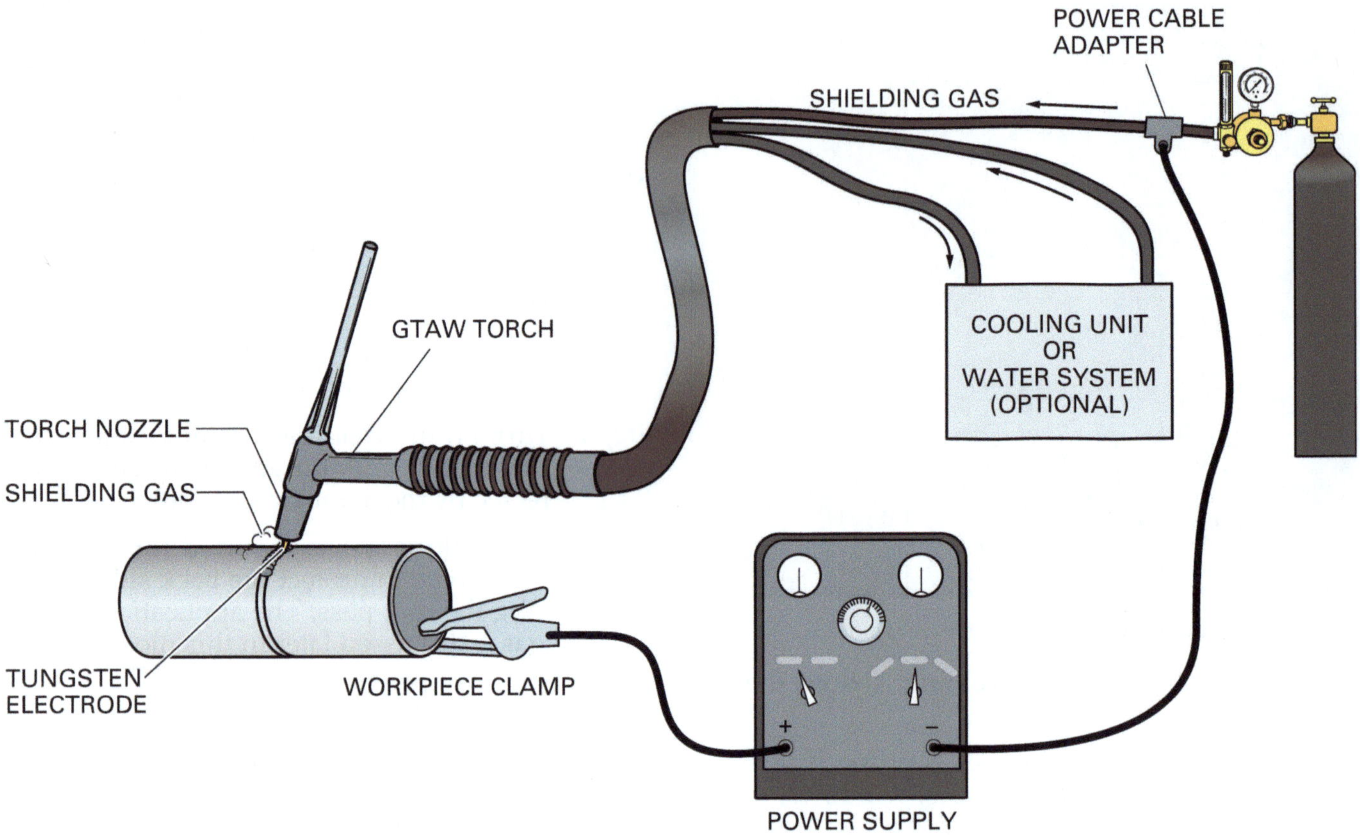

Figure 6 Typical GTAW configuration diagram.

80 low alloy or stainless steel pipe. Each welded joint will require two coupons of the same diameter and schedule. Typical ASME, API, and AWS pipe bevel specifications for bevel angles, root faces, and root openings are shown in *Figure 7.*

To prepare weld coupons for open-root V-groove weld joints, follow these steps:

Step 1 Clean all debris and mill scale from the inside and outside of the pipe with a grinder, wire brush, or other common method.

CAUTION

Ensure that grinding wheels, files, and similar tools used on stainless steel and low-alloy pipe have not been used on carbon steel materials or other metals. Wire brushes should also be stainless steel and must not have been used on other metals. Also confirm that cleaning and grinding tools have not been used for removing hydrocarbon contaminants.

CAUTION

There is little or no surface cleaning action resulting from the GTAW process. As a result, pipe surfaces must be thoroughly cleaned to prevent weld discontinuities.

Step 2 Bevel the end of the pipe to 30 to 37½ degrees by any acceptable beveling method, including manual, mechanical, and thermal methods.

Step 3 Cut off the end of the beveled pipe to the length specified by your instructor. Two coupons will be required for each weld.

The two coupons can now be further prepared for welding:

Step 1 Check the bevel. There should be no dross, and the bevel angle should be 30 to 37½ degrees.

Step 2 Grind or file a 0" to ⅛" (0 to 3 mm) root face on the bevel.

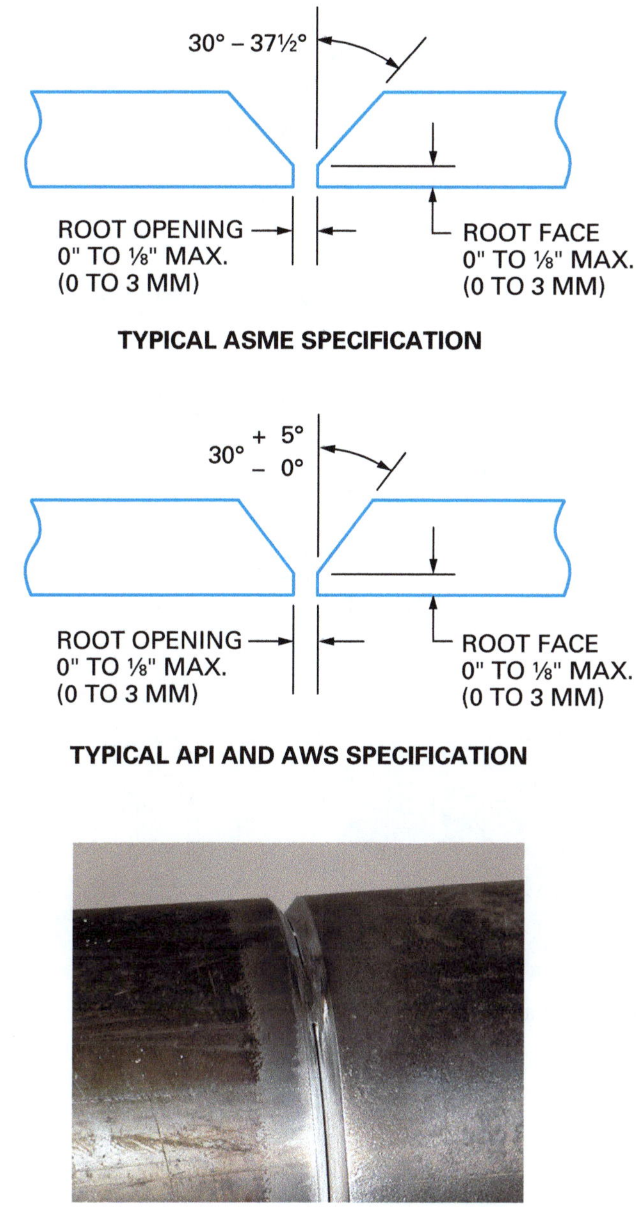

Figure 7 ASME, API, and AWS pipe bevel specifications.

NOTE

Welding codes allow the root face for open-root welds to be from 0" to ⅛" (0 to 3 mm) wide and the root opening to be from 0" to ⅛" (0 to 3 mm). Select and adjust the root face and opening as directed by the instructor as welding practice begins.

Step 3 Align the two pipe sections. The inside surfaces of the two coupons should be even all around and evenly separated by the required root opening. Pieces of filler rod, metal shims, or bare welding electrode can be used to achieve the desired root opening between the two pipe sections until they are tack welded. Small-diameter pipe coupons can be aligned by clamping both pieces to a piece of angle. Large-diameter pipe coupons can be aligned using a pipe alignment jig or by holding a straightedge across the joint, parallel to the pipe axis. The straightedge must be used to check all around the pipe in case one or both sections are distorted. Once aligned, secure the two pipe sections with a clamp.

Step 4 Provide for root backing using backup flux or gas backing.

CAUTION

The same root backing required for welding must be used for tack welding coupons because the tack welds become part of the finished root pass. Failure to provide root backing for tack welds results in unacceptable welds.

NOTE

If a consumable insert is used, make sure that it is secured to the groove side of the joint, and to both sides of the groove.

Step 5 When the root opening is correct and the pipe coupons are aligned, make the first of four tack welds. Longer tacks may be required on heavy-wall or large-diameter pipe.

Step 6 After the first tack weld, check the root opening on the opposite side. Adjust the root opening space if necessary, and make the second tack weld.

CAUTION

Before making the second tack, remove any spacer that was used. This will prevent the spacer from getting caught as the heated metal expands.

Step 7 Verify that the root opening remains correct before welding the third tack midway between the first and second tacks.

Step 8 Weld the fourth tack opposite the third tack and midway between the first and second tacks. There should now be four tack welds evenly spaced (90 degrees) around the pipe coupon (*Figure 8*).

Step 9 Feather the tack welds with the edge of a grinding wheel. Feathering the ends of the tack welds with a grinder helps fuse the root pass into the tack welds.

CAUTION

Some WPS or site quality standards may prohibit grinding stainless steel welds, so check before doing so. When using a grinding wheel, make sure that it is a type approved for use on stainless steel and that it has not been used to grind any other metals or a hydrocarbon-contaminated surface.

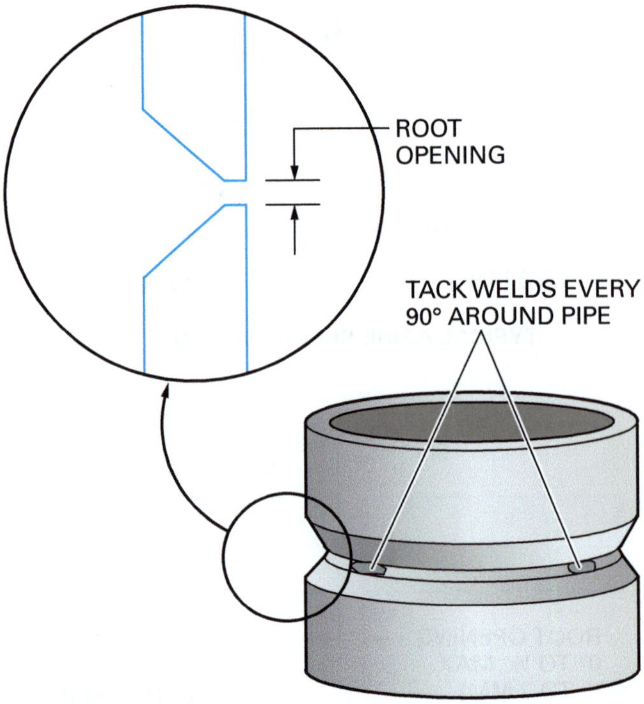

Figure 8 Tacked open-root V-groove weld coupon.

Additional Resources

AWS F4.1: Recommended Safe Practices for Preparation for Welding and Cutting of Containers and Piping. Miami, FL: American Welding Society.

Modern Welding Technology. Current Edition. Howard B. Cary and Scott Helzer. Englewood Cliffs, NJ: Prentice Hall.

OSHA 1926.351, Arc Welding and Cutting. Current Edition. Washington, DC: Occupational Safety & Health Administration.

The Procedure Handbook of Arc Welding. Current Edition. Cleveland, OH: The James F. Lincoln Arc Welding Foundation.

Welding Handbook, Volumes 1 through 4. Current Edition. Miami, FL: American Welding Society.

Welding Pressure Pipelines and Piping Systems (PDF). The Lincoln Electric Company, Cleveland, OH, USA. **www.lincolnelectric.com**. Last accessed May 29, 2015.

1.0.0 Section Review

1. When comparing low alloy and carbon steel welding, which of the following statements is correct?
   a. Low alloy steel is less likely to develop underbead cracking.
   b. Low alloy steel is more hardenable than carbon steel, which can cause cracks during cooling.
   c. The heating cycle for carbon steel must be managed more precisely.
   d. Carbon steel requires more attention to detail in the root opening and root face preparation.

2. The fuel gas that is heavier than air and tends to collect in low areas is _____.
   a. oxygen
   b. nitrogen
   c. acetylene
   d. propane

3. Because the tack welds become an integral part of the finished root pass, _____.
   a. the same root backing required for welding must be used for tack welding
   b. they must be done using the same process as the remaining passes
   c. no filler metal is used when applying the tack welds
   d. the common number of tack welds should be doubled

Section Two

2.0.0 GTAW Techniques and Positions

Objective

Describe open-root V-groove pipe welding positions and GTAW pipe welding techniques.

a. Describe the techniques used to apply GTAW to low alloy and stainless steel pipe.
b. Describe how to make the root pass with a gas backing.
c. Describe the techniques required to produce open-root GTAW low alloy and stainless steel pipe welds in various positions.

Performance Tasks

1. Make GTAW open-root V-groove low alloy or stainless steel pipe welds in the 2G position using the appropriate filler metal and a gas backing.
2. Make GTAW open-root V-groove low alloy or stainless steel pipe welds in the 5G position using the appropriate filler metal and a gas backing.
3. Make GTAW open-root V-groove low alloy or stainless steel pipe welds in the 6G position using the appropriate filler metal and a gas backing.

This section describes torch and filler rod handling techniques used in GTAW. It explains how to set up and manage the backing gas required when welding low alloy and stainless steel pipe. Instruction for the use of GTAW techniques to weld low alloy and stainless steel pipe in multiple positions is also provided.

2.1.0 GTAW Techniques

GTAW weld bead characteristics and quality are affected by several factors, each of which is influenced by the way the torch is handled. These factors include the following:

- Torch travel speed and arc length
- Torch angles
- Torch and filler rod handling techniques

2.1.1 Torch Travel Speed and Arc Length

Torch travel speed and/or arc length may affect the GTAW weld puddle and weld penetration. Slow travel speeds allow more heat to concentrate and form a larger, more deeply penetrating puddle. Faster travel speeds prevent heat buildup and form smaller, shallower puddles. Torch arc length is a significant control for bead width. As the torch is raised, arc length, voltage, and bead width increase.

2.1.2 Torch Angles

The two basic torch angles that must be controlled when performing GTAW are the work angle and the travel angle. The definition of these angles is common to all types of arc welding.

The torch work angle (*Figure 9*) is an angle of 90 degrees or less between a line perpendicular to the major workpiece surface at the point of electrode contact and a plane determined by the electrode axis and the weld axis. For pipe, the plane is determined by the electrode axis and a line tangent to the pipe surface at the same point. As shown in *Figure 9*, the work angle is adjusted up to 20 degrees as beads are added to the weld, depending on how they are positioned.

The torch travel angle is an angle of 90 degrees or less between the electrode axis and a line perpendicular to the weld axis at the point of electrode contact in a plane determined by the electrode axis and the weld axis. For pipe, the plane is determined by the electrode axis and a line tangent to the pipe surface at the same point. The travel angle used for GTAW is a push angle (*Figure 10*). The push angle is created when the torch is tilted back, away from the direction of travel, so that the electrode tip precedes the torch in the direction of the weld. In this position, the

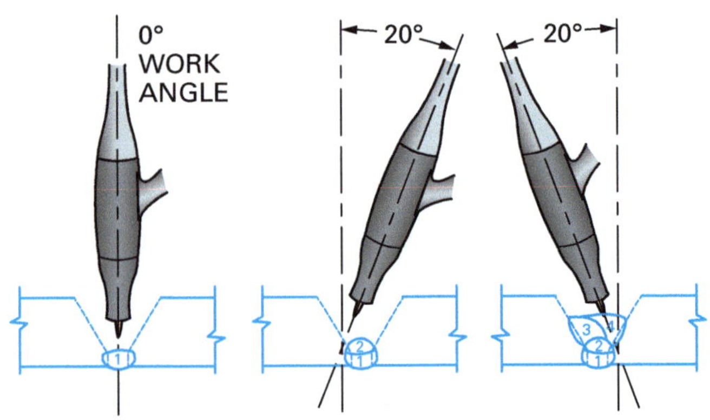

Figure 9 Typical GTAW torch work angles.

electrode tip and shielding gas are directed just ahead of the weld bead. Push angles of 15 to 20 degrees are normally used for GTAW.

A push angle that is too great (more than 20 degrees) will tend to draw air from under the back edge of the torch nozzle where it will mix with the shielding gas stream and contaminate the weld. In *Figure 10*, note that the angle of the filler rod is 90 degrees to the axis of the torch. As the torch position changes while progressing around the pipe, the angle of the filler rod changes with it.

2.1.3 Torch and Filler Metal Handling Techniques

The manual handling of the filler rod for GTAW makes the process unique and requires that the correct technique be learned and practiced. The two basic torch and filler metal handling methods used to perform GTAW are the freehand and walking-the-cup techniques. Try both, and use the one that gives the best results.

In the freehand technique (*Figure 11*), the torch electrode tip is held just above the weld puddle or base metal. The torch is supported by the welder's hand, which is usually steadied by resting some part of it on or against the base metal to maintain the proper arc length. If required, the welder can move the torch tip in a small circular motion within the molten puddle to maintain its size and advance the puddle. Filler metal is added as needed.

CAUTION

Do not insert the end of the filler rod into the molten puddle under the electrode in an attempt to melt it off. This can cause hard spots and weld defects.

The GTAW filler rod is held in the opposite hand. The tip of the rod is always kept within the shielding gas envelope, even when it is not being fed into the weld puddle, to protect the rod from atmospheric contamination and to keep it preheated. Note that this is important regardless of the welding technique used.

Insert the filler metal into the leading edge of the weld puddle, using extreme care not to touch the tungsten electrode with the end of the filler metal. If it touches the filler metal or weld puddle, it will become contaminated with filler metal. The electrode must then be removed and cleaned by grinding or chemical cleaning before

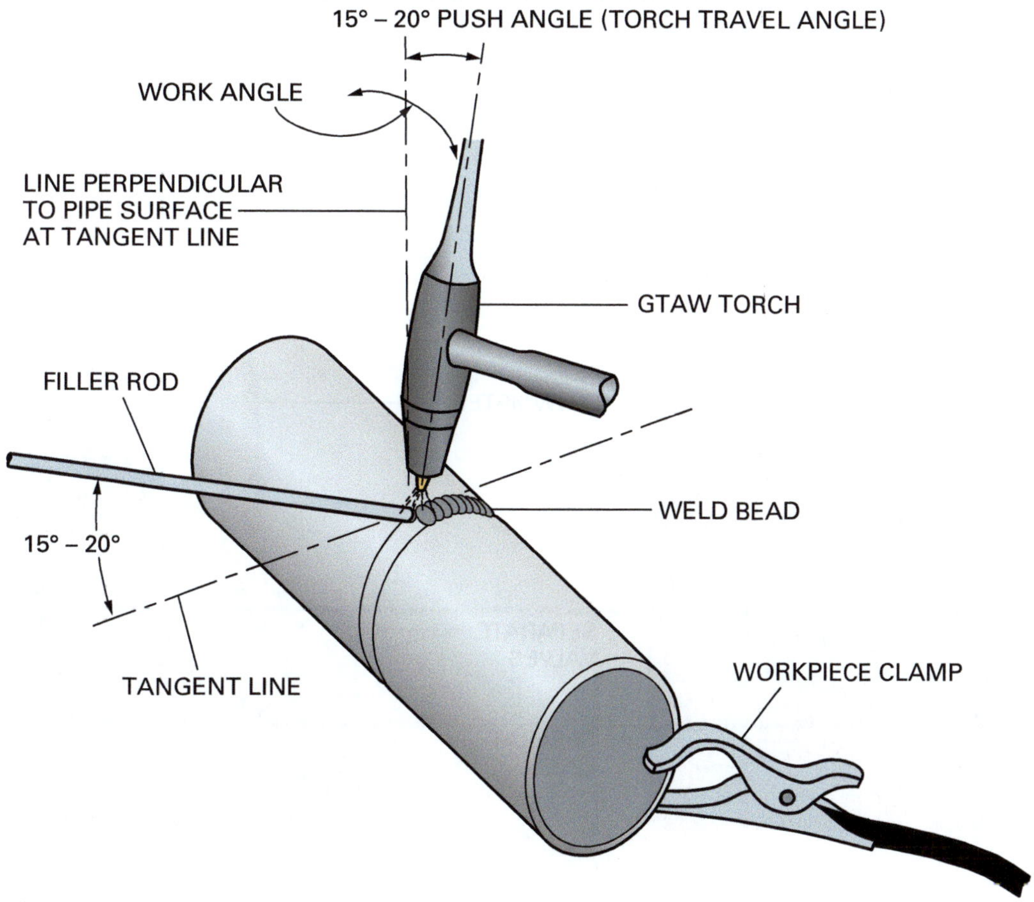

Figure 10 GTAW torch travel angle and filler rod position.

Figure 11 Freehand technique.

proceeding. Also, do not insert the end of the filler metal rod into the molten puddle under the electrode and then attempt to melt it off. Even if it does not touch the tungsten, this technique can cause hard spots and weld defects. A technique for preventing contamination of the electrode is to coordinate the movement of the torch toward the back edge of the weld puddle as the filler metal is inserted into the leading edge of the weld puddle.

For the walking-the-cup method, rest the edge of the torch nozzle (cup) against the base metal or groove edges to steady the torch and maintain a constant arc length. Rock the torch from side to side on the edge of the cup as it is advanced to maintain the puddle size and heat both sides of

The Use of a Y-Adapter

When using the same gas for backing as well as shielding, connect a Y-adapter with valves in each branch to the output of a flow regulator. Connect one branch to the backing gas device, and use the other branch for the torch supply. With the branch valves turned off, open the cylinder valve slowly at first, and then open it all the way. At the torch, turn on the torch valve or purge control and, using the torch branch valve on the Y-adapter, adjust for the proper torch flow rate as indicated on the flow regulator. At the torch, turn off or deactivate the gas flow. Then, open and adjust the backing gas branch valve on the Y-adapter for the proper backing gas flow, as indicated on the flow regulator. Once the branch valves are adjusted for the proper flow rates, if welding must be interrupted for a time, use the cylinder valve to shut off the gas.

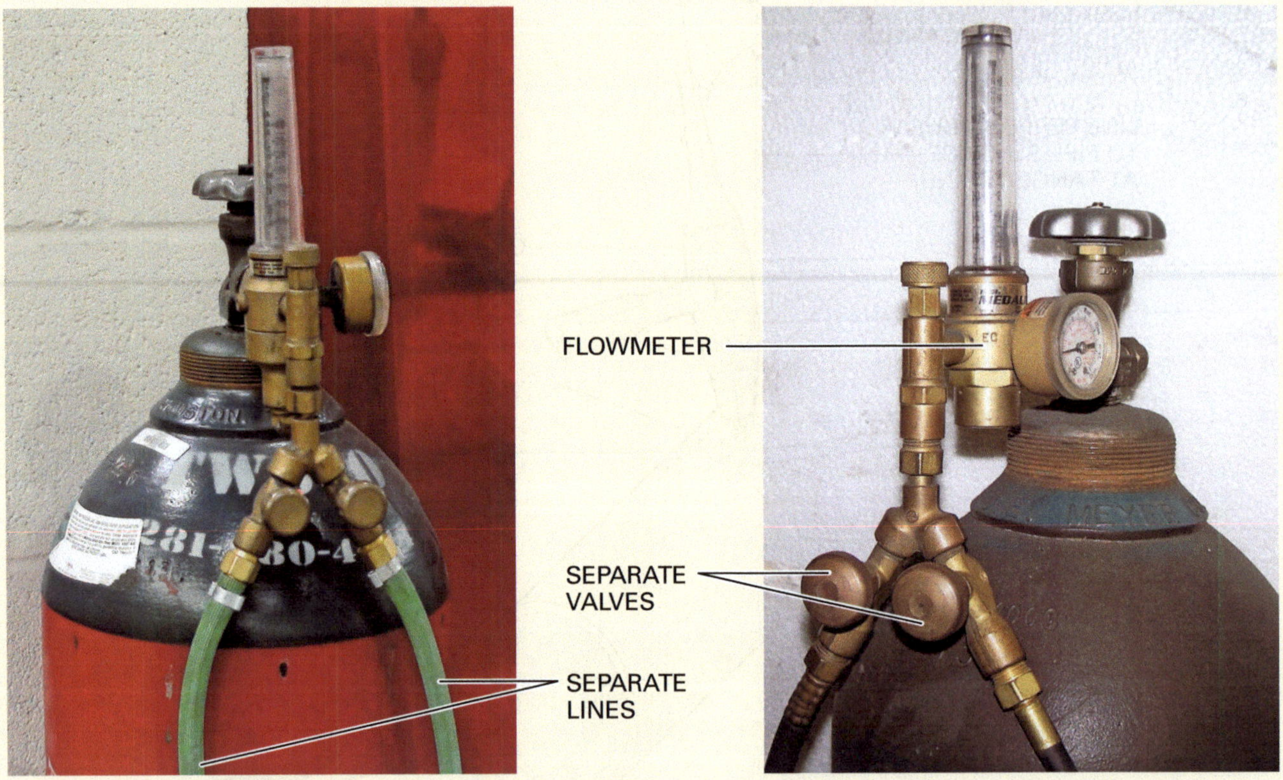

the groove. The filler metal is added during the fill passes in the same manner as in the freehand technique, using care not to contaminate the electrode. Feed the filler rod consistently as the weld progresses to ensure that the weld puddle develops evenly and produces an even bead.

Like all welding techniques, this filler rod technique requires practice to master, but is well worth the effort. Cradle the filler rod between the pointer finger and the middle finger, similar to the way you might hold a pencil. Then use the thumb to consistently push the rod forward in small motions. Once the thumb is extended pushing the rod forward, quickly bend and reposition the thumb to begin a new forward push. Consistent motion is important, so the thumb must be repositioned quickly to maintain the pace. The thumb can also be used to slowly roll the electrode as well during the pushing motion. This technique is easily practiced without actively welding; simply practice with a rod to become comfortable with the technique before trying it on a weld.

Another technique to consider is referred to as the *finger-pull technique*. The rod is secured between the pointer and middle fingers, which are used to pull the filler rod forward. At the end of the forward stroke, the fingers are slipped backwards along the rod to reposition, without pulling it backwards and out of the weld puddle in the process. Once the two fingers are repositioned, the pulling motion begins again.

2.2.0 The Root Pass and Back-Side Protection

Welding processes that use a shielding gas to protect the molten puddle on the face of the weld from the atmosphere do not provide any protection for the back side of the weld. Back-side protection is usually not required for carbon steel and some low alloy steels, but it essential for stainless and many other low alloy steels. This is because elements used in alloys are more prone to combine with the oxygen in the atmosphere to form oxides, which decrease the corrosion-resistance to an unacceptable level. Oxides can also be pulled into the weld puddle during subsequent welding, causing defects. The most common methods used to provide back-side protection are as follows:

- Backing gas (purge)
- Backup flux
- Consumable inserts

2.2.1 Backing Gas

Backing gas provides the highest degree of protection to the back side of the weld when welding open-root joints. Argon provides better protection than nitrogen, but nitrogen typically costs 25 to 33 percent less. Because it costs less, nitrogen is used when it will not adversely affect the weld. However, the gas specified by the WPS must always be used.

To be effective, the backing gas must reduce the presence of atmospheric gases around the back side of the weld to an acceptable level. The gas must be allowed to vent in a controlled manner from the weld area; no significant pressure can be allowed to develop inside the pipe as welding progresses. As a result, a continuous flow of gas must be provided.

For short and small-diameter piping, plug or cap the ends and any openings (*Figure 12*). This will provide containment and reduce the amount of gas needed (and the expense). However, this method of containment is not cost-effective for large piping or long sections. To decrease the volume of piping requiring backing gas replacement, plug the piping near the weld on both sides with removable devices called dams. Dams can be inflatable bladders such as those shown in *Figure 13*, inflated plastic bags, or foam rubber. Dams made of water-soluble paper can be fabricated to fit the pipe; prefabricated versions are also available. This material is designed to fully break down in water and reportedly leaves no debris.

When dams are used, they must be removed after welding has been completed. If argon is

Flux-Cored GTAW Wire

Although it is rarely used and many welders have never tried it, there is an alternative to the use of a purge gas for the root pass on stainless pipe. A limited selection of flux-cored filler metal for the GTAW process is available from several suppliers. This wire is only used for the root pass. Although it may eliminate the need for a purge gas, some slag will develop on the back side of the weld. If the back side is not accessible, flushing or chemical agents would be required to remove it. This is a primary reason for the continued use of purge gases in GTAW of stainless steel piping.

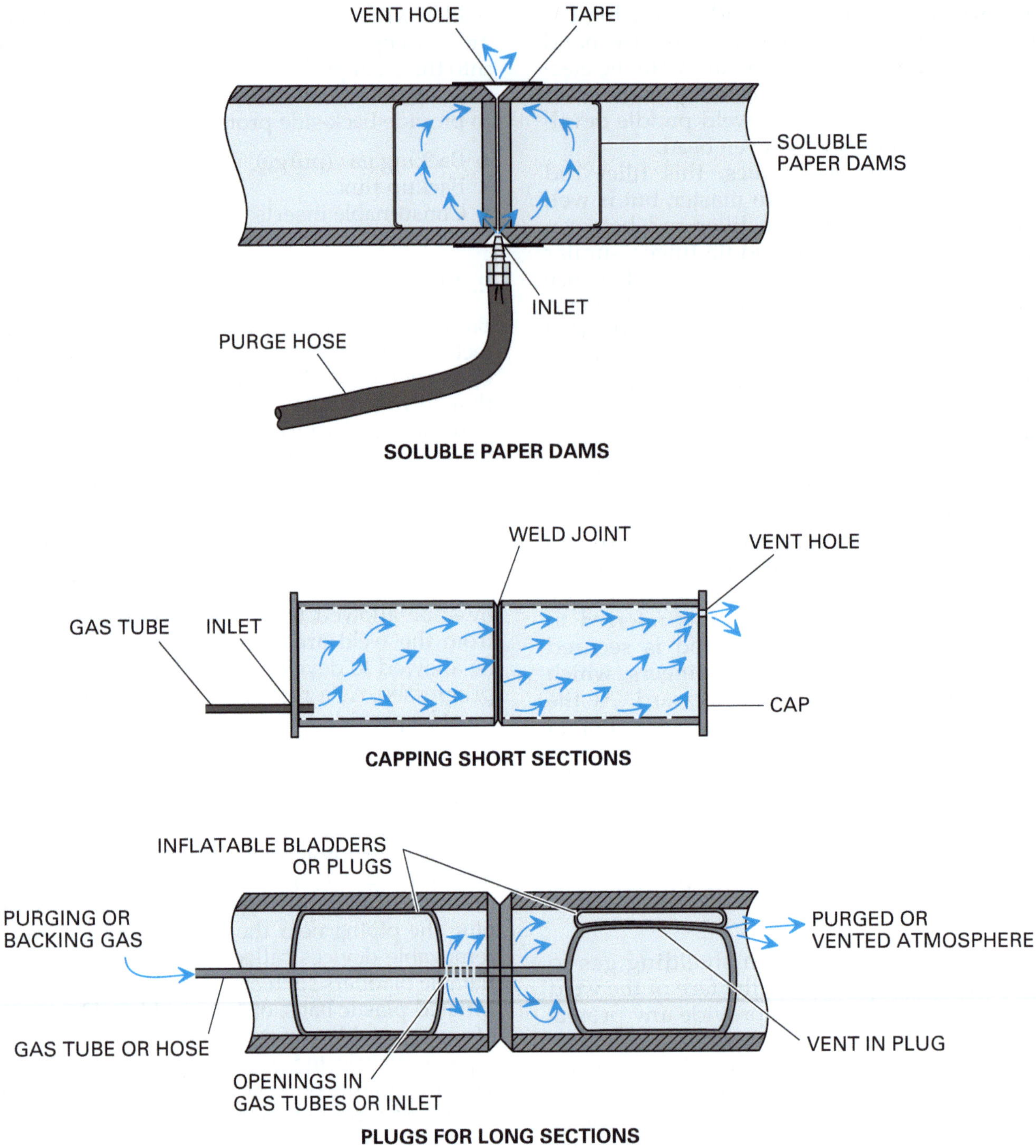

Figure 12 Typical pipe-purging arrangements.

required as the backing gas, another approach that reduces costs is to first purge the pipe with nitrogen to push out the air, and then begin a slow argon purge for welding.

As the weld progresses around a pipe and the root nears closure, the backing gas flow must be reduced to a minimum to prevent blowing the molten weld puddle out of the joint and causing pinholes in the root pass. No significant pressure can be allowed to build inside the pipe, or the escaping gas will blow through the weld. The backing gas flow should be kept low and should not exceed the capacity of the vent openings so that pressure does not build. The goal is to simply create a shielding-gas atmosphere inside the pipe near the weld area. Once the air has been pushed out of the weld zone, only the slightest positive pressure is required to keep it from re-entering. The purge should be maintained at least until the hot pass has been placed over the root pass and the root pass has been completely covered from view.

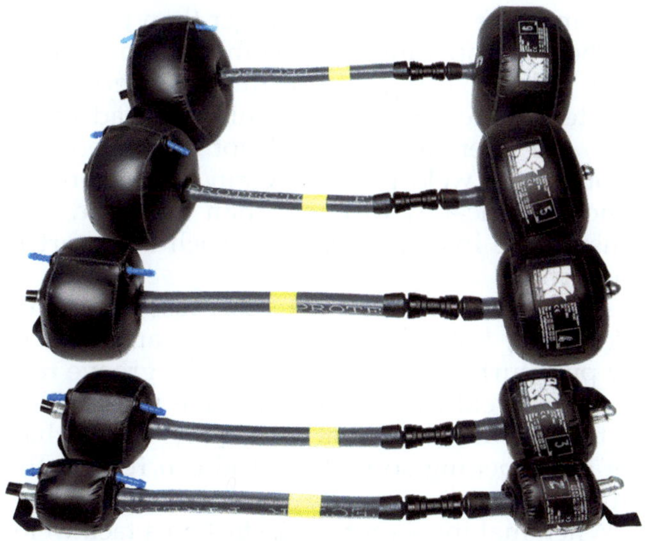

Figure 13 Inflatable bladder system for pipe purging.

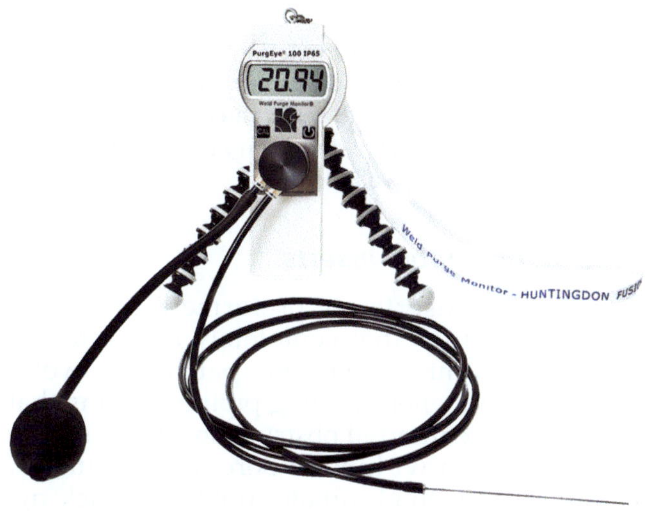

Figure 14 Purge gas analyzer.

When setting up the backing gas, establish the containment method and then seal the weld joint. Sealing is often done with masking tape. Introduce the backing gas through a tube or hose that passes through or past one of the containment devices. The backing gas can also be brought in through a nozzle that directs the gas through the root opening of the weld. Argon backing gas should always be introduced at the bottom, as it is heavier than the air it is displacing. As a result, there should be a vent located near the top as an exit for the atmospheric gases. If the vent hole is at the bottom, the argon gas will drop out and more air will be drawn in at the top. When nitrogen is used as the backing gas, the vent should be at the bottom, because nitrogen is lighter than air and will rise to the top.

Check the WPS to determine the flow rate range of the backing gas. The flow rate range is used for maintaining the backing gas atmosphere during welding, after the pipe has been purged. Higher flow rates can be used for the initial purge to reduce the time it takes to push out all of the air. An oxygen analyzer (*Figure 14*) is often used to confirm that the atmosphere inside the pipe is correct for the welding being performed. When the oxygen analyzer is used, a maximum of 1 percent oxygen is the usual standard for an acceptable welding atmosphere inside the pipe.

2.2.2 Backup Flux

Backup flux is another method used to provide root protection for open-root welds when weld quality requirements are less critical. A powdered flux (*Figure 15*) is mixed with alcohol or acetone to a thick, creamy consistency. It is left to set for several minutes to allow the alcohol or acetone to react with the flux. An even coat of the mixture is then brushed onto the back side of the joint. The mixture dries quickly and leaves a coating of flux. During welding, the flux powder melts and forms a barrier that prevents air from coming in contact with the molten metal. To remove the flux after welding, use a wire brush and hot water. Before using a backup flux, ensure that it is permitted by the WPS.

WARNING!

Backup flux may contain silica, fluorides, or other toxic materials. Alcohol and acetone used for the mixture are flammable. Weld only in well-ventilated spaces. Failure to provide proper ventilation could result in personal injury or death.

Figure 15 Powder flux for GTAW of stainless and low-alloy steels.

CAUTION

Backup fluxes are formulated for welding specific alloy types. Check the manufacturer's recommendations to select the correct flux for the metal being welded. Using the wrong flux could result in weld defects.

2.2.3 Consumable Inserts

Consumable inserts are preformed, preplaced filler metal rings that completely fuse into the root of the joint and become part of the weld. They are used when welding piping systems that cannot tolerate internal obstructions. Inserts keep flow restriction within the bore to a minimum. They also help to eliminate root-bead cracking. Consumable inserts can also be used when welding pipe, tube, and other components that require uniformity and a smooth root bead.

Since the insert becomes an integral part of the root pass, complete fusion and smooth blending of the insert and adjacent pipe or fitting results. A backing gas is used with inserts to prevent oxidation of the weld root when welding stainless or low alloy steel pipe.

To ensure complete penetration when using a consumable insert, look for a small swirling dot at the center of the silica puddle before advancing the torch. If this dot is not present and swirling, completely penetration is unlikely.

2.2.4 Making the Root Pass

The greatest challenge of an open V-groove pipe weld is making the root pass. The root must have complete penetration, but not have an excessive amount of penetration or melt-through root reinforcement. The penetration is controlled by using the correct amperage, filler metal application, and torch travel speed for the root opening and root face being used. Root reinforcement should be flush to ⅛" (3.2 mm), depending on the site specifications.

The two techniques for running the root pass are the on-the-wire technique and running a keyhole. Regardless of the technique used, always start the root pass from a tack weld.

To perform the on-the-wire root pass technique, secure or hold the filler metal in the bottom of the root opening of the V-groove parallel to the edge of the groove. Fuse the filler metal in place by moving the torch along the groove using either the freehand or walking-the-cup torch technique. Advance the torch slowly enough to thoroughly fuse the filler metal with the pipe edges, but not so slowly that it can burn through. This technique is used with a filler metal that is the same size as or slightly larger than the width of the root opening; the method is similar to pipe welding using a consumable insert. With this technique, to ensure a uniform bead thickness, the filler metal should not be moved into or out of the weld puddle. Again, this technique is not significantly different than that used as a consumable insert, except that such an insert is pre-formed.

To perform the keyhole root pass technique (*Figure 16*), hold the filler metal at an angle to the pipe surfaces and feed it into the groove root from the weld side. This technique uses a filler metal rod diameter that is slightly larger than the root opening so that the rod can rest on the groove.

When making a root pass, start on a tack weld. When a keyhole weld puddle has been established by melting the root faces of both pipe ends together, move the arc to the back of the puddle and insert the end of the filler metal into the leading edge of the weld puddle. Then move the filler metal away from the weld puddle. Advance the torch, using either the walking-the-cup or the freehand technique, maintaining the keyhole. Insert the filler metal back into the leading edge of the molten puddle and forward out of the molten puddle to add filler metal as needed. Be sure the filler metal is always kept in the shielding gas stream to keep the filler metal preheated and protected from atmospheric contamination.

When an insert is not used, clean and inspect the root pass after it has been run. Remove any excess buildup or undercut with the edge of a narrow grinding wheel (not a cutoff wheel). This will make it easier to apply the hot pass. Be careful not to widen the groove or grind through the root pass.

Figure 16 Root pass and keyhole.

Cutting Filler Rod

Snip the rod at a 45-degree angle. Then insert the rod into the puddle with the angled tip at the top of the feathered tack. This results in better penetration of the tie-in and a smoother transition from the previous weld into the new weld.

When a consumable insert is used, hold the torch perpendicular to the pipe surface and point it radially toward the center of the pipe. Use an arc length of about ⅛" (3 mm) to melt the insert. Forward movement is controlled by the melting rate of the insert and the properties of the weld puddle. Increased fluidity and a rising weld puddle indicate that there is sufficient melting.

2.3.0 Practicing Open-Root V-Groove Welds

Groove welds may be made in all positions on pipe. The weld position is determined by the axis of the pipe (*Figure 17*). Four standard weld test positions are used with pipe:

- *Flat (1G-ROTATED)* – The pipe axis is horizontal, and the pipe is slowly rotated while it is being welded on the top. The weld beads are flat.
- *Horizontal (2G)* – The pipe axis is vertical and the pipe rotation is fixed. The weld beads are horizontal.
- *Multiple (5G)* – The pipe axis is horizontal and the pipe rotation is fixed. The weld beads are flat, vertical, and overhead.
- *Multiple inclined (6G)* – The pipe axis is inclined at approximately 45 degrees ±5 degrees. Pipe rotation is fixed. Weld beads are horizontal, vertical, and overhead. If required for training purposes, a restricting ring may be added to the 6G position coupon to form a 6GR position coupon.

The pipe axis positions for 1G-ROTATED, 2G, and 5G can vary ±15 degrees from the basic position. The pipe axis position for 6G can vary ±5 degrees from the nominal 45-degree angle.

To determine the quality of a pipe weld, specific sections of pipe are removed for destructive testing. For positions 1G-ROTATED and 2G, two specimens are removed. They must be taken from opposite sides of the pipe, as shown in *Figure 18* (*A*). Once removed, the specimens are put through different bend tests, which reveal the quality of the welds. Alternatively, specimens may be tested by radiography, a non-destructive testing (NDT) method.

When 5G or 6G welds will be destructively tested, the test specimens must be cut from the four regions shown in *Figure 18* (*B*). These regions are midway between the twelve o'clock and three o'clock; three o'clock and six o'clock; six o'clock and nine o'clock; and nine o'clock and twelve o'clock positions. The twelve o'clock position is the top of the coupon groove; the six o'clock position is at the bottom.

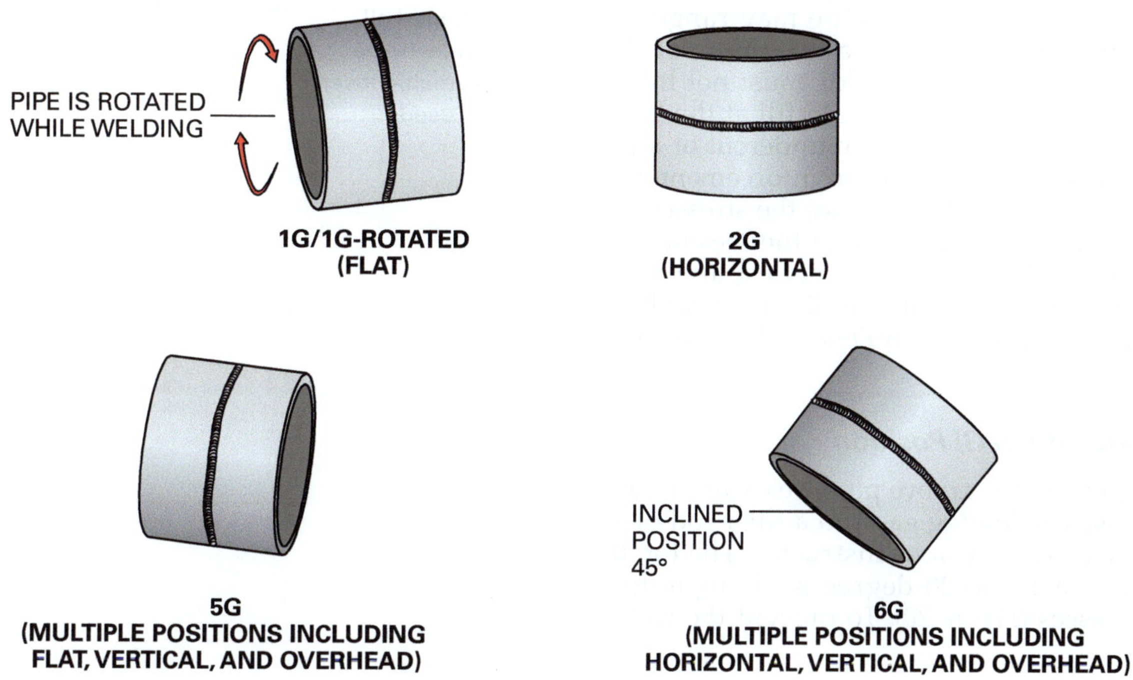

Figure 17 Four standard pipe groove-weld test positions.

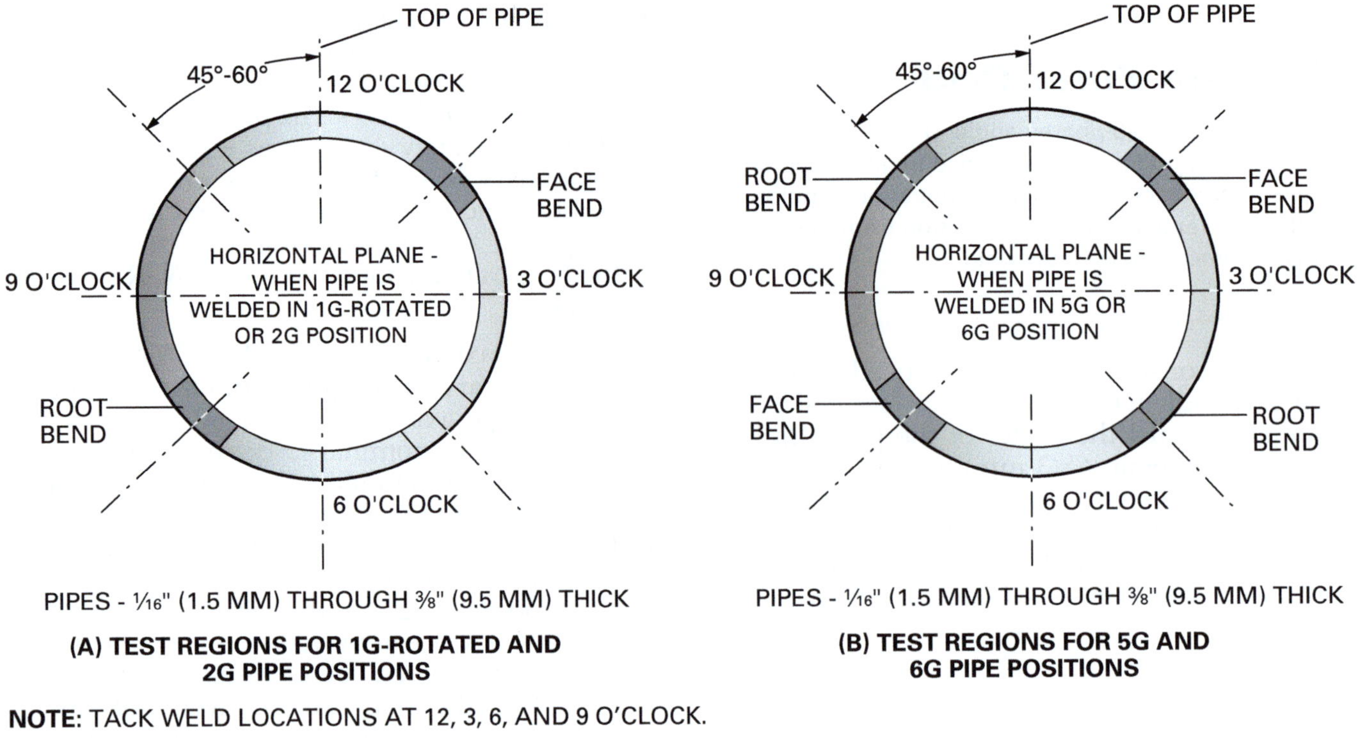

Figure 18 Test specimen regions for pipe positions 1G-ROTATED, 2G, 5G, and 6G.

2.3.1 Acceptable and Unacceptable Pipe Weld Profiles

Pipe groove welds should be made with slight face reinforcement (not exceeding ⅛" or 3 mm) and a gradual transition to the base metal at each toe. The root pass should have complete penetration along the entire circumference of the root. Any detectable incomplete penetration is typically rejected and should be repaired. The root reinforcement inside of the pipe may range in thickness from being flush to a maximum of ⅛" (3 mm) thick. Pipe groove welds must not have excess reinforcement or any underfill at the face or root. They must be free from undercut or any overlap. Excessively large face reinforcement reduces the pipe's strength. It causes the stresses in the pipe to be concentrated along the toes of the weld. As a result, the pipe will not expand and contract uniformly along its length. If a weld has any of the defects shown in *Figure 19*, it must be repaired.

2.3.2 Horizontal (2G) Position

Practice open-root V-groove pipe welds in the 2G position. Use a shielding gas and a filler rod of a diameter specified by your instructor. The torch should be at a 15- to 20-degree push angle for the weld passes (*Figure 20*). To prevent the weld

Hi-Lo Gauge

You can use a gauge, such as the one shown here, to check that internal surfaces are aligned all around when a consumable insert is not used.

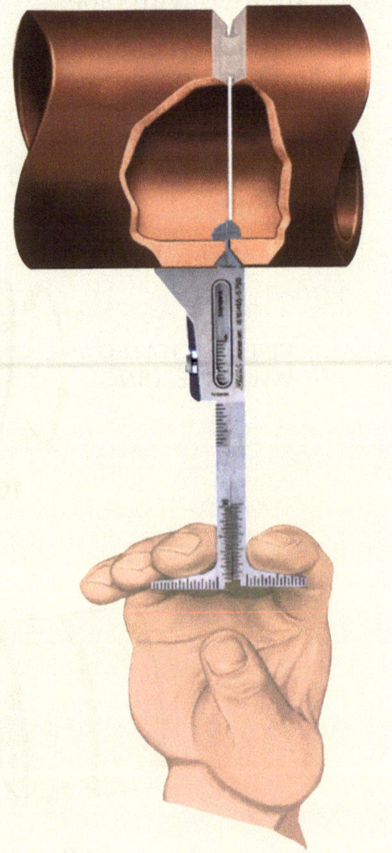

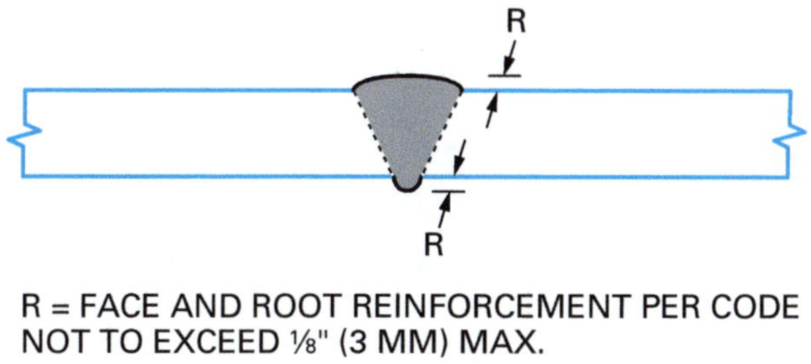

ACCEPTABLE WELD PROFILE

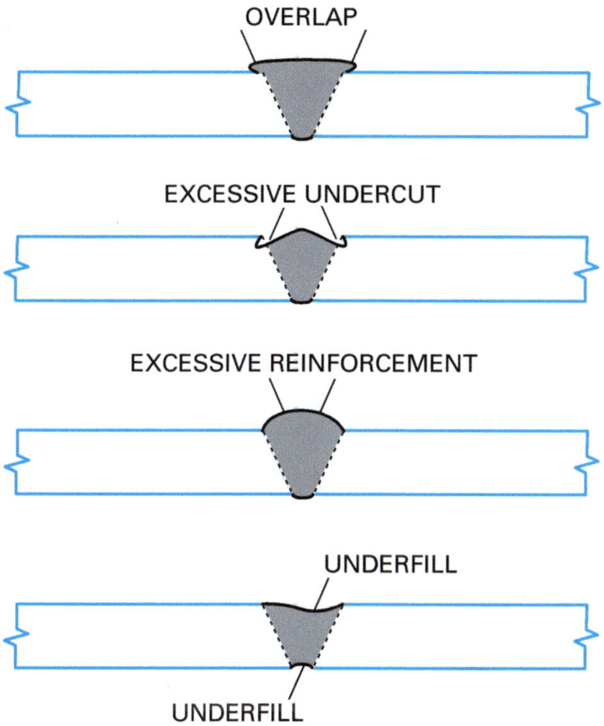

UNACCEPTABLE WELD PROFILES

Figure 19 Acceptable and unacceptable pipe groove-weld profiles.

puddle from sagging, the torch work angle can be dropped slightly, but not more than 10 degrees. For the root pass, use an oscillating motion. Be sure to achieve complete fusion into the tack welds.

When running the remaining passes, use stringer beads and either the walking-the-cup or the freehand torch-handling technique. Ensure proper fusion at the toes of the weld bead is achieved. Take special care to fill the crater at the termination of the weld, and stagger the location of the starts and stops for each pass.

Follow these steps to practice open-root V-groove pipe welds in the 2G position:

Step 1 Purge and tack weld together the practice pipe weld coupon as explained earlier.

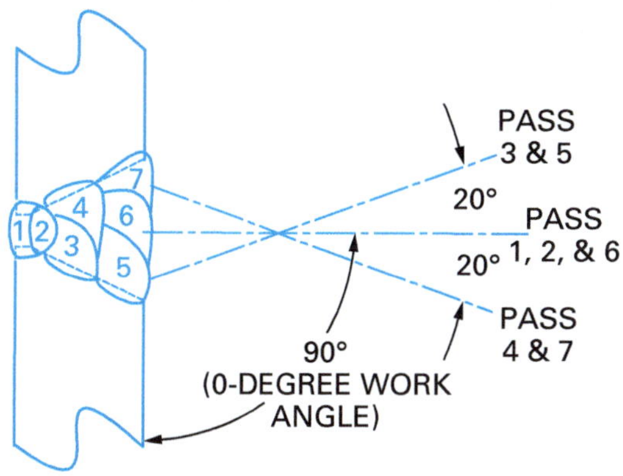

STRINGER BEAD SEQUENCE

Figure 20 Multiple-pass 2G bead sequence and work angles.

Step 2 Clamp or tack weld the pipe coupon with the axis of the pipe vertical.

Step 3 Prepare the purge system to be used for the duration of the weld. Set the backing gas flowmeter to the flow rate specified by the filler rod manufacturer, the WPS, or your instructor.

> **NOTE**
>
> If required by a WPS, site quality standard, or your instructor, observe any preheat/interpass temperatures or PWHT requirements.

Step 4 Run the root pass, shown as Bead 1 in *Figure 20*. Continue this procedure until the root pass has been completed.

Step 5 Brush the root pass to clean the weld.

Step 6 Run the remaining passes at the proper work angles using stringer beads. Overlap the passes, and clean the weld after each pass.

2.3.3 Multiple (5G) Position

Practice open-root V-groove pipe welds in the multiple (5G) position. Use a shielding gas and a filler rod of a diameter specified by your instructor. Start on a tack weld at the bottom of the pipe, and weld uphill toward the top. Use a torch push angle of 15 to 20 degrees to the pipe surface as shown in *Figure 21*.

The torch should be at approximately a 15- to 20-degree push angle for all passes as shown in *Figure 21*. Use stringer beads for the remaining passes (*Figure 22*), and either the walking-the-cup or the freehand torch-handling technique. Remember that weave beads are not used on stainless steel. Ensure proper fusion is achieved at the toes of the weld bead. Pay special attention to filling the crater at the termination of the weld, and stagger the location of the starts and stops for each pass.

Follow these steps to practice open-root V-groove pipe welds in the 5G position:

Step 1 Purge and tack weld together the practice pipe weld coupon as explained earlier.

Step 2 Clamp or tack weld the pipe coupon with the axis of the pipe horizontal.

Step 3 Prepare the purge system to be used for the duration of the weld. Set the backing gas flowmeter to the flow rate specified by the filler rod manufacturer, the WPS, or your instructor.

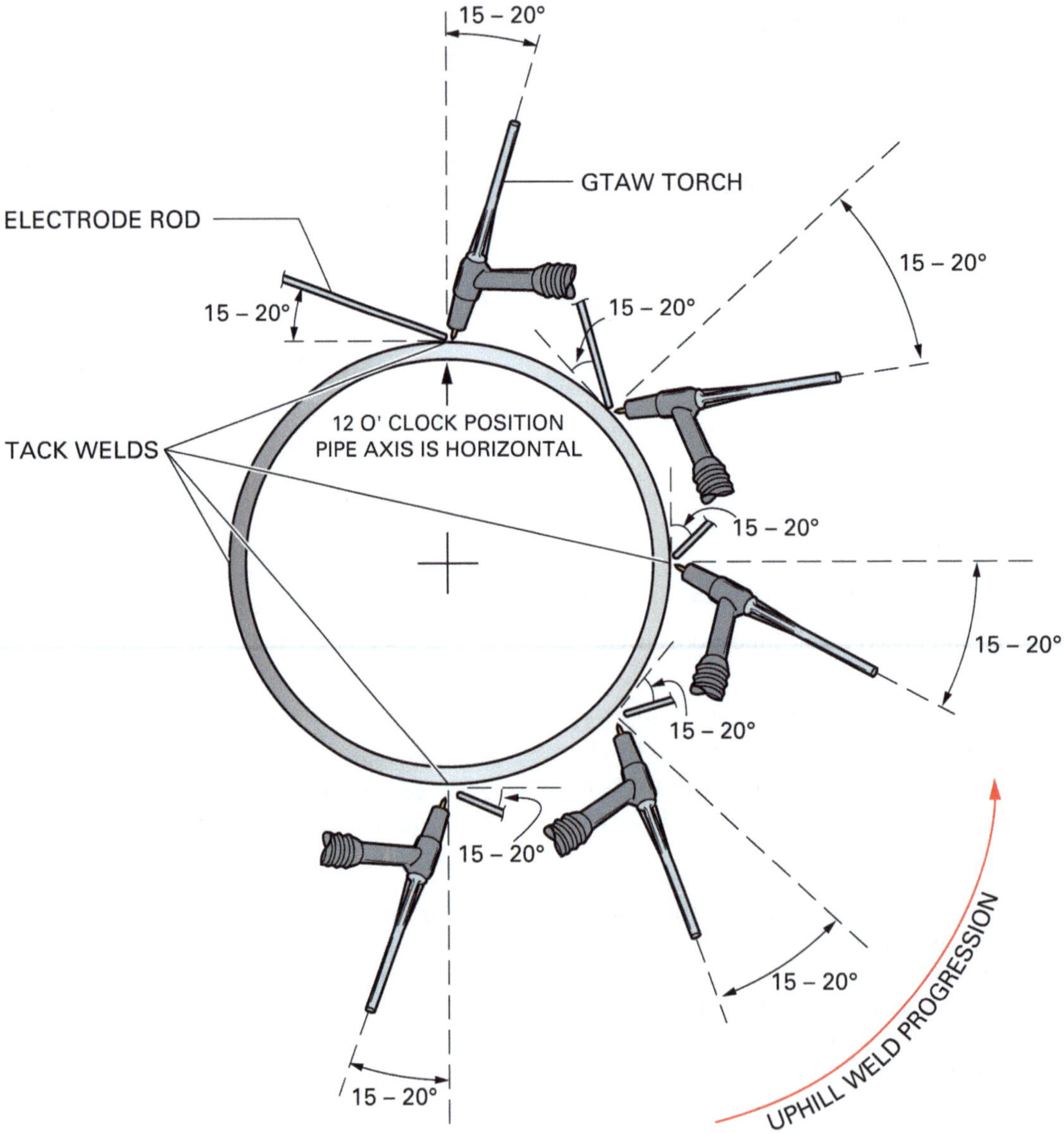

Figure 21 GTAW torch and filler rod angles for the 5G position.

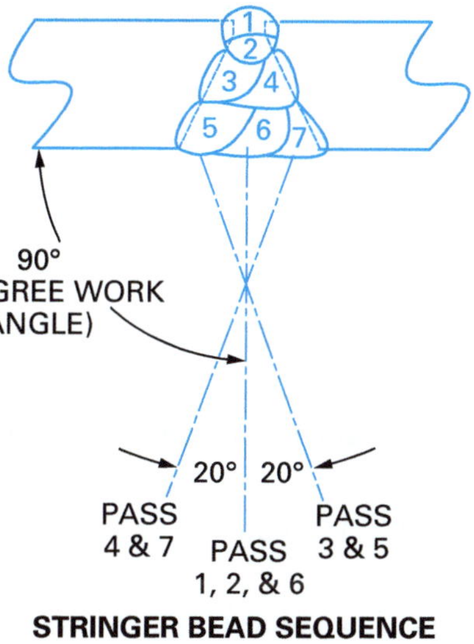

Figure 22 Multiple-pass 5G bead sequences and work angles.

> **NOTE**
> If required by a WPS, site quality standard, or your instructor, observe any preheat/interpass temperatures or PWHT requirements.

Step 4 Run the root pass uphill, as shown in *Figure 21*, using a stringer bead. Repeat the procedure for the opposite side of the pipe.

Step 5 Brush the root pass to clean the weld.

Step 6 Run the remaining passes at the proper work angles using stringer beads. Overlap the passes, and clean the weld after each pass.

2.3.4 Multiple Inclined (6G) Position

Practice open-root V-groove pipe welds in the multiple inclined 6G position. Use a shielding gas and a filler rod of a diameter specified by your instructor. Start at the bottom of the pipe, and weld uphill toward the top. For the root pass, the torch should be at approximately a 15- to 20-degree push angle and at 90 degrees (0-degree work angle) to the pipe (*Figure 23*).

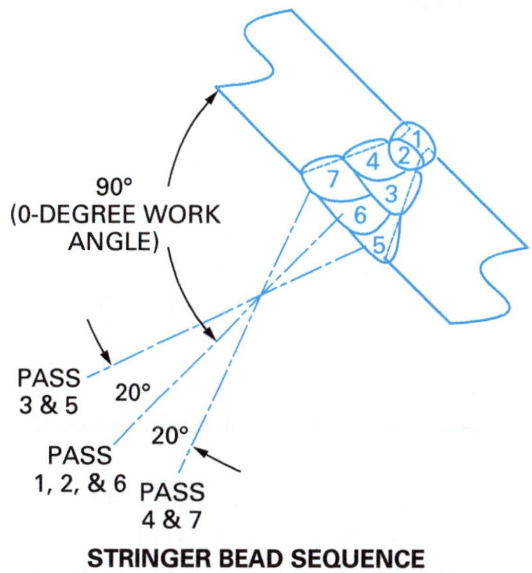

Figure 23 Multiple-pass 6G bead sequence and work angles.

When running the remaining passes, use stringer beads and either the walking-the-cup or the freehand torch-handling technique. Ensure proper fusion is achieved at the toes of the weld. Pay special attention to fill the crater at the termination of the weld, and stagger the location of the starts and stops for each pass.

Follow these steps to practice open-root V-groove pipe welds in the 6G position:

Step 1 Purge and tack weld together the practice pipe weld coupon as explained earlier.

Step 2 Clamp or tack weld the pipe coupon with the axis of the pipe inclined 45 degrees to the horizontal plane.

Step 3 Prepare the purge system to be used for the duration of the weld. Set the backing gas flowmeter to the flow rate specified by the filler rod manufacturer, the WPS, or your instructor.

NOTE

If required by a WPS, site quality standard, or your instructor, observe any preheat/interpass temperatures or PWHT requirements.

Step 4 Run the root pass uphill using a stringer bead. Repeat the procedure for the opposite side of the pipe.

Step 5 Brush the root pass to clean the weld.

Step 6 Run the remaining passes at the proper work angles using stringer beads. Overlap the passes, and clean the weld after each pass.

Additional Resources

AWS F4.1: Recommended Safe Practices for Preparation for Welding and Cutting of Containers and Piping. Miami, FL: American Welding Society.

Modern Welding Technology. Current Edition. Howard B. Cary and Scott Helzer. Englewood Cliffs, NJ: Prentice Hall.

OSHA 1926.351, Arc Welding and Cutting. Current Edition. Washington, DC: Occupational Safety & Health Administration.

The Procedure Handbook of Arc Welding. Current Edition. Cleveland, OH: The James F. Lincoln Arc Welding Foundation.

Welding Handbook, Volumes 1 through 4. Current Edition. Miami, FL: American Welding Society.

Welding Pressure Pipelines and Piping Systems (PDF). The Lincoln Electric Company, Cleveland, OH, USA. **www.lincolnelectric.com**. Last accessed May 29, 2015.

2.0.0 Section Review

1. With GTAW, faster torch travel speeds result in _____.
   a. an increased arc length
   b. larger, deeper weld puddles
   c. smaller, shallower weld puddles
   d. a greater concentration of heat

2. When using argon as a backing gas that is to be added through the root of the weld, it should be added _____.
   a. through the bottom of the weld
   b. through the top of the weld
   c. only after purging with nitrogen
   d. only after the root pass is complete

3. Which of the following is a *true* statement?
   a. The pipe axis for the 6G and 6GR positions should be at a 30-degree angle from horizontal.
   b. The GTAW torch work angle can be changed as much as 45 degrees as needed.
   c. GTAW root passes on stainless steel should be done with weave beads.
   d. The starting and stopping points for each weld pass should be staggered.

SUMMARY

Compared to other welding processes, GTAW is rather unique in that the welder must manually feed the filler metal into the weld. All welding requires good coordination, but GTAW requires coordinated movement from both hands. GTAW is excellent for use with thin materials such as sheet metal.

When working with low alloy and stainless steels, the welder must be mindful of potential weld contamination. Tools used on stainless steel must not have been used on carbon steel or other metals. Using grinding wheels and other tools that are contaminated will result in weld quality issues. A backing gas should be used when working with low alloy and stainless steel pipe. In some cases, the necessary weld-zone protection can be provided by a special flux.

Learning and mastering the GTAW process significantly increases a welder's value on the job and broadens the scope of opportunities. Continue practicing these welds and the handling of the filler rod until the needed skills have been acquired.

Review Questions

1. The welding process that provides greater control over root penetration and fill than most other process is _____.
   a. GTAW
   b. SMAW
   c. GMAW
   d. FCAW

2. When running a welding pass on low alloy steel, the oscillation width of the tungsten electrode should be _____.
   a. equal to the thickness of the workpiece
   b. less than or equal to the width of the filler rod
   c. twice the width of the electrode itself
   d. twice the width of the filler rod

3. When martensitic stainless steel contains 0 to 0.1 percent carbon, what preheat temperature is generally required?
   a. 1,600°F (871°C)
   b. 1,100°F (593°C)
   c. 500°F (260°C)
   d. None

4. Why is acetylene considered to be more dangerous than propane?
   a. Acetylene is heavier than air and collects in low places.
   b. Acetylene ignites at much lower concentrations when mixed with air or oxygen.
   c. Acetylene is a toxic gas.
   d. Acetylene auto-ignites on contact with oxygen.

5. The ASME specifications for pipe bevels call for a bevel that is _____.
   a. 30 ±5 degrees
   b. 30 to 37½ degrees
   c. 35 ±5 degrees
   d. 35 to 40 degrees

6. A significant control available to the welder for bead width in GTAW is _____.
   a. travel speed
   b. torch arc length
   c. torch angle
   d. filler rod diameter

7. The method that provides the highest degree of back-side protection when welding open-root joints is the use of _____.
   a. a flux-coated filler rod
   b. consumable inserts
   c. a backing gas
   d. a dry powder flux

8. Which of the following is a *true* statement about the use of a backing gas?
   a. The flow of backing gas should be increased as the root pass nears completion.
   b. Once the air has been pushed out of teh weld zone, only slight positive pressure is needed to keep it from re-entering.
   c. The backing gas purge can be stopped as soon as the root pass is completed.
   d. The maximum allowable percentage of oxygen in the weld zone is 3 percent.

9. Compared to air, argon gas is _____.
   a. lighter
   b. more toxic
   c. heavier
   d. less toxic

10. In the 5G position, the root pass should be applied _____.
    a. at a 75-degree work angle
    b. at a 45-degree work angle
    c. uphill
    d. downhill

Trade Terms Introduced in This Module

Austenitic steel: These stainless steels are nonmagnetic in the annealed condition and non-hardenable by heat treatment; they can be hardened significantly by cold working. They combine excellent corrosion- and heat-resistance with good mechanical properties over a broad temperature range. Austenitic stainless steels make up the largest of the three common stainless steel groups and are the most likely to be encountered by welders.

Carbide precipitation: The process of carbides combining with chromium to create chromium carbide. The result is a reduction of chromium and the corrosion-resistance it provides. Intergranular corrosion results from the loss of chromium.

Carburized: The process of iron or steel combining with carbon liberated from a metal when the metal is heated.

Ferritic steel: These stainless steels are always magnetic and can be hardened to some extent by cold working. They combine corrosion- and heat-resistance with fair mechanical properties over a narrower temperature range than austenitic steels. Ferritic stainless steels are straight chromium stainless steels containing 11.5 to 27 percent chromium, as well as other metals.

Intergranular corrosion: A selective corrosion attack of the crystalline grain boundaries of a stainless steel. It results from chromium depletion which occurs during carbide precipitation.

Low alloy steel: Steel that is alloyed with other elements, amounting to 1 to 8 percent of the steel by weight, to improve its mechanical properties.

Martensitic steel: These stainless steels comprise two groups: chromium-martensitic and martensitic. The chromium-martensitic stainless steels contain from 11.5 to 18 percent chromium, along with other metals. Martensitic stainless steels contain only 4 to 6 percent chromium; for this reason, they are not considered true stainless steels, although their corrosion-resistance is much greater than that of mild carbon steels. Martensitic stainless steels are magnetic and can be hardened by quenching and tempering.

Appendix

PERFORMANCE ACCREDITATION TASKS

Performance Accreditation Tasks (PATs) provide specific acceptable criteria for performance and help to ensure a true competency-based welding program for students.

The following tasks are designed to evaluate your ability to run open-root V-groove welds with GTAW equipment in three standard test positions using stainless and/or low alloy steel filler rod of the appropriate diameter and argon shielding gas. Perform each task when you are instructed to do so by your instructor. As you complete each task, show it to your instructor for evaluation. Do not proceed to the next task until told to do so. For AWS 2G and 5G certifications, refer to *AWS EG3.0-96 Guide for the Training and Qualification of Welding Personnel; Level II – Advanced Welder* for bend test requirements. For AWS 6G certifications, refer to *AWS EG4.0-96 Guide for the Training and Qualification of Welding Personnel; Level III – Expert Welder* for bend test requirements.

PATs 1 and 2 correspond to *AWS EG3.0-96, Section 3.3.6.4, Unit #4 GTAW, Learning Objective #15*. PAT 3 corresponds to *AWS EG4.0-96, Section 3.3.6.5, Unit #5 GTAW, Learning Objective #15* (partial).

GTAW OPEN-ROOT V-GROOVE PIPE WELD IN THE 2G POSITION

Using GTAW equipment and stainless and/or low-alloy steel filler rod of the appropriate diameter, argon shielding and backing gas, and stringer beads, make an open-root V-groove weld on stainless steel or carbon steel pipe in the 2G position.

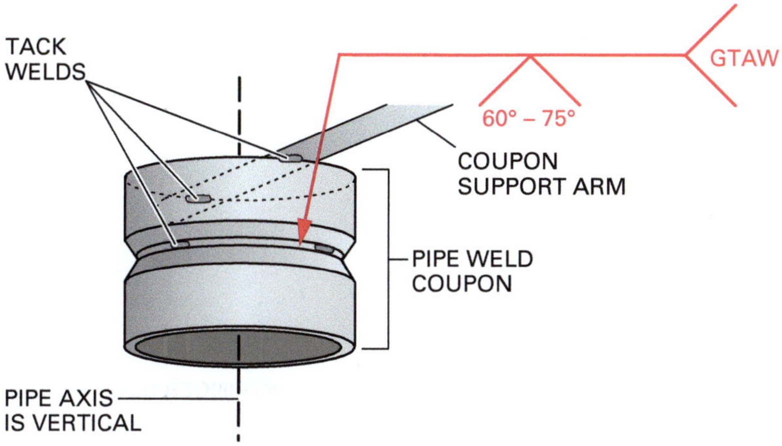

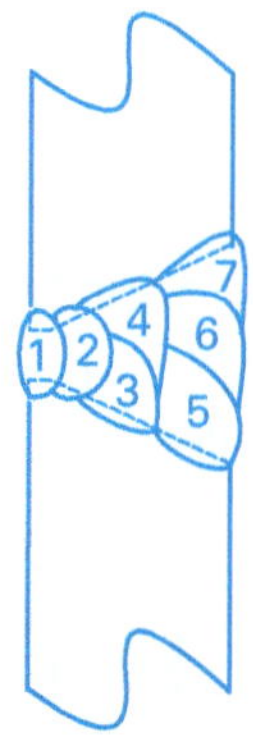

STRINGER BEAD SEQUENCE

Criteria for Acceptance:

- Uniform appearance on the bead face ________
- Craters and restarts filled to the full cross section of the weld ________
- Acceptable weld profile in accordance with the applicable codes and standards ________
- Smooth transition with complete fusion at the toes of the weld ________
- Complete joint penetration and face reinforcement with the root reinforcement at least flush with the inside of the pipe, to a maximum buildup of ⅛" (3 mm) ________
- No porosity ________
- No undercut ________
- No cracks ________
- No overlap ________
- No incomplete fusion ________

GTAW OPEN-ROOT V-GROOVE PIPE WELD IN THE 5G POSITION

Using GTAW equipment and stainless and/or low-alloy steel filler rod of the appropriate diameter, argon shielding and backing gas, and stringer beads, make an open-root V-groove weld on stainless steel or carbon steel pipe in the 5G position.

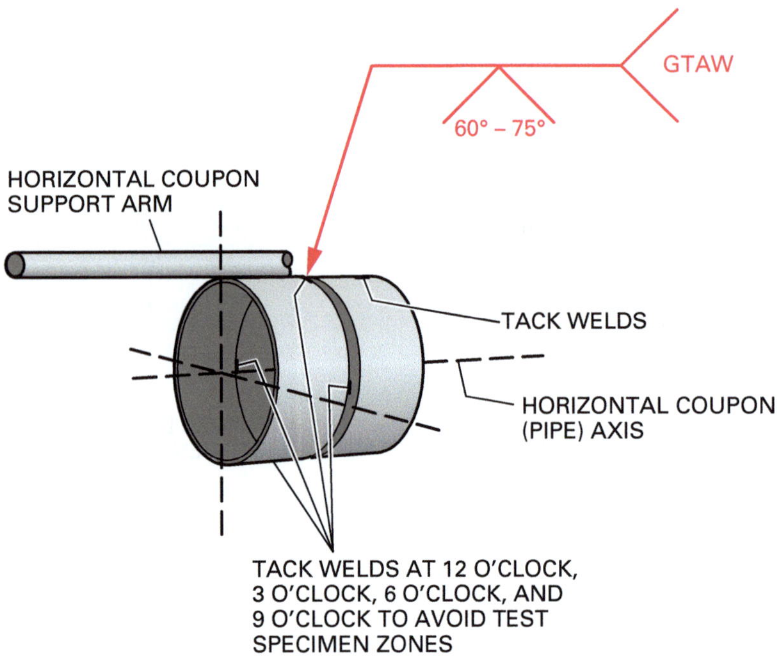

STRINGER BEAD SEQUENCE

Criteria for Acceptance:

- Uniform appearance on the bead face __________
- Craters and restarts filled to the full cross section of the weld __________
- Acceptable weld profile in accordance with the applicable codes and standards __________
- Smooth transition with complete fusion at the toes of the weld __________
- Complete joint penetration and face reinforcement with the root reinforcement at least flush with the inside of the pipe, to a maximum buildup of ⅛" (3 mm) __________
- No porosity __________
- No undercut __________
- No cracks __________
- No overlap __________
- No incomplete fusion __________

GTAW OPEN-ROOT V-GROOVE PIPE WELD IN THE 6G (OR 6 GR) POSITION

Using GTAW equipment and stainless and/or low-alloy steel filler rod of the appropriate diameter, argon shielding and backing gas, and stringer beads, make an open-root V-groove weld on stainless steel or carbon steel pipe in the 6G (or 6GR) position.

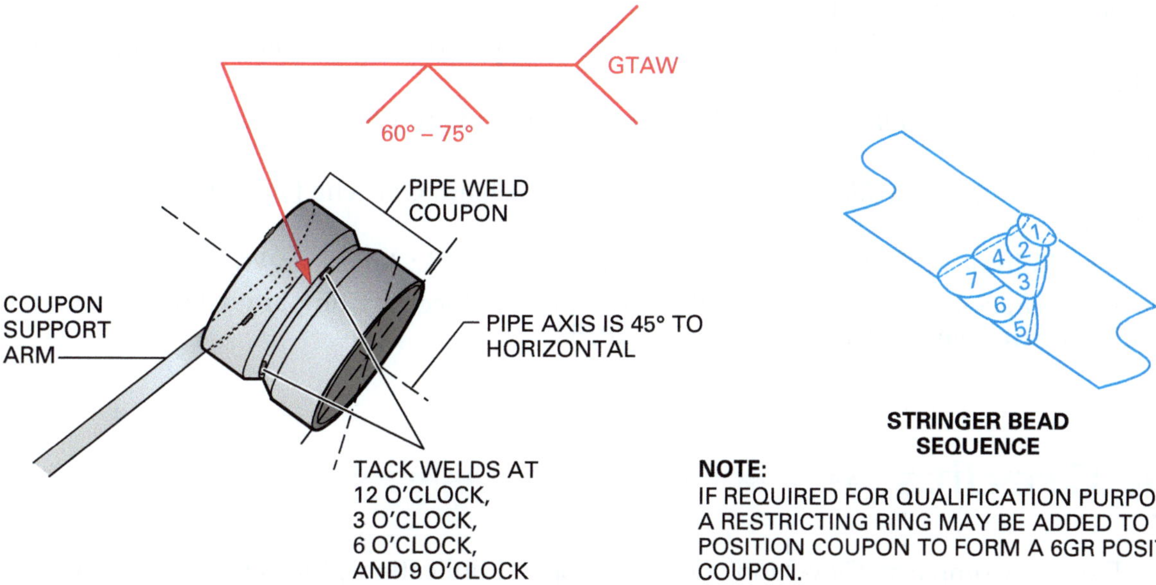

NOTE:
IF REQUIRED FOR QUALIFICATION PURPOSES, A RESTRICTING RING MAY BE ADDED TO THE 6G POSITION COUPON TO FORM A 6GR POSITION COUPON.

Criteria for Acceptance:

- Uniform appearance on the bead face ________
- Craters and restarts filled to the full cross section of the weld ________
- Acceptable weld profile in accordance with the applicable codes and standards ________
- Smooth transition with complete fusion at the toes of the weld ________
- Complete joint penetration and face reinforcement with the root reinforcement at least flush with the inside of the pipe, to a maximum of ⅛" (3 mm) ________
- No porosity ________
- No undercut ________
- No cracks ________
- No overlap ________
- No incomplete fusion ________

Additional Resources

This module presents thorough resources for task training. The following reference material is recommended for further study.

AWS F4.1: Recommended Safe Practices for Preparation for Welding and Cutting of Containers and Piping. Miami, FL: American Welding Society.

Modern Welding Technology. Current Edition. Howard B. Cary and Scott Helzer. Englewood Cliffs, NJ: Prentice Hall.

OSHA 1926.351, Arc Welding and Cutting. Current Edition. Washington, DC: Occupational Safety & Health Administration.

The Procedure Handbook of Arc Welding. Current Edition. Cleveland, OH: The James F. Lincoln Arc Welding Foundation.

Welding Handbook, Volumes 1 through 4. Current Edition. Miami, FL: American Welding Society.

Welding Pressure Pipelines and Piping Systems (PDF). The Lincoln Electric Company, Cleveland, OH, USA. **www.lincolnelectric.com**. Last accessed May 29, 2015.

Figure Credits

The Lincoln Electric Company, Cleveland, OH, USA, Module Opener, Figures 3, 4, 11, 15, SA01, SA02

Brett Schwaninger, @schwannman, Optimum Filter Inc., Figure 2

American Welding Society®, Figure 5

Topaz Publications, Inc., Figure 7 (photo)

Terry Lowe, Figures 8 (photo), 16, 20 (photo), 22 (photo), 23 (photo), SA03

G.A.L. Gage Co., SA04

Huntingdon Fusion Techniques Limited, Figures 13, 14

Section Review Answer Key

Answer	Section Reference	Objective
Section One		
1. b	1.1.1	1a
2. d	1.2.3	1b
3. a	1.3.4	1c
Section Two		
1. c	2.1.1	2a
2. a	2.2.1	2b
3. d	2.3.2	2c

NCCER CURRICULA — USER UPDATE

NCCER makes every effort to keep its textbooks up-to-date and free of technical errors. We appreciate your help in this process. If you find an error, a typographical mistake, or an inaccuracy in NCCER's curricula, please fill out this form (or a photocopy), or complete the online form at **www.nccer.org/olf**. Be sure to include the exact module ID number, page number, a detailed description, and your recommended correction. Your input will be brought to the attention of the Authoring Team. Thank you for your assistance.

Instructors – If you have an idea for improving this textbook, or have found that additional materials were necessary to teach this module effectively, please let us know so that we may present your suggestions to the Authoring Team.

NCCER Product Development and Revision

13614 Progress Blvd., Alachua, FL 32615

Email: curriculum@nccer.org
Online: www.nccer.org/olf

❑ Trainee Guide ❑ Lesson Plans ❑ Exam ❑ PowerPoints Other ______________

Craft / Level: Copyright Date:

Module ID Number / Title:

Section Number(s):

Description:

Recommended Correction:

Your Name:

Address:

Email: Phone:

SMAW – Stainless Steel Plate and Pipe Groove Welds

Overview

Welders working on stainless steel must understand the differences between welding this type of steel and other types. Further, the various grades of stainless steel require different welding techniques, filler metal composition, and heat treatments. This module covers stainless steel metallurgy, the required SMAW electrodes, and the techniques and equipment used to make open-root V-groove welds on stainless steel plate and pipe using the SMAW process.

Module 29306

Trainees with successful module completions may be eligible for credentialing through the NCCER Registry. To learn more, go to **www.nccer.org** or contact us at 1.888.622.3720. Our website has information on the latest product releases and training, as well as online versions of our *Cornerstone* magazine and Pearson's product catalog.

Your feedback is welcome. You may email your comments to **curriculum@nccer.org**, send general comments and inquiries to **info@nccer.org**, or fill in the User Update form at the back of this module.

This information is general in nature and intended for training purposes only. Actual performance of activities described in this manual requires compliance with all applicable operating, service, maintenance, and safety procedures under the direction of qualified personnel. References in this manual to patented or proprietary devices do not constitute a recommendation of their use.

29306 V5

29306
SMAW – Stainless Steel Plate and Pipe Groove Welds

Objectives

When you have completed this module, you will be able to do the following:

1. State special considerations for SMAW of various types of stainless steel and identify compatible electrodes.
   a. Describe various types of stainless steel.
   b. Describe methods for controlling carbide precipitation.
   c. Identify various types of SMAW electrodes compatible with stainless steel.
   d. Describe the selection and storage of stainless steel electrodes.
2. State the basic concepts of SMAW of stainless steel and how to prepare for welding.
   a. State the basic concepts of SMAW of stainless steel.
   b. Identify common welding safety practices.
   c. Explain how to prepare the area, materials, and equipment for SMAW of stainless steel.
3. Describe open-root V-groove plate and pipe welding positions and SMAW stainless steel welding techniques.
   a. Identify and describe plate and pipe welding positions and acceptable weld profiles.
   b. State general electrode handling considerations for SMAW of stainless steel.
   c. Describe how to make the root pass.
   d. Describe the techniques required to produce open-root V-groove SMAW stainless steel plate welds in four positions.
   e. Describe the techniques required to produce open-root V-groove SMAW stainless steel pipe welds in four positions.

Performance Tasks

Under the supervision of your instructor, you should be able to do the following:

1. Make open-root V-groove welds on stainless steel plate in the 1G position.
2. Make open-root V-groove welds on stainless steel plate in the 2G position.
3. Make open-root V-groove welds on stainless steel plate in the 3G position.
4. Make open-root V-groove welds on stainless steel plate in the 4G position.
5. Make open-root V-groove welds on stainless steel pipe in the 1G-ROTATED position.
6. Make open-root V-groove welds on stainless steel pipe in the 2G position.
7. Make open-root V-groove welds on stainless steel pipe in the 5G position.
8. Make open-root V-groove welds on stainless steel pipe in the 6G position.

Trade Terms

Annealing
Carbide precipitation
Heat-affected zone
Interpass temperature
Sensitization

Note

Module 29306 is an elective for *Welding Level Three*. It is not required for successful level completion.

Industry Recognized Credentials

If you are training through an NCCER-accredited sponsor, you may be eligible for credentials from NCCER's Registry. The ID number for this module is 29306. Note that this module may have been used in other NCCER curricula and may apply to other level completions. Contact NCCER's Registry at 888.622.3720 or go to **www.nccer.org** for more information.

Contents

Contents (continued)

Figures and Tables

SECTION ONE

1.0.0 STAINLESS STEELS AND ELECTRODES

Objective

State special considerations for SMAW of various types of stainless steel and identify compatible electrodes.

a. Describe various types of stainless steel.
b. Describe methods for controlling carbide precipitation.
c. Identify various types of SMAW electrodes compatible with stainless steel.
d. Describe the selection and storage of stainless steel electrodes.

Trade Terms

Annealing: A heat treatment process that reduces the brittleness of the work.

Carbide precipitation: The process of carbides combining with chromium to create chromium carbide. The result is a reduction of chromium and the corrosion-resistance it provides. Intergranular corrosion results from the loss of chromium.

Heat-affected zone: The portion of the workpiece/base metal whose mechanical properties or microstructure have been or will be altered by the heat from welding, brazing, soldering, or thermal cutting processes.

Sensitization: The process by which carbide precipitation takes place.

Each of the various types of stainless steel has special considerations that welders must take into account before using SMAW techniques on the metal. These considerations include the properties and characteristics of the different types of stainless steel; how to control carbide precipitation during SMAW of stainless steel; and how to select and store the appropriate stainless steel electrodes for a specific application.

1.1.0 Stainless Steel Metallurgy

Stainless steels are ferrous (iron-based) alloys that have a chromium content of at least 11 percent. Other alloying agents, including nickel and carbon, may be present in varying amounts, depending on the type of stainless steel. The primary physical characteristic of all stainless steels is their resistance to corrosion and heat, which varies with the category of stainless steel. Some stainless steels also possess good low- and high-temperature mechanical properties.

Stainless steel is categorized based on its grain structure. The type of grain structure depends on the specific alloy content of the stainless steel and its heat treatment during manufacture. The alloy content and heat treatment also affect the steel's mechanical and magnetic properties. Stainless steels are categorized into these three basic grades depending on their crystalline grain structure:

- Austenitic
- Martensitic
- Ferritic

Table 1 summarizes some of the AISI (American Iron and Steel Institute) stainless steels by category.

1.1.1 Austenitic Stainless Steels

Austenitic stainless steels are the largest of the stainless steel groups, and the group most often encountered by welders in the field. Austenitic stainless steels include the AISI 200 and 300 series stainless steels, all of which contain significant amounts of both chromium and nickel. The chromium content of austenitic stainless steels ranges from 16 to 26 percent. The nickel content varies from 3.5 to 23 percent, depending on the type of steel. The 200 series substitutes manganese and nitrogen for part of the nickel content.

Austenitic stainless steels range from nonmagnetic to very weakly magnetic. They remain stable in the austenite phase over a wide range of temperatures and do not transform to martensite when cooled. Austenitic stainless steels cannot be hardened by heat treatment, but they do harden somewhat when cold-worked. They have high resistance to oxidation and maximum resistance to corrosion. Austenitic stainless steels are extremely tough and ductile as welded and do not usually require annealing unless used under severely corrosive conditions. They are generally easy to weld.

1.1.2 Martensitic Stainless Steels

Martensitic stainless steels are made up of two groups: plain martensitic and chromium-martensitic. The plain martensitic stainless steels contain only 4 to 6 percent chromium and no nickel. They are not considered true stainless steels; however, their corrosion resistance is much greater than the mild carbon steels, even at high temperatures.

Table 1 Certain Basic AISI-Type Stainless Steels by Category

AUSTENITIC			
Type	% Carbon	% Chromium	% Nickel
201	0.15	16/18	3.5/5.5
202	0.15	17/19	4/6
301	0.15	16/18	6/8
302	0.15	17/19	8/10
303	0.15	17/19	8/10
304	0.08	18/20	8/12
304L	0.03	18/20	8/12
308	0.08	19/21	10/12
309	0.20	22/24	12/15
310	0.25	24/26	19/22
316	0.08	16/18	10/14
316L	0.03	16/18	10/14
321	0.08	17/19	9/12
347*	0.08	17/19	9/13
348*	0.08	17/19	9/13

* Stabilized with columbium and/or titanium

MARTENSITIC			
Type	% Carbon	% Chromium	% Nickel
403	0.15	11.5/13	—
410	0.15	11.5/13.5	—
414	0.15	11.5/13.5	1.25/2.5
416	0.15	12/14	—
420	0.15+	12/14	—
422	0.20/0.25	11/13	0.5/1.0
431	0.20	15/17	1.25/2.5
440A	0.60/0.75	16/18	—
440B	0.75/0.95	16/18	—
440C	0.95/1.20	16/18	—

FERRITIC			
Type	% Carbon	% Chromium	% Nickel
405	0.08	11.5/14.5	—
409	0.08	10.5/11.75	—
430	0.12	14/18	—
442	0.20	18/23	—
444	0.25	17.5/19.5	1.0
446	0.20	23/27	—

Examples of plain martensitic stainless steels include AISI types 501 and 502. In the chromium-martensitic steels, the austenite-to-martensite grain structure is formed by slow cooling. The chromium-martensitic stainless steels contain from 11.5 to 18 percent chromium. They also contain about 1 percent manganese and, in some cases, 0 to 2.5 percent nickel. Examples of chromium-martensitic stainless steels include AISI types 403, 410, 414, 416, 420, 422, 431, and 440.

All martensitic stainless steels are magnetic and can be hardened by heat treatment. They are highly heat-resistant and keep many of their properties up to 1,100°F (593°C). They are more corrosion-resistant than alloy steels, but they are not as corrosion-resistant as the other stainless steel types. Martensitic stainless steels are air-hardened. They must be cooled slowly in still air after welding to prevent cracking. They may also require heat treatment after welding.

1.1.3 Ferritic Stainless Steels

Ferritic stainless steels are straight chromium stainless steels. They contain 11.5 to 27 percent chromium, about 1 percent manganese, and little or no nickel. Their carbon content is 0.20 percent or less. The ferrite grain structure is stable at all temperatures and does not change to austenite. Examples of ferritic stainless steels include AISI types 405, 409, 430, 430F, 442, 444, and 446.

The ferritic stainless steels have a low coefficient of expansion and good resistance to corrosion. They are magnetic and cannot be hardened by heat treatment. They can be hardened somewhat by cold working. They are used for pressure vessel linings, automobile mufflers and trim, building trim, and low-cost equipment that must be corrosion-resistant. Ferritic stainless steels have high resistance to nitric acid and high temperature scaling. Ferritic stainless steels are typically the most difficult to weld of all the stainless steel types. AISI types 430F and 430FSe are generally considered unweldable or very difficult to weld.

1.2.0 Carbide Precipitation and Hot Cracking

When austenitic stainless steel containing more than 0.02 percent carbon is heated to between 800°F (427°C) and 1,600°F (871°C), carbide precipitation occurs in the heat-affected zone. The carbon in excess of 0.02 percent moves to the grain boundaries where it combines with chromium, forming chromium carbide. Chromium prevents corrosion in stainless steel. Consuming chromium by forming the chromium carbide compound reduces the corrosion-resistance in the stainless steel. This process leading to carbide precipitation is known as sensitization. It makes austenitic stainless steel subject to intergranular corrosion when exposed to a corrosive agent.

Carbide precipitation increases with the carbon content of the steel. No other property of the stainless steel is affected other than corrosion-resistance. The corrosion is often seen on the surface in the heat-affected zone after exposure to a corrosive agent. *Figure 1* shows a section of tubing from a heating-water preheater assembly that failed, along with a microscopic image of the carbide precipitation present along the surface.

(A) FAILED PREHEATER TUBE

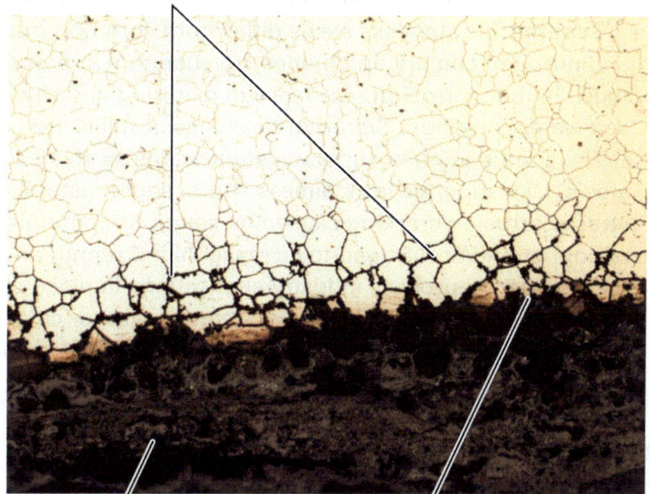

(B) CARBIDE PRECIPITATION (100X MAGNIFICATION)

Figure 1 Carbide precipitation.

Methods for controlling carbide precipitation include the following:

- Use of a stabilized base metal and filler metal
- Use of a low-carbon base metal and filler metal (L grades)
- Post-weld heat treatment

1.2.1 Stabilized Base Metal and Filler Metal

Stabilized stainless steels contain columbium or titanium to reduce carbide precipitation. Both columbium and titanium are used to stabilize base metals. Only columbium is used in electrodes, however, because it transfers across the arc more readily than titanium.

Both columbium and titanium have a greater attraction for carbon than chromium does. When a stabilized metal is heated to between 800°F (427°C) and 1,400°F (760°C), the columbium or titanium combines with the carbon. This leaves

Specialty Stainless Steels

Besides the three basic grades of stainless steel, four specialty grades are also available. These include duplex, precipitation-hardening, superaustenitic, and superferritic stainless steels.

Duplex stainless steels have roughly equal amounts of ferrite and austenite. They contain about 24 percent chromium and 5 percent nickel. They are the most recently developed of the stainless steels. Duplex stainless steels offer higher yield strength and greater resistance to corrosion and stress cracking than the 300 series steels. These steels may be welded using SMAW, GTAW, GMAW, FCAW, submerged arc welding (SAW), or plasma arc welding (PAW) techniques.

Precipitation-hardening stainless steels contain alloying additions, such as aluminum. These alloying additions allow the stainless steel to be hardened by a solution and aging heat treatment. Precipitation-hardening stainless steels are classified into subgroups as martensitic, semiaustenitic, and austenitic precipitation-hardening stainless steels. The weldable versions of these steels can be welded using GTAW, GMAW, or PAW techniques.

Superferritic stainless steels differ from common grade ferritic stainless steels in that they have lower carbon and nitrogen content and higher molybdenum and/or chromium content. Their chemical composition provides much higher corrosion resistance in oxidizing and chloride environments, and they have outstanding resistance to sea-water corrosion. Superferritic stainless steels are only welded with GTAW and GMAW techniques.

The composition of superaustenitic stainless steels lies between 300 series stainless steels and nickel-alloy steels. Like superferritic stainless steels, superaustenitic stainless steels are highly resistant to corrosion and stress-corrosion cracking caused by solutions containing chloride. Many of these alloys contain planned amounts of copper and/or nitrogen to increase their strength and corrosion resistance. These stainless steel types are widely used in process industry applications in place of common austenitic stainless steels where the common types may fail due to pitting, crevice corrosion, or chloride-induced corrosion cracking. Superaustenitic stainless steels can be welded using SMAW, GTAW, GMAW, or SAW techniques.

the chromium free to inhibit corrosion. Stabilized stainless steels can be used at temperatures up to 1,400°F (760°C) as welded, without heat treatment and without affecting their corrosion resistance. AISI 321 and 347 are examples of stabilized stainless steels.

1.2.2 Low-Carbon Base Metal and Filler Metal

Chromium carbide is formed by the carbon in excess of 0.02 percent. When the carbon content of the stainless steel base metal is kept closer to 0.02 percent, less carbide precipitation occurs. For the filler metal, keeping the carbon content below 0.03 percent will decrease carbide precipitation. Low-carbon stainless steels have a letter L in their AISI designation to signify low carbon. The designation ELC, for extra-low carbon, is also sometimes used. AISI 304L and 316L are examples of low-carbon stainless steels.

1.2.3 Postweld Heat Treatment (PWHT)

The chromium carbide formed in the heat-affected zone can be broken and the elements separated by postheating the weldment (for example, to 1,950°F, or 1,066°C) and then quenching it. While this method can solve the problem of carbide precipitation, it is not used often because of the size of the weldment and the problems that can occur when the weldment is heated to such a high temperature. Some of the problems that are difficult to control are sagging and loss of dimension control. Others, such as excessive scaling, can be controlled with an appropriate shielding gas.

1.2.4 Hot Cracking

Hot cracking of austenitic stainless steel occurs when metallic compounds of sulfur and phosphorus penetrate grain boundaries. These metallic compounds can be present in the weld or the heat-affected zone. When the weld cools, cracks appear as shrinkage stresses occur. Hot cracking is usually prevented by introducing a small amount of ferrite into the austenitic matrix from the base and filler material. The ferrite provides ferrite-austenite grain boundaries that control sulfur and phosphorous compounds and prevent hot cracking.

1.3.0 Stainless Steel Electrodes

SMAW electrodes for welding stainless steel are basically the same as SMAW electrodes for welding carbon steel. Most electrodes are a solid filler metal wire with an external flux coating. The wire core transfers the welding current from the electrode holder to the work. The arc at the end of the electrode melts the tip of the filler-metal core, the flux coating at the electrode tip, and the base metal under the arc. The molten electrode metal fuses with the base metal; the filler metal forms the majority of the weld. As the flux coating melts and vaporizes, it performs the following functions:

- It generates a cloud of carbon dioxide (CO_2) gas that shields the molten weld. It does this by pushing away the harmful oxygen and hydrogen present in the atmosphere.
- It acts as a fluxing agent to clean and deoxidize the molten metal.
- It stabilizes the arc and reduces spatter.
- It forms a slag covering over the weld. This further protects the weld from the atmosphere and slows its cooling rate.

The action of an SMAW electrode used for welding stainless steel is shown in *Figure 2*.

Some flux coatings contain alloying elements. These alloying elements change the chemical composition and strength of the deposited weld metal. Alloys are normally added to the electrode filler metal so that the deposited metal will better match the base metal in chemical composition and strength.

Some stainless electrodes, such as E430-16, have a carbon steel core. The chromium, nickel, and other alloys required for the deposited weld metal to become stainless steel are contained in the flux coating. As the arc melts the carbon steel core and flux coating, the deposited weld metal becomes stainless steel.

Coated tubular electrodes are specialty stainless electrodes. They have a thin carbon steel tube filled with powdered metal that contains chromium, nickel, and other alloys. The metal tube has an external flux coating. As the arc melts the tube, the steel and powdered metals melt and flow together to form stainless steel.

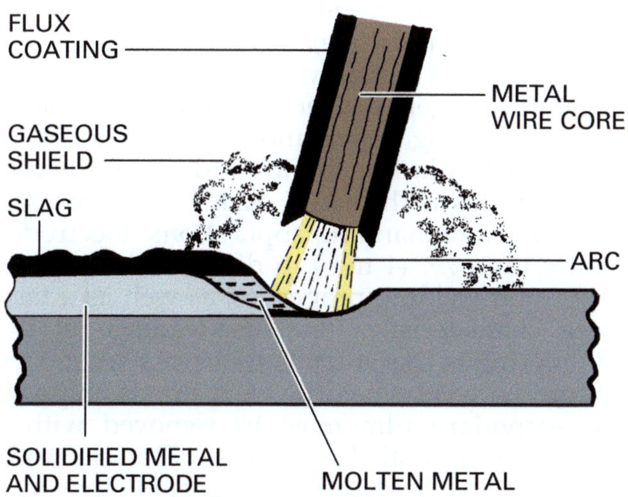

Figure 2 Stainless steel SMAW electrode action.

Coated tubular electrodes are most often used for welding specialty stainless alloys, dissimilar steels, manganese, and other wear-resistant alloys. Coated tubular electrodes can operate at lower amperage ranges than solid wire electrodes. This reduces the possibility of carbon precipitation and allows the welding of thin materials. At higher amperage ranges, the deposition rates of coated tubular electrodes are higher than solid wire electrodes. Most tubular electrodes can be used in all positions.

CAUTION

Do not use stainless steel electrodes if the flux coating is chipped off. Without the alloys in the flux, the deposited weld metal will be mild steel, not stainless steel.

1.3.1 AWS Filler Metal Specification System

The AWS writes specifications for welding consumables, such as the electrodes, filler rods, and fluxes used with various welding processes. There is another series of different specifications known as AWS A5.XX. These specifications are used in all major welding codes and are accepted in all industries except where other approvals are specified. Industries or agencies in which other approvals are specified include the American Bureau of Shipping (ABS) and the Coast Guard, Federal Highway Administration, Military Specifications (MIL), and Lloyds of London. The electrode specifications of the ASME Boiler and Pressure Vessel Code closely follow the AWS specifications.

The purpose of the AWS specifications is to set standards that all manufacturers must follow when making welding consumables. This ensures consistency for the user no matter what company makes the product. The specifications set standards for the following:

- Classification system, identification, and marking
- Chemical composition of the deposited weld metal
- Mechanical properties of the deposited weld metal

The specification that applies to covered SMAW stainless steel electrodes is *AWS A5.4, Specification for Stainless-Steel Electrodes for Shielded Metal Arc Welding*.

> **NOTE**
>
> The specification number is sometimes followed by a hyphen and a two-digit number, such as *A5.1-91* or *A5.5-96.* The hyphen and number indicate the year in which the specification was last revised. These documents are on a five-year revision cycle. Always use the latest version.

1.3.2 Classification System

The classification system used to identify *AWS A5.4* stainless steel electrodes is based on the numbering system developed by AISI. This numbering system uses a three- or four-digit number to identify the type of stainless steel alloy. The letter E is added as a prefix to indicate an electrode. The initial AISI digits are followed by a one- to four-letter alloy modification designation (if required). To the right of that is a two-digit usability designation number that is preceded by a hyphen or space. This two-digit number is based on the flux coating composition, welding position, and type of welding current that can be used with the electrode. The classification system number is printed on the flux coating of each electrode near the bare end. *Figure 3* shows markings typically found on stainless steel electrodes. *Figure 4* shows details of the marking system used to classify stainless steel electrodes.

All electrodes made according to AWS specifications must have the AWS classification number printed on them and/or on their shipping containers. In addition, electrode manufacturers usually print their own unique classification name or number on the electrode and on the shipping container. Manufacturers may make more than one electrode that fits within a specific AWS classification. They may also make electrodes that cover more than one AWS classification or usability designation.

For example, ESAB® (Elektriska Svetsnings Aktiebolaget) makes two different SMAW electrodes that cover AWS classifications E308 and E308L -15, -16, and -17. They are called Arcaloy® 308L and 308L Plus. Arcaloy® 308L is an AWS E308/308L -15 and -16 low-carbon electrode used to weld chemical plant equipment of type 304L

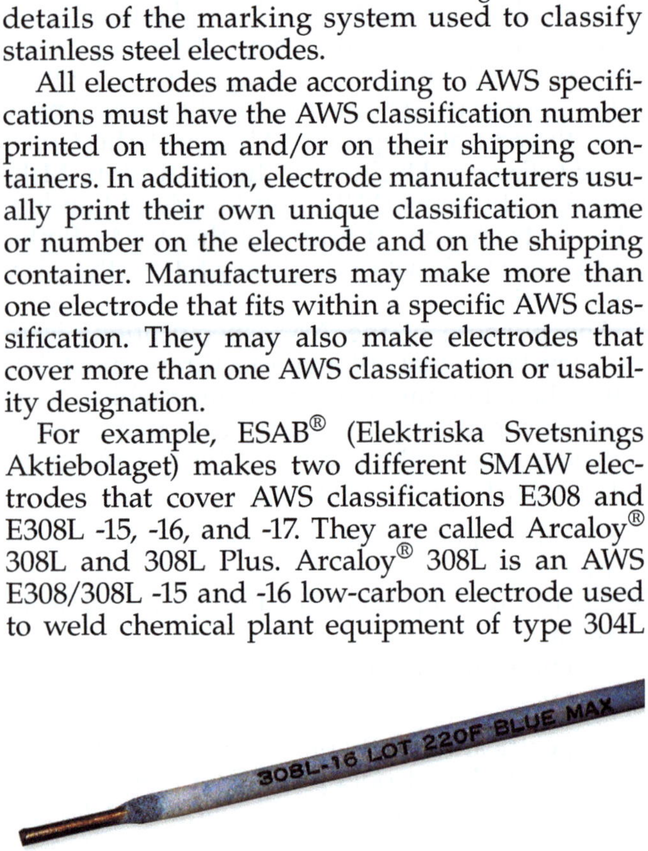

Figure 3 Stainless steel electrode marking.

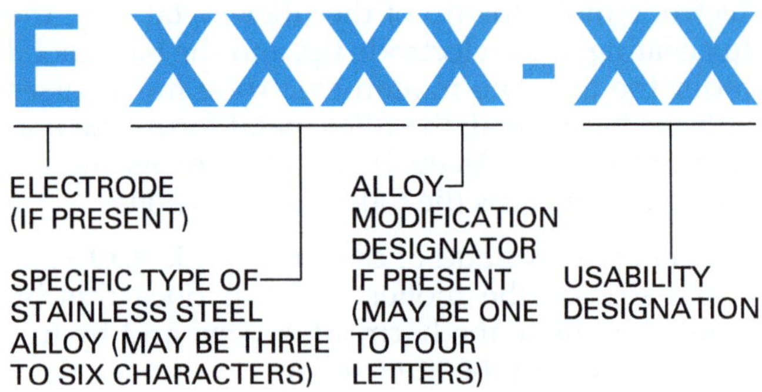

Figure 4 Stainless steel electrode classification marking system.

stainless steel, where carbon in the weld must be under 0.04 percent. Arcaloy® 308L is also used to weld stainless steel Types 321 and 347. Arcaloy® 308L Plus is an AWS E308/208L -17 electrode that contains less carbon than Arcaloy® 308L, but it contains more silicon. It is used to weld stainless steel Types 304, 304L, 308, and 308L when bead appearance is important. When there is more than one electrode in an AWS classification and you are unsure which electrode to use, check with your instructor or supervisor.

Many electrode manufacturers offer applications suitable for smartphones and similar devices that include information about their electrodes, their uses, and recommended amperage settings. Most also continue to publish paperback welding guides. Local distributors usually provide these booklets free of charge.

1.3.3 Usability Designations

The four primary usability designations for stainless steel SMAW electrodes are -15, -16, -17, and -26.

The -15 usability designation indicates that the coating of the electrode is typically lime (lime-fluorspar or basic lime). A lime coating gives the purest weld metal, which is also usually the lowest in nitrogen, oxygen, and other inclusions. The lime coating is also most resistant to moisture. As a result, it tends to produce less porosity. Electrodes with the -15 designation are used for the most demanding applications. Electrodes smaller than 5/32" (4 mm) in diameter have good all-position welding properties, as well as a fast-freeze characteristic. One disadvantage of the lime coating is that it tends to leave a secondary film on the weld when the weld flux is chipped. This secondary film must be removed with a stainless steel-bristled wire brush. This electrode can be difficult to manipulate. Coated electrodes with the -15 designation are used with direct current electrode positive (DCEP) current. This electrode is designed to produce a convex bead that is ideal for root passes on heavy plate.

Electrodes with a usability designation of -16 usually have a rutile (titania) coating, which also contains some limestone. They are also known as basic-rutile coatings. This is the most popular coating type. It has excellent arc characteristics. The slag can be easily and completely removed without leaving a secondary film. Electrodes smaller than 5⁄32" (4 mm) in diameter can be used in all positions. However, vertical and overhead welds are somewhat more difficult than those performed with 15 electrodes. This is because of the slower freeze characteristics of the -16 electrode. The -16 electrodes can be used with either AC or DCEP welding current, but preferably DCEP. These electrodes are designed to produce smooth, flat beads. Penetration should be complete and free of porosity. Ductility and strength can be better than with lime-coated electrodes.

Electrodes with a usability designation of -17 typically use a modified version of the -16 designation coating. The coating they use is a titania-silica combination. The electrodes produce more of a spray arc, resulting in a more finely rippled weld bead. They also have a slower-freezing slag that has better handling qualities during drag applications. Like the -16 electrodes, -17 electrodes can be used with AC or DCEP welding current, but preferably with DCEP. They are typically used in flat or horizontal positions, although they can also be used in all other positions but with greater difficulty. When used for vertical welds, -17 electrodes require wider weave beads in uphill travel than are needed with -16 electrodes. Electrodes 3⁄16" (4.8 mm) or larger are not recommended for overhead or vertical welding.

Electrodes with a usability designation of -26 are similar in coatings and usage to -17 and some -16 electrodes. The core wire of a -26 electrode can be different, such as plain carbon steel with the required alloys contained in the coating. This makes the electrode larger in diameter than a -16 or -17 type. Note that -26 electrodes are to be used for flat and horizontal fillet welds only.

Special -16 and -17 Electrodes

Some manufacturers market specially coated solid or tubular electrodes that are specifically designed for vertical or overhead welding and for 5G and 6G pipe positions.

Some classification numbers have additional letters that follow the classification digits. These letters indicate an alloy modification designation. The following is a list of those extra letters and their meanings:

- *L – Low carbon (0.03 percent or less)* – Less than 0.04 percent carbon is deposited in the weld, commonly used to weld chemical plant equipment.
- *H – High carbon (0.04 to 0.08 percent)* – Used for creep-resistant welds in 304H stainless steel.
- *HC – Very high carbon (0.08 to 0.15 percent)* – Used to limit magnetic permeability to 2.0 and where corrosion is not a major factor.
- *Cb – Columbium (0.8 percent)* – Added to electrode alloys to stabilize the chromium and inhibit carbide precipitation.
- *Mo – Molybdenum (2.2 to 2.5 percent)* – Added to electrode alloys used to weld stainless steel containing molybdenum.
- *LR – Low residuals* – Have lower carbon, silicon, phosphorus, and sulfur content and very controlled columbium and manganese content.

1.4.0 Electrode Selection and Storage

Welders must be able to distinguish between the many types of electrodes that are available and select the correct electrode for the job. The two most important considerations when selecting stainless steel electrodes are the WPS or site quality standard, and the base metal composition.

If there is a WPS or site quality standard that may include specific work packages, the electrode type will be specified. The electrode type will be given as an AWS classification and/or manufacturer's standard, which will sometimes include a manufacturer's name. If there is a WPS, the specified filler metal must be used.

CAUTION

To maintain weld quality, always follow the filler metal specifications in the WPS or site quality standards.

Stainless steel electrodes are usually selected to match the chemical composition of the base metal when welding stainless steel. Because there are so many different types of stainless base metals and electrodes, always refer to a table or chart when making a selection. The manufacturers of stainless steel electrodes provide tables or charts that show the type of electrode (filler metal) that works best with specific types of stainless steel.

Table 2 is an example of a chart used to identify the correct filler metal for the base metal to be welded. Note that this is only an example. Unless specified, electrodes with usability designations of types -15 (EXXX-15), -16, -17, and -26 can be used. The intended uses of various classes of stainless steel electrodes as identified in *ANSI/AWS 5.4* are provided in a series of tables found in *Appendix A*.

CAUTION

The recommendations shown in *Table 2* and in the tables found in *Appendix A* may not be suitable for some applications. Always refer to the WPS for specific electrode requirements.

1.4.1 Storing Stainless Steel Electrodes

The flux coating on stainless steel electrodes is low-hydrogen (low in moisture, which produces hydrogen during welding). For this reason, stainless steel electrodes must be kept dry. The dry mineral flux coatings on stainless steel electrodes start to absorb moisture as soon as they are exposed to the atmosphere. Moisture in the flux causes porosity and other defects in welds.

Stainless steel electrodes are shipped in hermetically sealed containers to keep the electrodes dry. Immediately after opening the sealed containers, the electrodes must be stored in special ovens (*Figure 5*). The ovens are kept at temperatures of

Table 2 Manufacturer's Chart for Welding Stainless Steels

		AUSTENITIC		
Base Metal	**% Carbon**	**% Chromium**	**% Nickel**	**Filler Metal**
201	0.15	16/18	3.5/5.5	308
202	0.15	17/19	4/6	308
301	0.15	16/18	6/8	308
302	0.15	17/19	8/10	308
303	0.15	17/19	8/10	308[1]
304	0.08	18/20	8/12	308
304L	0.03	18/20	8/12	308L
308	0.08	19/21	10/12	308
309	0.20	22/24	12/15	309
310	0.25	24/26	19/22	310
316	0.08	16/18	10/14	316
316L	0.03	16/18	10/14	316L
321	0.08	17/19	9/12	347[2]
347[3]	0.08	17/19	9/13	347[2]
348[3]	0.08	17/19	9/13	347[2]

1 EXXX-15 flux coating only
2 Stabilized with columbium
3 Stabilized with columbium and/or titanium

		MARTENSITIC		
Base Metal	**% Carbon**	**% Chromium**	**% Nickel**	**Filler Metal**
403	0.15	11.5/13	–	410
410	0.15	11.5/13.5	–	410
414	0.15	11.5/13.5	1.25/2.5	410
416	0.15	12/14	–	410
420	0.15+	12/14	–	410
431	0.20	15/17	1.25/2.5	430

		FERRITIC		
Type	**% Carbon**	**% Chromium**	**% Nickel**	**Filler Metal**
405	0.08	11.5/14.5	–	410
430	0.12	14/18	–	430
442	0.20	18/23	–	309/310
446	0.20	23/27	–	309/310

Figure 5 Electric drying oven.

200°F to 250°F (93°C to 120°C), or as recommended by the manufacturer. After stainless steel electrodes have been removed from the oven, they can only be exposed to the atmosphere for about four hours. Electrodes exposed for longer than four hours may require redrying at higher temperatures. Portable electrode holders (*Figure 6*) are recommended to minimize exposure.

The precise exposure times and procedures for handling and drying filler metal at a specific site are usually written down in a quality standard. General precautions that normally apply to handling electrodes include the following:

- Never handle hot electrodes from the oven with bare hands or dirty gloves; use clean gloves only.
- After removing electrodes from the oven for use, store them in a pouch or rod holder to avoid contamination.

Figure 6 Portable electrode holder.

Welders must follow procedures exactly. If unsure of the requirements at a site, check with your supervisor.

> **CAUTION**
>
> Site quality standards are very strict. Be sure that you are familiar with them before handling electrodes. The information provided in this module is general in nature. Always follow the specific quality standards for your site.

Additional Resources

AWS A5.4, Specification for Stainless-Steel Electrodes for Shielded Metal Arc Welding. Miami, FL: American Welding Society.

AWS F4.1: Recommended Safe Practices for Preparation for Welding and Cutting of Containers and Piping. Miami, FL: American Welding Society.

Modern Welding Technology. Current Edition. Howard B. Cary and Scott Helzer. Englewood Cliffs, NJ: Prentice Hall.

OSHA 1926.351, Arc Welding and Cutting. Current Edition. Washington, DC: Occupational Safety & Health Administration.

The Procedure Handbook of Arc Welding. Current Edition. Cleveland, OH: The James F. Lincoln Arc Welding Foundation.

Welding Handbook, Volumes 1 through 4. Current Edition. Miami, FL: American Welding Society.

Welding Pressure Pipelines and Piping Systems (PFD). The Lincoln Electric Company, Cleveland, OH, USA. **www.lincolnelectric.com**. Last accessed May 29, 2015.

1.0.0 Section Review

1. The primary physical characteristic of all stainless steels is their resistance to corrosion and heat, which varies depending on their _____.
   a. size
   b. shape
   c. nitrogen content
   d. category

2. When austenitic stainless steel containing more than 0.02 percent carbon is heated to between 800°F (427°C) and 1,600°F (871°C), the phenomenon that occurs in the heat-affected zone is called _____.
   a. warping
   b. hot cracking
   c. corrosion resistance
   d. carbide precipitation

3. When welding stainless steel, the majority of the weld is formed by the _____.
   a. base metal
   b. electrode filler metal
   c. flux coating of the electrode
   d. equal mix of electrode metal and base metal

4. Two important considerations when selecting stainless steel electrodes are the WPS or site quality standard and the _____.
   a. base metal composition
   b. AISI type
   c. filler rod composition
   d. crystalline grain structure

SECTION TWO

2.0.0 STAINLESS STEEL SMAW PREPARATIONS

Objective

State the basic concepts of SMAW of stainless steel and how to prepare for welding.

a. State the basic concepts of SMAW of stainless steel.
b. Identify common welding safety practices.
c. Explain how to prepare the area, materials, and equipment for SMAW of stainless steel.

Trade Terms

Interpass temperature: The temperature of the base metal in the general area of the weld before the next pass is started. The definition of the area of the weld varies by the applicable welding code.

Before performing SMAW on any type of stainless steel, welders must make sure that the job will be done safely and correctly. They need to be knowledgeable about the properties and welding characteristics of the type of stainless steel that will be welded. They must also follow the appropriate safety practices and properly prepare the area, materials, and equipment that will be used for the job.

2.1.0 Basic Welding Variations for Stainless Steel

Welding stainless steel is slightly more difficult than welding carbon steel. Compared to carbon steel, stainless steel has a lower melting point, conducts heat more slowly, expands more when heated, and has a higher resistance to electrical current flow. These characteristics are similar for all stainless steels, but they vary greatly with the metallurgical structure of the steel. Stainless steels in the same metallurgical group (austenitic, martensitic, or ferritic) have similar welding characteristics.

> **NOTE**
> Information in this module is provided as a general guideline only. Refer to your WPS or site quality standards for the specific requirements for a weld.

2.1.1 Welding Austenitic Stainless Steels

Austenitic stainless steel is generally easier to weld than the other types of stainless steels. It normally does not require preheating or postheating. However, the **interpass temperature** must be controlled to prevent carbide precipitation. The interpass temperature should typically not exceed 350°F (177°C). Refer to the site WPS for specific information.

2.1.2 Welding Martensitic Stainless Steels

Martensitic stainless steels air harden in proportion to their chromium and carbon content. This makes the weld and weld zone very hard and brittle unless it is heat treated.

Martensitic heating requirements are based on carbon content. Approximate heating requirements are as follows:

- *0 to 0.1 percent carbon* – No preheating or postheating is required.
- *0.1 to 0.2 percent carbon* – Preheat to 500°F (260°C) and cool slowly.
- *0.2 to 0.5 percent carbon* – Preheat to 500°F (260°C), and then anneal after welding by heating to 1,500°F to 1,600°F (816°C to 871°C), furnace cool to 1,100°F (593°C), and then air cool the workpiece.

When preheating and postheating cannot be used, weld martensitic stainless steels with austenitic type filler metal (stainless or nickel-alloy), such as E308 or E309 (if allowed by the procedure). The heat-affected zone will still be hard and brittle. However, the filler metal will be ductile and able to absorb some stresses to protect the hard, heat-affected zone.

2.1.3 Welding Ferritic Stainless Steels

Ferritic stainless steel has the poorest weldability of the stainless types. This characteristic can be traced to the loss of toughness that comes from severe grain growth. Welding certain ferritic stainless steels (such as 430, 430F, 430FSe, 434, 442, and 446) causes embrittlement and cracking that result from martensite formation. When heat treatment is possible, the recommended procedure includes preheating followed by annealing.

The welding of ferritic stainless steel is usually done using austenitic E308, E309, E310, or E312 electrodes. This filler metal is ductile and will resist cracking better than a less ductile filler metal. The only ferritic stainless steel electrode for SMAW that is listed in the AWS specifications is E430.

2.1.4 Welding Carbon Steels to Stainless Steels

When welding carbon steels to stainless steels, use a high-alloy electrode, such as E309, E312, or the nickel-alloy electrodes. This will reduce the effect that mixing carbon steel with stainless steel has on the deposited weld metal. E309 or E312 electrodes are sometimes used for welding medium- and high-carbon steels when preheating and postheating cannot be performed. E309 and E312 electrodes provide a strong, ductile weld deposit. E308 electrodes are recommended for welding austenitic manganese steel or carbon steel to manganese steel.

2.2.0 Safety Practices

Safety considerations are a critical part of preparing for a welding operation. The following is a summary of safety procedures and practices that must be observed while cutting or welding. Be advised that this is only a summary. Complete safety coverage is provided in the *Welding Level One* module, "Welding Safety." That module should be studied, if not already completed, before continuing. Above all, be sure to wear appropriate protective clothing and equipment at all times when welding or cutting.

2.2.1 Protective Clothing and Equipment

Protective clothing and equipment are a necessary part of every welding task. Follow the guidelines below as a general rule, and ensure that you are familiar with any unique requirements your employer may have regarding personal protective equipment (PPE).

- A welding hood equipped with a properly shaded lens is best for all forms of welding. The welding hood lens must be an appropriate shade for the application.
- Wear proper leather or flame-retardant clothing along with welding gloves for protection from flying sparks, molten metal, and heat.
- Wear high-top safety shoes or boots. Make sure that the tongue and lace area of the footwear is covered by a pant leg. If the tongue and lace area is exposed or if the footwear must be protected from burn marks, wear spats under the pant leg or leather chaps over the front and top of the footwear.
- Wear a 100-percent cotton cap without any mesh material. If a hard hat is required for the environment, use one that allows the attachment of rear deflector material and a face shield. A hard hat with a rear deflector is generally preferred when working overhead, and may be required by some employers and job sites.
- Wear earmuffs, or at least earplugs, to prevent sparks from entering the ear canals.

2.2.2 Housekeeping and Fall Protection

Welding is often performed outside in the weather and in the construction environment. Production work done inside large, open shops exposes a welder to equipment moving across the floor and overhead. Welders often have to crawl under or reach over objects to complete their work.

At times, welders must work at height. Fall protection equipment is required for all work at height (*Figure 7*). Always keep fall protection devices untangled and properly attached. Combined with the other equipment a welder must carry, there is a great deal to manage, but fall protection equipment must always be the priority. Fall protection equipment saves lives. Never attach any kind of welding gear to the harness that could interfere with its ability to perform its intended function.

Regardless of where you are working, always keep the work area as neat as possible. This makes the area safer and helps to reduce distractions. Mindful actions, such as making sure that welding cables are not tripping hazards, contribute to a safer environment. Electrode stubs and unused electrodes must not be left on the floor. They can cause someone to lose their footing and fall. Parts waiting to be welded or left over from

Austenitic Stainless Steel Welding

Rapid cooling of austenitic stainless steel is beneficial and can help prevent distortion. Chill bars can be used near the weld, both at the top and under the weld. Copper chill bars should be nickel-plated to prevent contamination of the weld. In applications requiring high duty cycles, water-cooled chill bars can be very beneficial. Using skip-welding and back-step welding techniques will also help to minimize distortion. Make sure that interpass temperatures do not exceed 350°F (177°C) to reduce sensitization of the stainless steel.

Figure 7 Working at height.

previous work are a tripping hazard and should be kept out of the walking area. If the parts have sharp edges, they could cut welding cable insulation or torch gas hoses.

2.2.3 Fire and Explosion Prevention

Flammable gases are commonly found around welding and cutting operations, as are sparks and other significant sources of ignition. For that reason, and because many fire-related incidents over the years have been traced back to welding operations, it is crucial that fire and explosion prevention be given consistent attention. Follow these guidelines to maintain a safe environment:

- Never carry matches or gas-filled lighters. Sparks can cause the matches to ignite or the lighter to explode in your pocket, causing serious injury.
- Never use oxygen to blow dust or dirt from clothing. Traces of the oxygen can remain trapped in the fabric for some time. If a spark hits the fabric during this time, the clothing can burn rapidly and violently out of control.
- Make sure that any flammable material in the work area is removed or shielded by a fire-resistant covering. Approved fire extinguishers must be available before attempting any heating, welding, or cutting operations. If a hot-work permit and a fire watch are required, be sure these are in place before beginning.
- Ensure that all site safety requirements are being observed.
- Never release a large amount of oxygen or use oxygen in place of compressed air. Its presence around flammable materials or sparks can cause rapid and uncontrolled combustion. Keep oxygen away from oil, grease, and other petroleum products.
- Never release a large amount of fuel gas, especially acetylene. Propane tends to concentrate in and along low areas and can ignite at a considerable distance from the release point. Acetylene is lighter than air, but is even more dangerous than propane. When mixed with air or oxygen, acetylene will ignite at much lower concentrations than any other fuel gas.
- Even though shielding gases are not toxic, any gas released in a sufficient quantity can displace the oxygen in a small room or enclosed work area, creating an atmosphere with little or no oxygen. Avoid large releases of any type of gas, regardless of its characteristics.
- To prevent fires, maintain a neat and clean work area, and make sure that any metal scrap or slag is cold before disposal.
- Before cutting containers such as tanks or barrels, check to see if they have contained any explosive, hazardous, or flammable materials, including petroleum products, citrus products, or chemicals that decompose into toxic fumes when heated. As a standard practice, always clean and then fill any tanks or barrels with water, or purge them with a flow of inert gas such as nitrogen to displace any oxygen. Containers must be cleaned by steam cleaning, flushing with water, or washing with detergent until all traces of the material have been removed. Proper procedures for cutting or welding hazardous containers are described in the *American Welding Society (AWS) F4.1, Safe Practices for the Preparation of Containers and Piping for Welding and Cutting.*

WARNING!

Welding or cutting must never be performed on drums, barrels, tanks, vessels, or other containers until they have been emptied and cleaned thoroughly, eliminating all flammable materials and all substances (such as detergents, solvents, greases, tars, or acids) that might produce flammable, toxic, or explosive vapors when heated. Clean containers only in well-ventilated areas. Vapors can accumulate during cleaning, causing explosions and serious personal injury.

2.2.4 Work Area Ventilation

Many gases are introduced to the atmosphere through welding processes. Some of these gases are harmful, and all can displace or combine with atmospheric oxygen in a confined space. Follow these guidelines, as well as those of your employer, for work area ventilation:

- Make sure to follow all confined space procedures before conducting any welding or cutting in a confined space. Use powered extraction systems as needed.
- Never use oxygen to provide ventilation in confined spaces.
- Always perform cutting or welding operations in a well-ventilated area.

Cutting or welding operations involving zinc or cadmium materials or coatings result in toxic fumes. Another significant concern is Chromium 6, commonly referred to as *hexavalent chromium*, which is used in stainless steel. It releases toxic fumes when heated. Manganese is another hazardous metal used in stainless steel. It is also used in some welding rods, increasing the risk of exposure.

For long-term cutting or welding of materials containing these components, always wear an approved full-face, supplied-air respirator that uses breathing air supplied from outside the work area. For occasional short-term exposure, a HEPA-rated or metal-fume filter may be used on a standard respirator.

Powered fume-extraction equipment must be used when welding materials that contain zinc, cadmium, hexavalent chromium, manganese, and similar materials that produce hazardous fumes if the area is not naturally well-ventilated. *Figure 8* shows examples of both mobile and portable units.

2.3.0 Preparations for Welding

Using the appropriate equipment in a clean, well-maintained work area is fundamental for safe, efficient work in any type of welding. Performing SMAW on stainless steel requires that welders use a properly prepared work area and welding machine. They must also take special care to clean and shape the plate welding coupons and pipe welding coupons that will be used for the job.

2.3.1 Preparing the Welding Area

To practice welding, a welding table, bench, or stand is needed. The welding surface must be steel. Provisions must be made for placing welding coupons out of position.

To set up the area for welding, follow these steps:

Step 1 Make sure that the area is properly ventilated. Use doors, windows, and fans.

Step 2 Check the area for fire hazards. Remove any flammable materials before proceeding.

Step 3 Know the location of the nearest fire extinguisher. Do not proceed unless the extinguisher is charged and you know how to use it.

Step 4 Set up flash shields around the welding area.

NOTE

The following sections explain how to prepare welding coupons with backing (V-groove with backing) and without backing (open-root V-groove). All welding objectives in this module involve open-root V-groove welds. Depending on site conditions, V-groove welds with backing can be substituted using a wider root opening of up to ¼" (6.4 mm). The techniques for both types of welds are identical except for the open-root V-groove root pass.

2.3.2 Preparing the Welding Machine

Identify an appropriate welding machine to use. *Figure 9* shows portable welding machines and their controls. Follow these steps to set up the welding machine for use:

Step 1 Verify that the welding machine can be used for DC welding.

Step 2 Make sure that the area is properly ventilated.

Step 3 Verify the location of the power disconnect. This may be a circuit breaker, disconnect switch, or even the power cord for corded models.

Step 4 Set the polarity to DCEP.

Classes of Fire Extinguishers

Fire extinguishers have different class ratings based on the types of fires they are used to extinguish. For example, Class A fire extinguishers are used for common materials like paper, wood, and plastic. Class B fire extinguishers are used for liquids such as gasoline and oil. Class C fire extinguishers are used for electrical equipment fires, and Class D extinguishers are used for combustible metals. Always make sure the fire extinguishers available on site are suitable for the types of fires that might occur.

(A) MOBILE UNIT

(B) PORTABLE UNIT

Figure 8 Mobile and portable extraction equipment.

Step 5 Connect the clamp of the workpiece lead to the workpiece.

Step 6 Set the amperage for the electrode type and size to be used. *Table 3* provides examples of common settings.

> **NOTE**
> Amperage settings vary by manufacturer, position, current type, and electrode brand. For specific information, refer to the WPS, the manufacturer's recommendations, or your instructor for the electrode being used.

Step 7 Check to be sure that the electrode holder is not grounded.

Step 8 Turn on the welding machine.

2.3.3 Preparing the Plate Coupons

When possible, the weld coupons should be ⅜" (10 mm) thick stainless steel. If this size is not readily available, use steel that is ¼" to ¾" (6 to 20 mm) thick for practice welds. Open-root welds on stainless steel plate generally have a bevel angle of 30 degrees.

> **NOTE**
> For performance qualification under some codes, the base metal is not an essential variable. In these cases, carbon steel can be used in place of the more expensive stainless steel. For practice, use carbon steel if stainless steel is not available.

Clean a carbon plate before welding by using a wire brush or grinder to remove all heavy mill scale and corrosion. If stainless steel is used, clean it with a stainless steel wire brush to prevent contaminating the base metal with carbon steel.

Table 3 Electrode Types and Amperage Settings

Electrode Diameter	300 Series			400 Series		
	F	V	OH/HOR	F	V	OH/HOR
3⁄32" (2.4 mm)	40–70	35–50	40–60	50–75	40–55	45–65
⅛" (3.2 mm)	65–100	55–70	60–85	80–115	65–80	75–100
5⁄32" (4.0 mm)	100–145	85–100	95–120	15–160	95–110	105–130

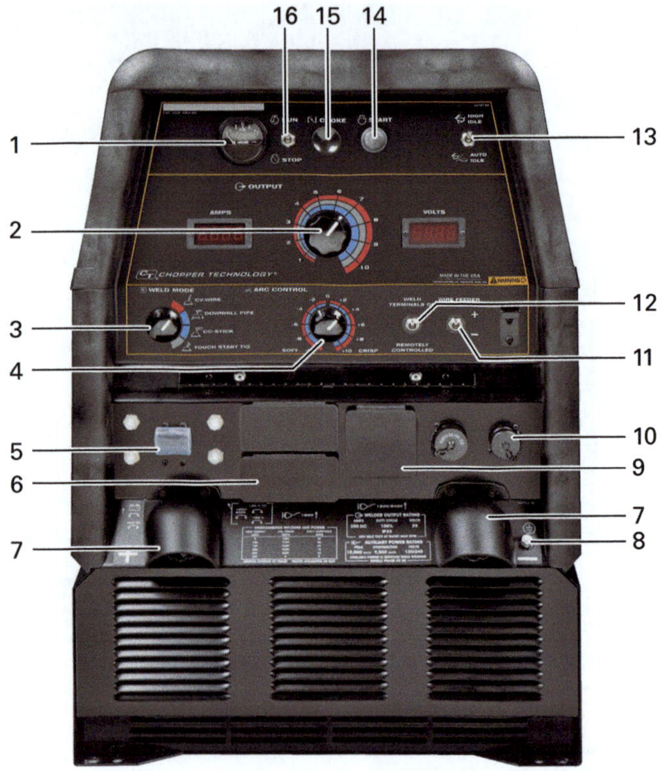

KEY CONTROLS
1. Fuel Gauge/Hour Meter
2. Output Control Dial
3. Weld Mode Switch
4. Arc Force & Inductance/Pinch Control
5. Circuit Breakers
6. 120 Volt AC Receptacles (NEMA 5-20R)
7. Output Stud Covers
8. Ground Stud
9. 120/240 Volt AC Full KVA Receptacle (NEMA 14-50R)
10. Remote Control & Feeder Connectors
11. Wire Feeder Voltmeter Polarity Switch
12. Weld Terminals Control Switch
13. Idler Control Switch
14. Pushbutton Start Switch
15. Engine Choke Control
16. Run/Stop Switch

(A) GAS ENGINE-DRIVEN 300-AMP PORTABLE DC WELDER

KEY CONTROLS
1. Coarse Current Dial
2. Engine Hour Meter
3. Battery Charging Ammeter
   Engine Temperature Gauge
   Oil Pressure Gauge
4. Idler Control
5. Run/Stop Switch
6. Start Button
7. Glow Plug Button
8. Engine Protection Light
9. 230V Circuit Breaker at 15 Amps
10. Sealed GFCI
11. Local/Remote Control Switch
12. Local/Remote Receptacle
13. 115V Circuit Breaker at 20 Amps
14. 115V AC Receptacles and Cover
15. 230V AC Receptacles and Cover
16. Fine Current and OCV Adjustment for Stick Welding

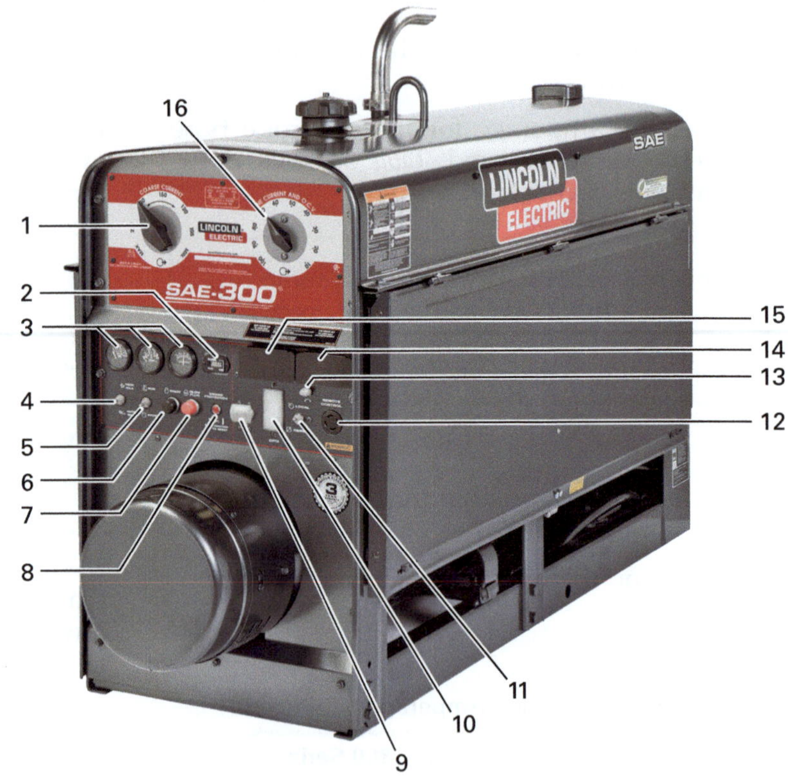

(B) DIESEL ENGINE-DRIVEN 300-AMP DC GENERATOR WELDER

Figure 9 Portable welding machines.

CAUTION

Ensure that grinding wheels, files, and similar tools used on stainless steel and low-alloy pipe have not been used on carbon steel materials or other metals. Wire brushes should also be stainless steel and must not have been used on other metals. Also confirm that cleaning and grinding tools have not been used for removing hydrocarbon contaminants.

Cut the metal into 3" × 5" (76 mm × 127 mm) rectangles with one (or both) of the 5" (127 mm) lengths beveled at 30 degrees. Grind a 0" to ⅛" (0 to 3 mm) root face on the bevel. The 5" (127 mm) length is used to save material. If backing strips will be used, cut them 1" × 6" (25 mm × 152 mm) long. The backing strips should be cut from the same type of base metal as the coupons, but they can be made of a thinner material than the coupons.

NOTE

The welding codes allow the root face and root opening on open-root welds to be from 0" to 1/8" (0 to 3 mm). Select and adjust the root face and root opening as needed when beginning your welding practice.

Follow these steps to prepare each V-groove weld coupon:

Step 1 Check the bevel face. No dross should be present. There should be a 0" to ⅛" (0 mm to 3 mm) root face for open-root welds, or no root face with a backing. If a backing is to be used, the root opening should be ¼" (6.4 mm). The bevel angle should be 22½ or 30 degrees (*Figure 10*). This will form a 45 or 60 degree groove angle.

Step 2 Center the beveled strips with the proper root opening as shown in *Figure 10*, and tack weld them in place. Place the tack welds on the back side of the joint. Use three to four ¼" (6 mm) tack welds.

Amperage Control

Amperage control is extremely important when welding open-root passes on stainless steels. Different amperage settings may be required as the weld progresses around a 5G or 6G pipe joint.

2.3.4 Preparing Pipe Coupons

Pipe coupons should be cut from 2" to 12" (DN50 to DN300) Schedule 40 or Schedule 80 stainless steel pipe. Carbon-steel pipe may be used for practice if necessary, but note that the Performance Tasks for this module require stainless steel materials. Each welded joint requires two coupons of the same diameter and schedule.

Figure 11 shows typical ASME, American Petroleum Institute (API), and AWS pipe bevel specifications for bevel angles, root faces, and root openings.

To clean and prepare weld coupons for open-root V-groove weld joints, follow these steps:

Step 1 If carbon steel is used, clean all heavy rust and mill scale from the inside and outside of the pipe with a grinder, wire brush, or other common method. Clean for a distance of ½" (13 mm) back from the weld joint. If stainless steel is used, clean with a stainless steel wire brush.

CAUTION

Ensure that grinding wheels, files, and similar tools used on stainless steel and low-alloy pipe have not been used on carbon steel materials or other metals. Wire brushes should also be stainless steel and must not have been used on other metals. Also confirm that cleaning and grinding tools have not been used for removing hydrocarbon contaminants.

GOING GREEN

Conserve and Protect

For practice welding, stainless steel is expensive and may be difficult to obtain. To conserve resources and reduce waste, completely use weld coupons until all surfaces have been welded upon. Cutting coupons apart and reusing the pieces conserves materials. Use material that cannot be cut into weld coupons to practice running beads. All materials that are beyond use for welding practice should be recycled.

Because of the potentially hazardous materials used in stainless steel, be sure to manage and properly dispose of any waste cleaned from the base metals. Also, ensure that any used electrode stubs are disposed of properly, as they may also contain hazardous elements.

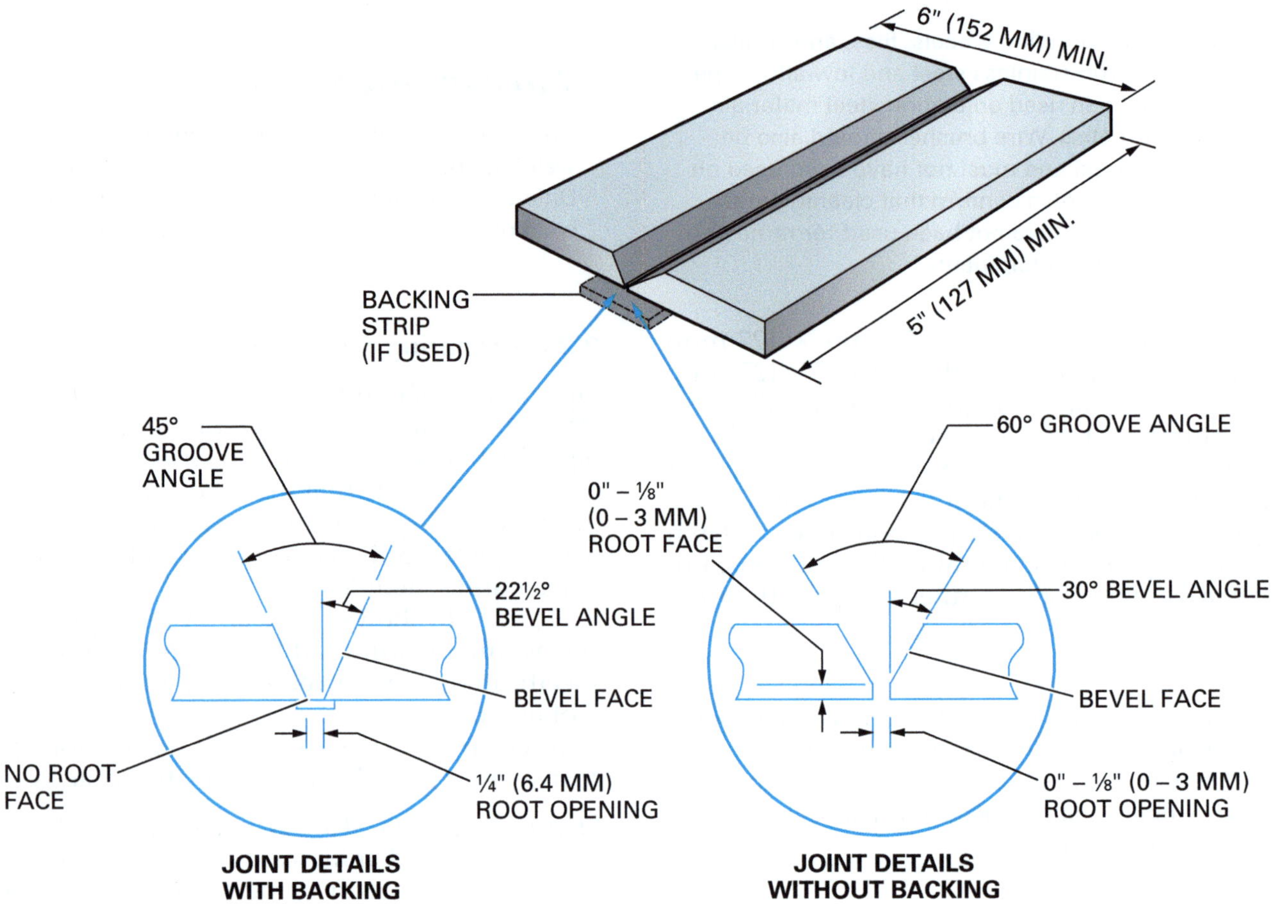

Figure 10 Open-root V-groove weld coupon.

LASH

To ensure good welding results, remember the LASH acronym:

- *Length of arc* (**L**) –The distance between the electrode and the base metal (usually one times the electrode diameter).
- *Angle* (**A**) –Two angles are critical:
  - Travel angle – The longitudinal angle of the electrode in relation to the axis of the weld joint
  - Work angle –The traverse angle of the electrode in relation to the axis of the weld joint
- *Speed* (**S**) –Travel speed is measured in inches per minute (IPM). The width of the weld will determine if the travel speed is correct.
- *Heat* (**H**) – Is controlled by the amperage setting. Heat is dependent upon the electrode diameter, base metal type, base metal thickness, welding position, and the interpass temperature.

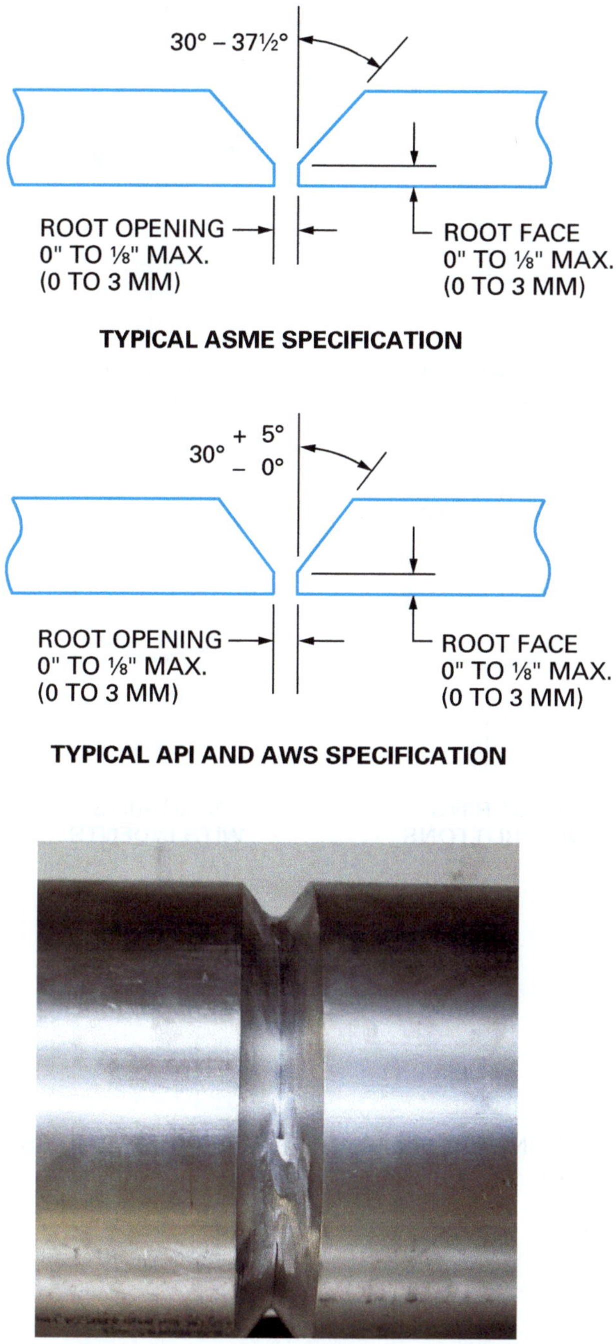

Figure 11 ASME, API, and AWS pipe bevel specifications.

Step 2 Bevel the end of the pipe to 30 to 37½ degrees. Use any acceptable beveling method. Techniques include manual cutting, mechanical cutting, and flame cutting.

Step 3 Cut off a 1½" (4 cm) section of beveled pipe end; this is the minimum length that should be used. The practice coupons for large-diameter pipe should be cut longer; they require more metal to help absorb the increased heat needed to make the welds.

Follow these steps to shape and align the open-root V-groove pipe weld coupons:

Step 1 Check the bevel. There should be no dross, and the bevel angle should be 30 to 37½ degrees.

Step 2 Grind or file a 0" to ⅛" (0 to 3 mm) root face on the bevel as specified by your instructor.

Step 3 Align the two pipe sections so that the inside surfaces are flush all around. Small-diameter pipe can be aligned by clamping both pieces to a piece of angle iron. Large-diameter pipe can be aligned with the aid of a pipe alignment jig or by holding a straightedge across the joint, parallel to the pipe axis. The straightedge must be used all around the pipe in case one or both sections are distorted.

Step 4 Set the root opening with pieces of filler wire, metal shims, or pieces of bare welding electrode of the correct diameter.

Step 5 When the root opening is correct and the pipe ends are aligned, make the first of four tack welds.

Step 6 After the first tack weld, check the root opening opposite the first tack weld. Adjust the root opening space if necessary. Make the second tack weld on the opposite side from the first tack.

Step 7 Check the root opening again. Weld the third tack midway between the first and second tacks.

Step 8 Weld the fourth tack opposite the third tack and midway between the first and second tacks. There should now be four tack welds evenly spaced at 90 degrees around the pipe coupon (*Figure 12*).

Step 9 Clean and feather the tack welds with the edge of a grinding wheel. Feathering the ends of the tack welds with a grinder helps to fuse the root pass into the tack weld.

Backing rings (*Figure 13*) are flat metal strips that have been rolled to fit inside a pipe. They can be ordered in a variety of base metal types to match the base metal being welded. Some rings are split, making them easier to insert and adjust to slightly out-of-round piping. Backing rings that have three or more nubs around the outside of the ring are also available. The selected diameter of the nub is the root opening required for the

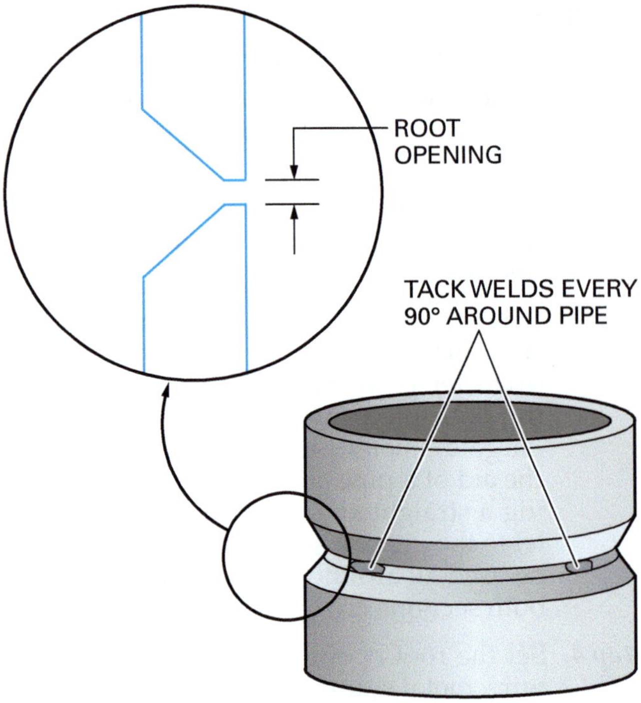

Figure 12 Tacked open-root V-groove weld coupon.

diameter of the pipe being welded. Nubs can be pegs, buttons, or indents punched into the ring. To remove the pegs, simply strike them with a hammer after the joint has been tack welded. Lower-profile buttons can be removed or left in position to be melted during the root pass.

The pipe joint preparation for backing rings is shown in *Figure 14*. The groove angle should be 60 or 75 degrees with a root opening of ¼" (6.4 mm) and no root face, or as specified on the WPS. If backing rings with nubs are used, the root opening will be set automatically when the beveled ends of the pipe are positioned firmly against the nubs.

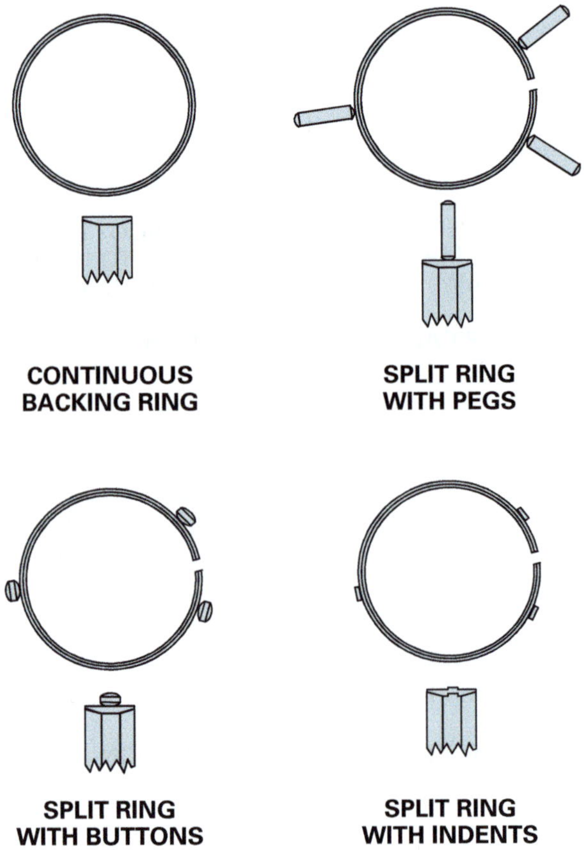

Figure 13 Examples of backing rings.

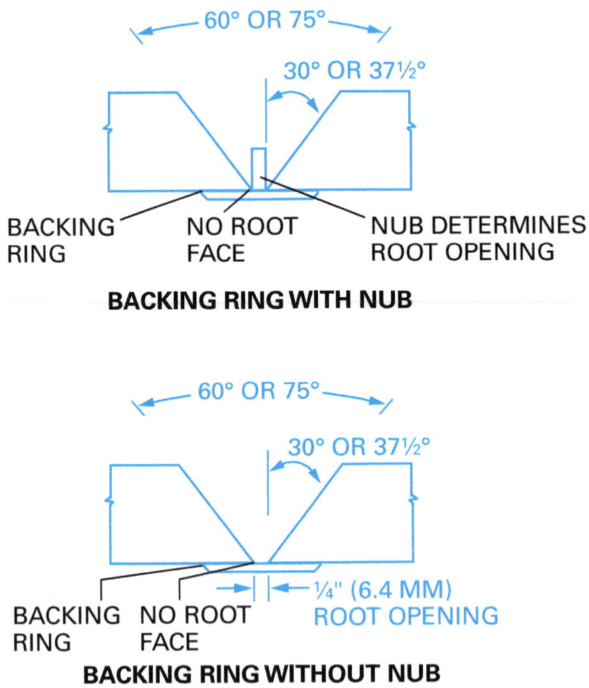

Figure 14 Stainless steel pipe joint with backing ring.

Additional Resources

AWS A5.4, Specification for Stainless-Steel Electrodes for Shielded Metal Arc Welding. Miami, FL: American Welding Society.

AWS F4.1: Recommended Safe Practices for Preparation for Welding and Cutting of Containers and Piping. Miami, FL: American Welding Society.

Modern Welding Technology. Current Edition. Howard B. Cary and Scott Helzer. Englewood Cliffs, NJ: Prentice Hall.

OSHA 1926.351, Arc Welding and Cutting. Current Edition. Washington, DC: Occupational Safety & Health Administration.

The Procedure Handbook of Arc Welding. Current Edition. Cleveland, OH: The James F. Lincoln Arc Welding Foundation.

Welding Handbook, Volumes 1 through 4. Current Edition. Miami, FL: American Welding Society.

Welding Pressure Pipelines and Piping Systems (PDF). The Lincoln Electric Company, Cleveland, OH, USA. **www.lincolnelectric.com**. Last accessed May 29, 2015.

2.0.0 Section Review

1. The type of stainless steel that normally does *not* require preheating or postheating is _____.
   a. austenitic
   b. ferritic
   c. crystalline
   d. martensitic

2. Before cutting tanks or barrels welders must clean the containers by washing them with detergent, steam cleaning them, or _____.
   a. purging them with oxygen
   b. flooding them with a fluxing agent
   c. scrubbing them with an abrasive
   d. flushing them with water

3. The two allowed groove angles for a pipe joint when using a backing ring are _____.
   a. 30 and 45 degrees
   b. 45 and 60 degrees
   c. 45 and 75 degrees
   d. 60 and 75 degrees

SECTION THREE

3.0.0 WELDING POSITIONS AND TECHNIQUES

Objective

Describe open-root V-groove plate and pipe welding positions and SMAW stainless steel welding techniques.

a. Identify and describe plate and pipe welding positions and acceptable weld profiles.
b. State general electrode handling considerations for SMAW of stainless steel.
c. Describe how to make the root pass.
d. Describe the techniques required to produce open-root SMAW stainless steel plate welds in four positions.
e. Describe the techniques required to produce open-root SMAW stainless steel pipe welds in four positions.

Performance Tasks

1. Make open-root V-groove welds on stainless steel plate in the 1G position.
2. Make open-root V-groove welds on stainless steel plate in the 2G position.
3. Make open-root V-groove welds on stainless steel plate in the 3G position.
4. Make open-root V-groove welds on stainless steel plate in the 4G position.
5. Make open-root V-groove welds on stainless steel pipe in the 1G-ROTATED position.
6. Make open-root V-groove welds on stainless steel pipe in the 2G position.
7. Make open-root V-groove welds on stainless steel pipe in the 5G position.
8. Make open-root V-groove welds on stainless steel pipe in the 6G position.

In order to make quality welds on stainless steel, it is critical to use the correct welding positions and proper techniques for the specific application. When performing SMAW on stainless steel, welders must use the appropriate welding position and weld profile for the job. To ensure quality welds, welders must follow the correct techniques for making the root pass and for producing open-root SMAW stainless steel plate welds and pipe welds. They must also use, store, and dispose of stainless steel electrodes properly.

3.1.0 Welding Positions and Acceptable Weld Profiles

The appropriate welding positions and acceptable weld profiles for SMAW stainless steel welding vary depending on the job. The following sections present standard requirements for several basic types of stainless steel plate groove welds and pipe groove welds.

3.1.1 Plate Groove Weld Positions

Groove welds can be made in all positions. The positions for plate are shown in *Figure 15*. The weld position for plate is based on the axis of the weld and the orientation of the workpiece. The groove weld positions for plate are flat (1G), horizontal (2G), vertical (3G), and overhead (4G).

In the 1G and 2G positions, the weld axis can be inclined up to 15 degrees. Any weld axis inclination for the other positions varies with the rotational position of the weld face as specified in the AWS standards.

3.1.2 Pipe Groove Weld Test Positions

The pipe welding positions are determined by the axis of the pipe (*Figure 16*). Four standard weld test positions are used with pipe:

- *Flat (1G-ROTATED)* – The pipe axis is horizontal. The pipe is slowly rotated while it is being welded on the top. Weld beads are flat.
- *Horizontal (2G)* – The pipe axis is vertical, and the pipe rotation is fixed. Weld beads are horizontal.
- *Multiple (5G)* – The pipe axis is horizontal, and the pipe rotation is fixed. Weld beads are flat, vertical, and overhead.
- *Multiple inclined (6G)* – The pipe axis is inclined nominally at 45 degrees ±5 degrees. The pipe rotation is fixed. Weld beads are horizontal, vertical, and overhead.

The pipe axis positions for 1G-ROTATED, 2G, and 5G can vary ±15 degrees from the basic position. The pipe axis position for 6G can vary ±5 degrees from the nominal 45-degree angle. See *Table 4* for information about all four weld test positions.

To determine the quality of a pipe weld, specific sections of pipe are removed for destructive testing. For positions 1G-ROTATED and 2G, two specimens are removed at random. They must be

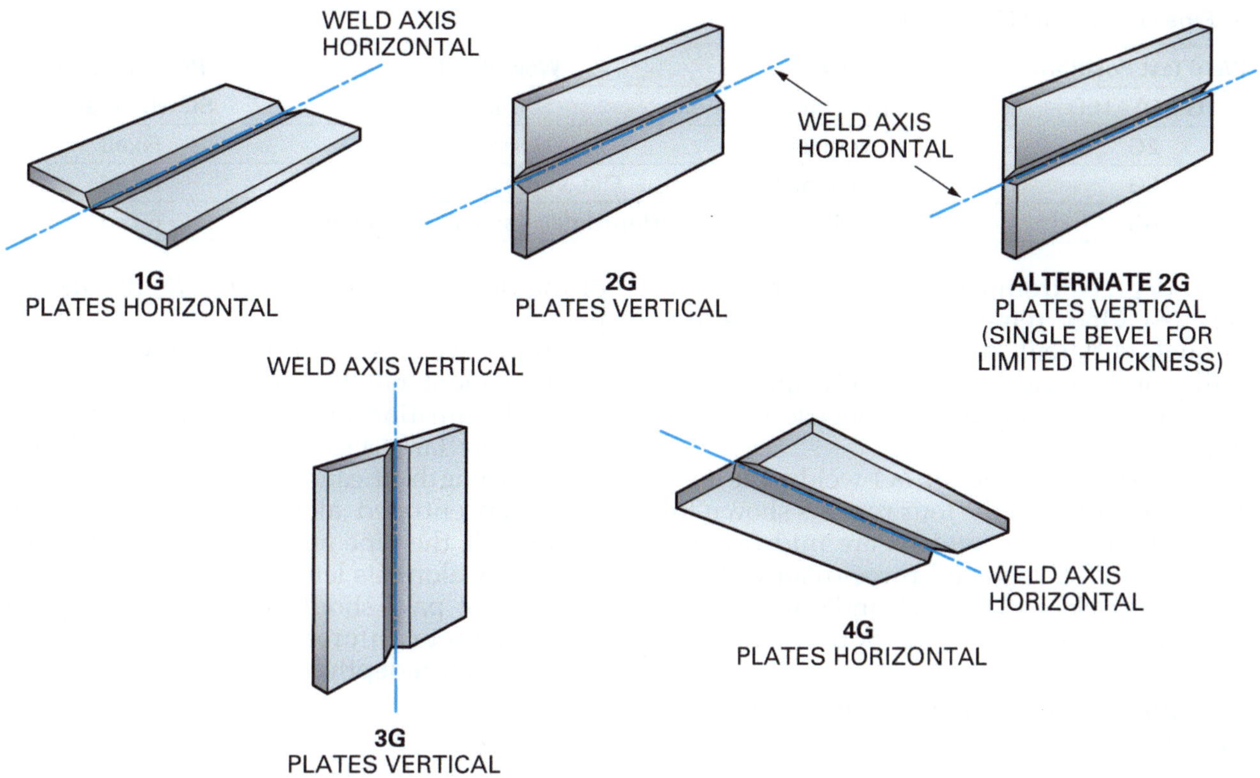

Figure 15 Groove weld positions for plate.

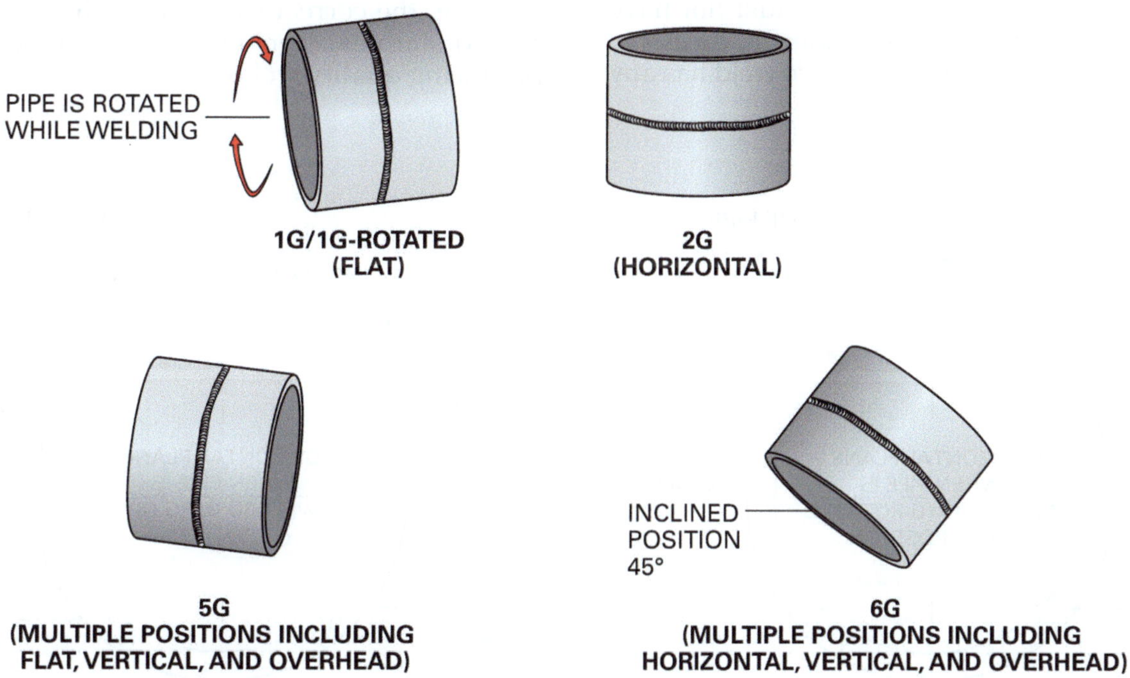

Figure 16 Four basic pipe groove-weld test positions.

Table 4 Pipe Groove-Weld Test Positions

Weld Test Position	Pipe Axis	Weld Beads	Pipe Position
1G-ROTATED	Horizontal	Flat	Slowly rotated
2G	Vertical	Horizontal	Fixed
5G	Horizontal	Flat, vertical, overhead	Fixed
6G	Inclined 45 degrees	Horizontal, vertical, overhead	Fixed

taken 180 degrees from each other, as shown in *Figure 17 (A)*. Once removed, the specimens are put through different bend tests, which reveal the quality of the welds. Alternatively, specimens may be tested by radiography, a non-destructive testing (NDT) method.

To destructively test 5G or 6G welds, the test specimens are cut from the four regions shown in *Figure 17 (B)*. These four regions are midway between the twelve o'clock and three o'clock; three o'clock and six o'clock; six o'clock and nine o'clock; and nine o'clock and twelve o'clock positions.

3.1.3 Acceptable and Unacceptable Groove Weld Profiles

Groove welds should be made with slight reinforcement and a gradual transition to the base metal at each toe. Groove welds must not have excessive reinforcement, any underfill, excessive undercut, or any overlap. If a groove weld has any of the defects shown in *Figure 18*, it must be repaired.

Groove welds should be made with slight face reinforcement not to exceed ⅛" (3 mm), and a gradual transition to the base metal at each toe. Excessively large face reinforcement reduces the pipe's strength. It causes the stresses in the pipe to be concentrated along the sides of the weld. As a result, the pipe will not expand and contract uniformly along its length.

The root pass should have complete penetration. The root reinforcement can range from flush to a maximum depth of ⅛" (3 mm).

3.2.0 Electrodes

In addition to appropriate welding positions and acceptable weld profiles for SMAW stainless steel welding, the correct use and handling of the required stainless steel electrodes is essential for producing quality welds.

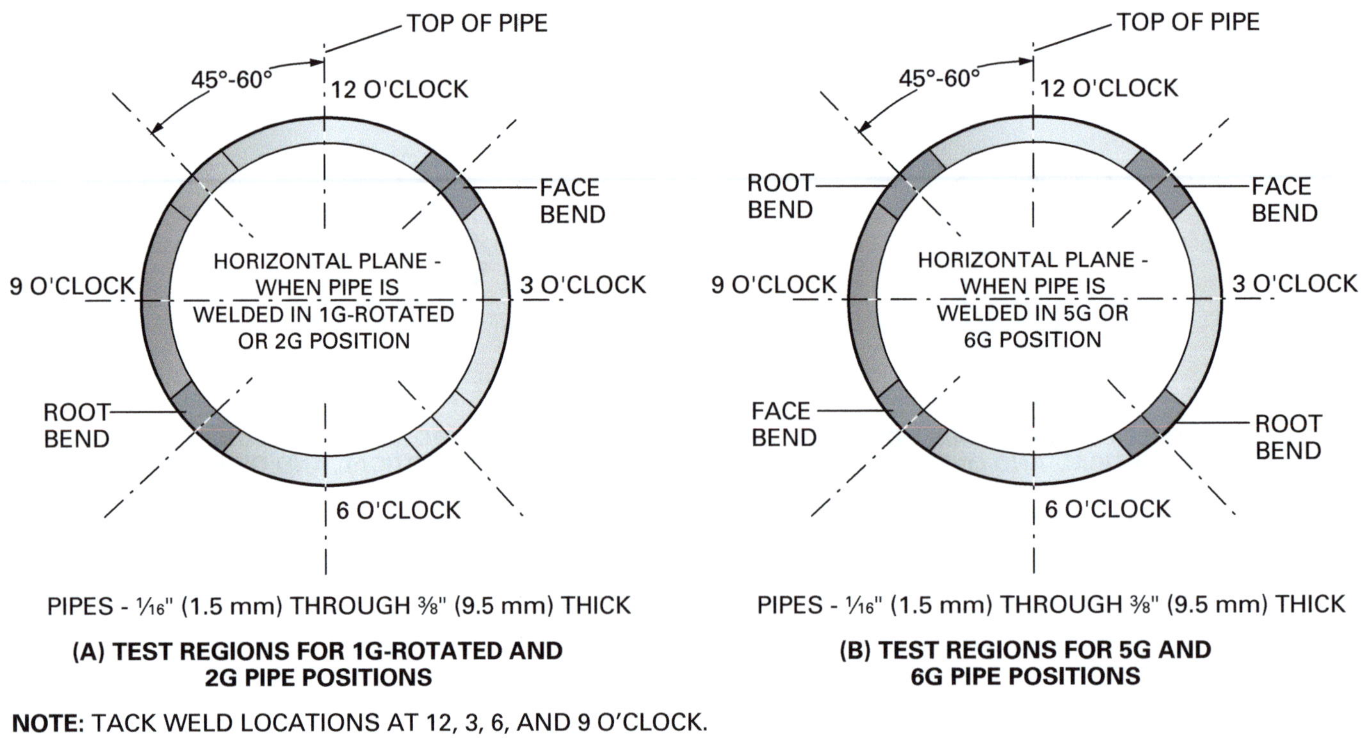

Figure 17 Test specimen regions for pipe positions 1G-ROTATED, 2G, 5G, and 6G.

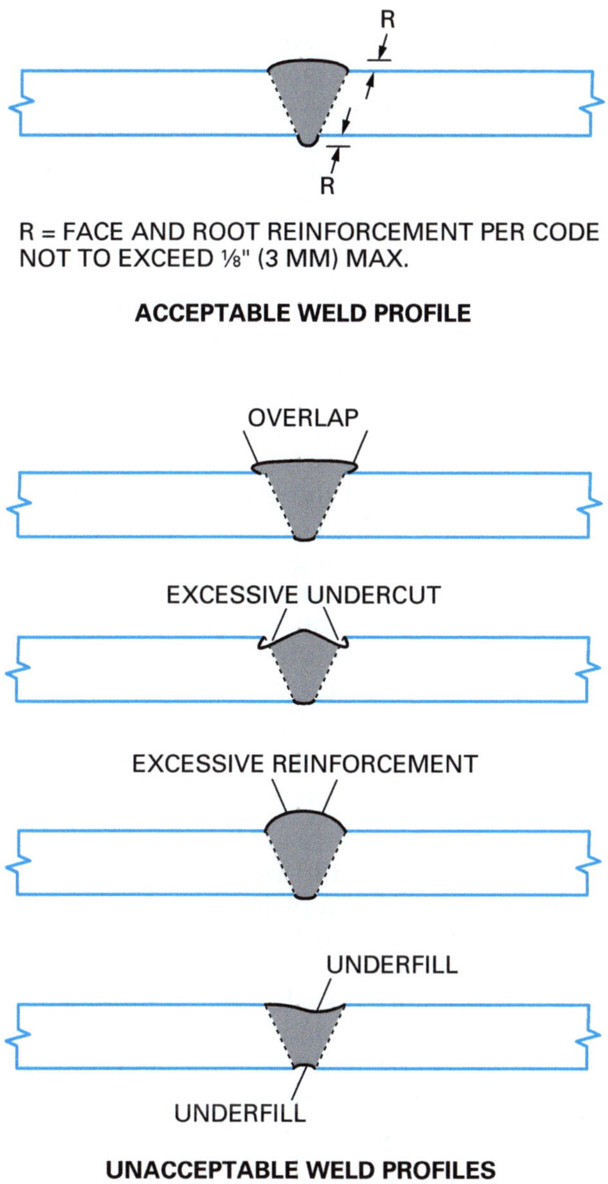

Figure 18 Acceptable and unacceptable groove weld profiles.

Obtain a small quantity of the electrodes that will be used. For the welding exercises in this module, electrodes compatible with the base metal being welded must be used. Keep these general rules in mind when running stainless steel electrodes:

- Never whip stainless steel electrodes. Keep the arc in the weld puddle at all times.
- Note that certain electrodes from some manufacturers are designed to be dragged on the base metal in downhill and horizontal travel directions.
- Maintain a short arc because stainless steel is very susceptible to atmospheric contamination. A longer arc will allow the atmosphere to contaminate the weld, causing porosity.

6GR Test Position

Welding pipe in the 6G position is a common welding test. A ring may be added to test for restricted accessibility (6GR).

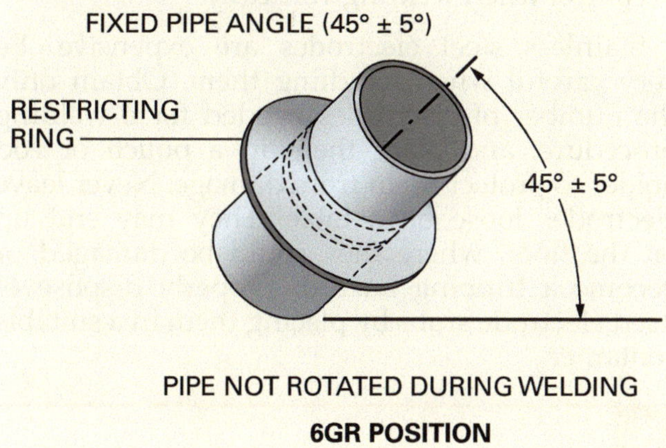

Back Welds

On plate only, single V-groove welds can be finished with a back weld. Back welds are performed on the opposite side of a V-groove weld after the V-groove weld passes have been completed. This method can save time because a melt-through root pass is not required. The edge of a grinding disk is used to cut through the root face to sound metal from the opposite side of the V-groove weld. After back-grinding is complete, the ground side of the joint should resemble a U-groove. Then the back-weld pass is made.

- Fill craters before breaking the arc to prevent crater cracks.
- Clean the slag thoroughly from the finish of a bead before starting another bead.
- Completely remove all slag between passes.
- Remove only a small number of electrodes from the oven at a time. Stainless steel electrodes pick up moisture quickly.
- Restart by striking the arc in the leading edge of the crater, move back into the crater to achieve proper fusion, and then proceed as usual. This technique welds over the arc strike, making a smoother restart.
- Stringer beads should be no more than three times the diameter of the electrode.
- When running stringer beads, use a slight side-to-side motion with a small pause at the toe to achieve proper fusion and flatten the bead.
- Never use stainless steel electrodes with chipped or missing flux.
- Use the lowest current possible to achieve sufficient penetration while reducing heat input to the work. This will decrease distortion and carbide precipitation, and allow better puddle control when welding vertically.

Stainless steel electrodes are expensive. Be very careful when handling them. Obtain only the number of electrodes needed for a welding procedure, and place them in a pouch or rod holder to protect them from damage. Never leave electrodes loose on a table. They may end up on the floor, where they could be damaged or become a tripping hazard. Properly dispose of used electrode stubs by placing them in a suitable container.

WARNING!

Never throw electrode stubs on the floor. They roll easily and could cause trips, falls, or other job-related accidents.

3.3.0 The Root Pass

The greatest challenge of an open-root V-groove weld is making the root pass. The root pass must have complete penetration, but not an excessive amount of root reinforcement. The amount of penetration is very important. If the penetration is too little, the joint will be weak. When there is too much penetration and excessive root reinforcement occurs, the inside diameter of the pipe is reduced and the flow is restricted.

CAUTION

When welding pipe, avoid making arc strikes outside of the weld groove on the surface of the pipe. An arc strike can cause a hardened spot that can crack as the pipe expands and contracts. An arc strike on the pipe surface is considered a defect, which will require repair or reworking.

After the root pass has been applied, it should be cleaned and inspected. Check the face of the weld for excess buildup and undercut, which could trap slag when the fill passes are run.

Use a hand grinder to remove any excess buildup or undercut. Grind the face of the root pass with the edge of the grinding disk. Most of the slag along the sides of the weld bead will be removed by the grinding action. This will make it easier to add the remaining passes. Be careful not to grind through the root pass or widen the groove. *Figure 19* shows the effect of grinding a root pass. A narrow grinding wheel (not a cutoff wheel) is effective for this purpose.

3.4.0 Open-Root V-Groove Plate Welds

The open-root V-groove weld is a common groove weld made on plate and pipe. Practicing open-root V-groove welds on plate is good preparation for making the more difficult pipe welds, which will be covered later.

The performance qualification requirements of this module for visual and destructive testing are based on the *ASME Boiler and Pressure Vessel Code, Section IX: Welding and Brazing Qualifications.*

This section explains how to perform the following open-root V-groove welds on plate:

- Flat (1G) position
- Horizontal (2G) position
- Vertical (3G) position
- Overhead (4G) position

3.4.1 Practicing Flat (1G) Position Open-Root V-Groove Welds

Practice flat open-root V-groove welds on plate. Use stringer beads and stainless steel electrodes. The bead sequence and work angles are shown in *Figure 20*. Keep the electrode angle at 90 degrees to the plate surface (0-degree work angle) with a 10- to 15-degree drag angle and a very short arc. Adjust the electrode work angle to achieve fusion and fill the bead. Pay special attention to fill the crater at the termination of the weld.

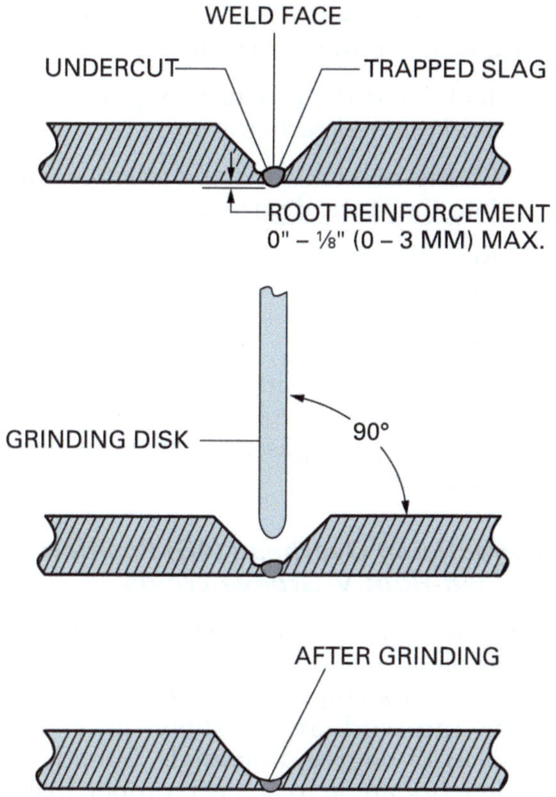

Figure 19 Grinding a root pass on pipe.

> **CAUTION**
> The interpass temperature should typically not exceed 350°F (177°C) when welding stainless steels. Refer to the WPS or site specifications for the exact temperature.

> **WARNING!**
> Always use pliers when handling a hot practice coupon. Wear gloves when placing the coupon in water. The steam that rises off the coupon can burn or scald unprotected hands.

Follow these steps to practice open-root V-groove welds in the flat position:

Step 1 Tack weld the practice coupons together as explained earlier.

Step 2 Clamp or tack weld the practice coupon in a flat position above the surface of the welding table.

Step 3 Run the root pass using 3⁄32" (2.4 mm) stainless steel electrodes.

Step 4 Clean the root pass, and grind it if required.

Step 5 Run the remaining passes at the proper work angles using 3⁄32" (2.4 mm) stainless steel electrodes. Overlap each pass. Clean the weld after each pass.

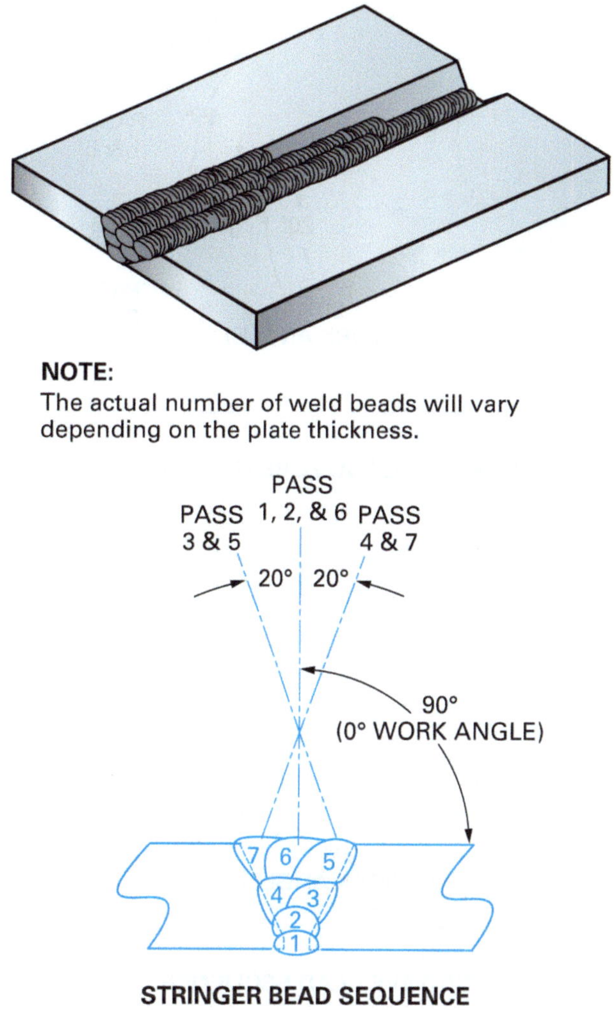

Figure 20 Multiple-pass 1G bead sequence and work angles.

3.4.2 Practicing Horizontal (2G) Position Open-Root V-Groove Welds

Practice horizontal open-root V-groove welds on plate. Use stainless steel electrodes and stringer beads. The proper bead sequence and work angles are shown in *Figure 21*. For the electrode angles, use a 10- to 15-degree drag angle, and adjust the work angle as required. Pay special attention to fill the crater at the termination of the weld.

Follow these steps to practice welding open-root V-groove welds in the horizontal position:

Step 1 Tack weld the practice coupons together as explained earlier. Use the standard or alternate horizontal weld coupon as directed by your instructor.

Step 2 Clamp or tack weld the flat plate weld coupon into the horizontal position.

Step 3 Run the root pass using 3⁄32" (2.4 mm) electrodes.

Step 4 Clean the weld.

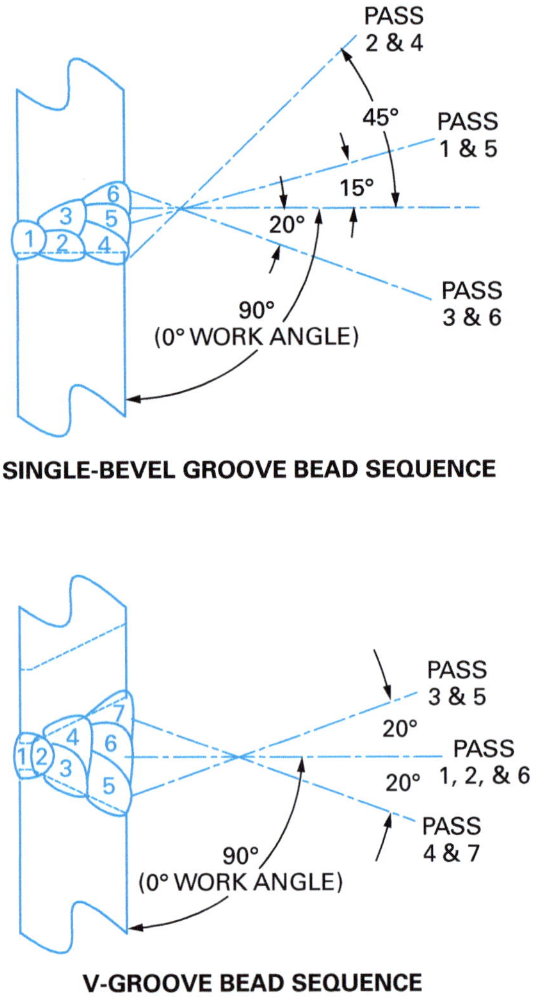

Figure 21 Multiple-pass 2G bead sequences and work angles.

Step 5 Run the remaining passes at the proper work angles. Use ³⁄₃₂" (2.4 mm) electrodes and stringer beads. Overlap each pass. Clean the weld after each pass.

3.4.3 Practicing Vertical (3G) Position Open-Root V-Groove Welds

Practice vertical open-root V-groove welds on plate. Use ³⁄₃₂" (2.4 mm) stainless steel electrodes and stringer beads. The appropriate bead sequence and work angles are shown in *Figure 22*. Set the amperage in the lower part of the range for better puddle control. Use an upward travel direction with a 0- to 10-degree push angle. Adjust the electrode work angle to achieve proper fusion on one side or the other as needed. To control the weld puddle for the fill passes, first establish a shelf of weld metal. Maintain a short arc, and be careful not to remove the arc from the weld puddle. Pay special attention to fill the crater at the termination of the weld.

Follow these steps to practice open-root V-groove welds in the vertical position:

Step 1 Tack weld the practice coupons together as explained earlier.

Step 2 Clamp or tack weld the practice coupon into the vertical position.

Step 3 Run the root pass using ³⁄₃₂" (2.4 mm) electrodes.

Step 4 Clean the weld.

Step 5 Run the remaining passes at the proper work angles. Use ³⁄₃₂" (2.4 mm) electrodes. Overlap each pass. Clean the weld after each pass.

3.4.4 Practicing Overhead (4G) Position Open-Root V-Groove Welds

Practice running overhead open-root V-groove welds on plate. Use ³⁄₃₂" (2.4 mm) stainless steel electrodes and stringer beads. Follow the bead sequence and work angles shown in *Figure 23*. Use a 10- to 15-degree drag angle, but adjust the electrode work angle to achieve proper fusion as needed. Pay special attention to fill the crater at the termination of the weld.

Follow these steps to practice welding open-root V-groove welds in the overhead position:

Step 1 Tack weld the practice coupons together as explained in the section related to plate weld-coupon preparation.

Step 2 Clamp or tack weld the practice coupon into the overhead position.

Step 3 Run the root pass using ³⁄₃₂" (2.4 mm) electrodes.

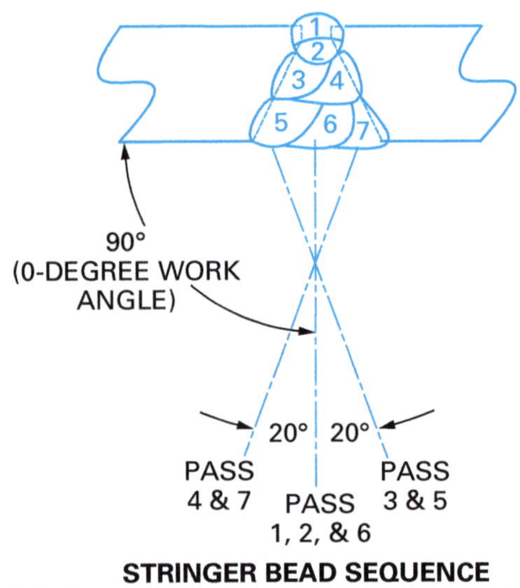

Figure 22 Multiple-pass 3G bead sequence and work angles.

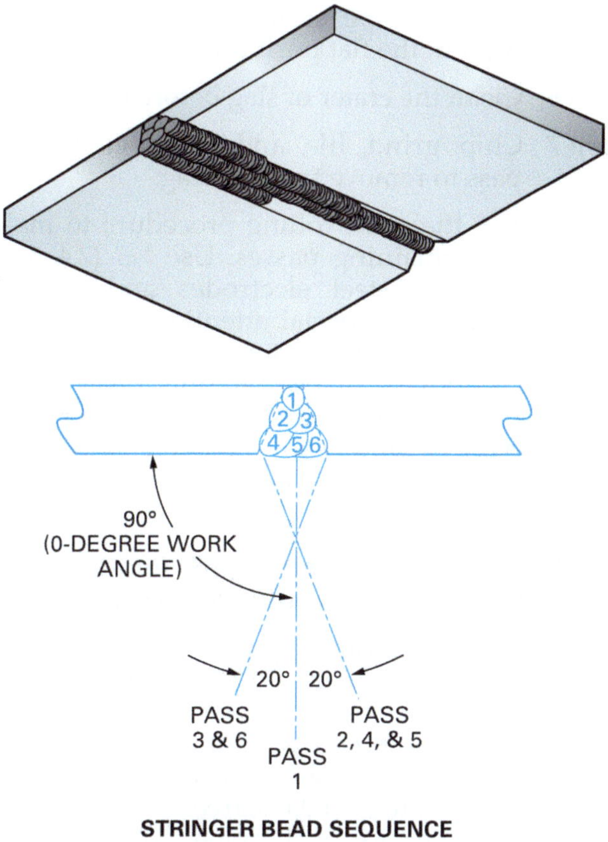

Figure 23 Multiple-pass 4G bead sequence and work angles.

Step 4 Clean the weld.

Step 5 Run the remaining passes at the proper work angles. Use 3⁄32" (2.4 mm) electrodes. Overlap each pass. Clean the weld after each pass.

3.5.0 Open-Root V-Groove Pipe Welds

The open-root V-groove joint is the most common groove weld made on pipe. As noted earlier, the ends of the pipe are beveled. This gives the welder some additional room to reach the root opening where the first weld, the root pass, will be made.

This section explains how to perform open-root V-groove pipe welds in all four standard positions.

3.5.1 Practicing Flat (1G-ROTATED) Position Open-Root V-Groove Pipe Welds

Practice 1G-ROTATED position open-root V-groove pipe welds using 3⁄32" (2.4 mm) stainless steel electrodes.

Keep the electrode angle at 90 degrees (0-degree work angle) to the pipe axis. Use a 10- to 15-degree drag angle and a very short arc for the root pass. A side-to-side oscillation is used to run a keyhole for the root pass. Be sure to achieve proper fusion at the tack welds during each restart. Clean the face of the root pass with a file, grinder, or stainless steel wire brush.

When running the remaining passes, use stringer beads with a slight side-to-side motion. This will help to ensure proper fusion is achieved at the toes of the weld bead. Take special care to fill the crater at the termination of the weld. Run all passes at or near the top of the pipe as it is rotated.

Follow these steps to practice welding open-root V-groove welds in the 1G-ROTATED position:

Step 1 Tack weld the practice coupons together as explained earlier.

Step 2 Position the pipe coupon horizontally on two sets of rollers at a comfortable welding height. A welding positioner or pipe rotator is also required. The positioner shown in *Figure 24 (A)* is very popular. The pipe rotator, shown in *Figure 24 (B)*, is another option. This model can handle up to 16" (DN400) pipe and weights up to 1,200 pounds (544 kg). Adjustable roller stands can be used to help support longer stock.

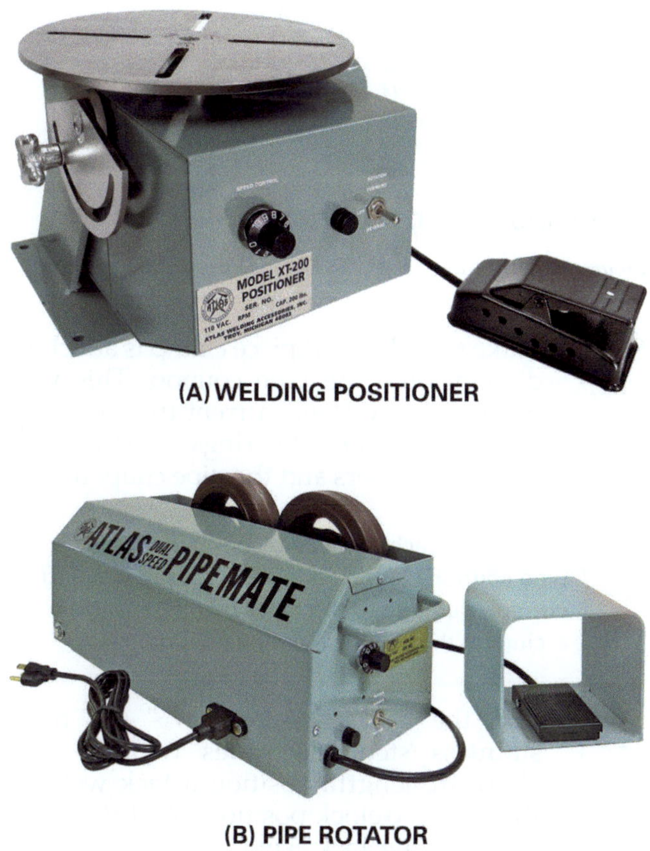

Figure 24 A welding positioner and a pipe rotator.

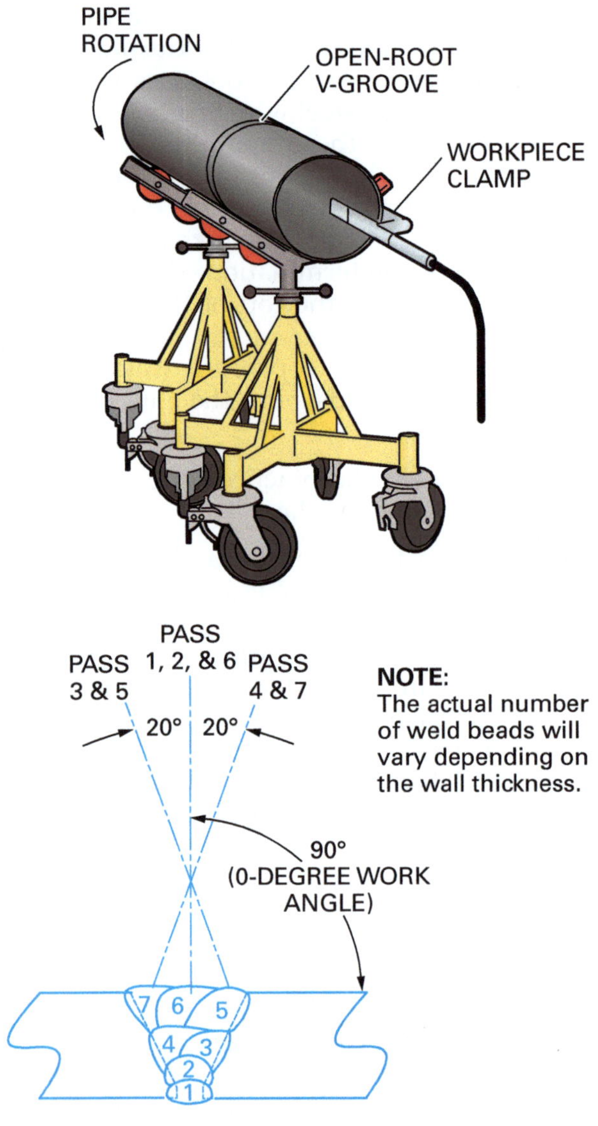

Figure 25 Multiple-pass 1G-ROTATED bead sequence and work angles.

Step 3 Make sure the workpiece clamp is attached directly onto the pipe coupon. This will prevent the welding current from passing through the roller bearings or arcing between the rollers and the pipe coupon.

> **CAUTION**
>
> Failure to attach the workpiece clamp to the pipe coupon can result in variations in welding current, damage to the roller bearings, and arcing on the rollers and pipe coupon.

Step 4 Run the root pass using 3⁄32" (2.4 mm) stainless steel electrodes with a very short arc length. Position a tack weld at the eleven o'clock position. Start the weld bead on the tack weld, and advance toward the twelve o'clock position.

Step 5 Roll the pipe as necessary to keep the weld in the flat position.

Step 6 Clean the crater of slag before restarting.

Step 7 Chip, grind, file, and/or brush the root pass to remove trapped slag.

Step 8 Use the same rolling procedure to make the remaining passes. Use 3⁄32" (2.4 mm) stainless steel electrodes and stringer beads. Pay special attention to the areas where fusion occurs to prevent slag inclusions or excess buildup. Overlap the passes, and clean the weld after each pass. *Figure 25* shows the related bead sequence and work angles.

3.5.2 Practicing Horizontal (2G) Position Open-Root V-Groove Pipe Welds

Practice 2G (horizontal) position open-root V-groove pipe welds using 3⁄32" (2.4 mm) stainless steel electrodes.

The electrode should be at a 10- to 15-degree drag angle with a very short arc. For the root pass, position the electrode at 90 degrees (0-degree work angle) to the surface of the pipe. To prevent the weld puddle from sagging, the electrode work angle can be dropped slightly, but no more than 10 degrees. Use a side-to-side oscillation to run a keyhole for the root pass. Be sure to achieve proper fusion at the tack welds during each restart.

When running the remaining passes, use stringer beads and a slight side-to-side motion. This will help to ensure fusion is achieved at the toes of the weld bead. Take special care to fill the crater at the termination of the weld. *Figure 26* shows the 2G pipe weld coupon position and the bead sequence and work angles.

Follow these steps to practice open-root V-groove pipe welds in the 2G position:

Step 1 Tack weld the practice coupons together as explained earlier.

Step 2 Clamp or tack weld the pipe coupon with the axis of the pipe vertical.

Step 3 Run the root pass using 3⁄32" (2.4 mm) stainless steel electrodes with a very short arc length. Start the root pass on a tack weld. Clean and chip the crater between restarts. Continue this procedure until the root pass has been completed.

Step 4 Chip, grind, file, and/or brush the root pass to remove trapped slag.

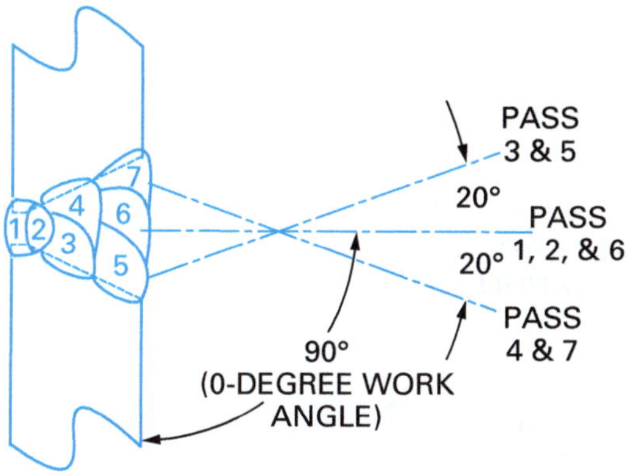

STRINGER BEAD SEQUENCE

Figure 26 Multiple-pass 2G bead sequence and work angles.

Step 5 Run the remaining passes at the proper work angles. Use ³⁄₃₂" (2.4 mm) stainless steel electrodes. Overlap the passes and clean the weld after each pass.

3.5.3 Practicing Multiple (5G) Position Open-Root V-Groove Pipe Welds

Practice multiple (5G) position open-root V-groove pipe welds using ³⁄₃₂" (2.4 mm) stainless steel electrodes.

Weld uphill by starting at the bottom of the pipe and welding toward the top. Use a side-to-side oscillation to run a keyhole for the root pass. Be sure to achieve proper fusion at the tack welds during each restart.

When running the remaining passes, use a slight side-to-side motion. This will help to ensure fusion is achieved at the toes of the weld bead. Pay special attention to fill the crater at the termination of the weld. *Figure 27* shows the 5G pipe weld coupon position and the bead sequence and work angles.

Follow these steps to practice open-root V-groove pipe welds in the 5G position:

Step 1 Tack weld the practice coupons together as explained earlier.

Step 2 Clamp or tack weld the pipe weld coupon into position with the pipe axis horizontal and with a tack weld located at the bottom.

Step 3 Run the root pass using ³⁄₃₂" (2.4 mm) electrodes. Start the root pass at the bottom of the pipe and weld uphill. When welding at the top and bottom of the coupon (flat and overhead positions), the electrode should usually point toward the pipe axis. When welding up the coupon sides (vertical welding), use a slight electrode push angle (a maximum of 10 degrees) with a very short arc length to control the weld puddle.

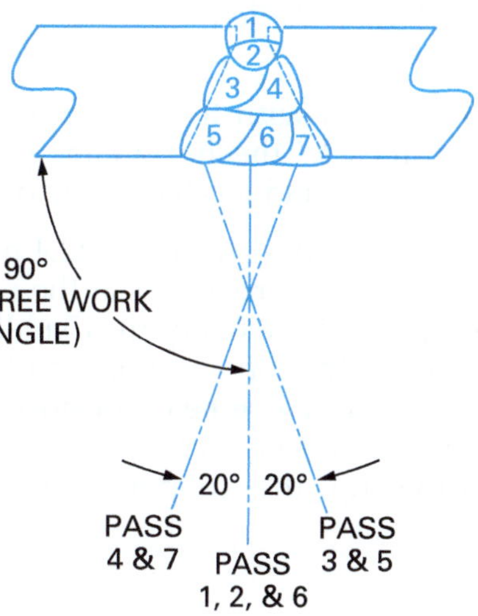

STRINGER BEAD SEQUENCE

Figure 27 Multiple-pass 5G bead sequence and work angles.

Step 4 Chip, grind, file, and/or brush the root pass to remove trapped slag.

Step 5 Run the remaining passes at the proper work angles using 3⁄32" (2.4 mm) stainless steel electrodes. Overlap the passes and clean the weld after each pass.

3.5.4 Practicing Multiple-Inclined (6G) Position Open-Root V-Groove Pipe Welds

Practice 6G multiple-inclined (45-degree) position open-root V-groove pipe welds using 3⁄32" (2.4 mm) stainless steel electrodes.

Weld uphill by starting at the bottom of the pipe and welding toward the top. Use a side-to-side oscillation to run a keyhole for the root pass. Be sure to achieve proper fusion at the tack welds during each restart.

When running the remaining passes, use stringer beads with a slight side-to-side motion. This will help to ensure fusion is achieved at the toes of the weld bead. Take special care to fill the crater at the termination of the weld. *Figure 28* shows the 6G pipe weld coupon position and the bead sequence and work angles.

Follow these steps to practice open-root V-groove pipe welds in the 6G position:

> **NOTE**
> If required for training purposes, a restricting ring may be added to the 6G position coupon to form a 6GR position coupon.

Step 1 Tack weld the practice coupons together as explained earlier.

Step 2 Clamp or tack weld the pipe weld coupon into position with the pipe axis inclined 45 degrees to the horizontal plane and with a tack weld located at the bottom.

Step 3 Run the root pass using 3⁄32" (2.4 mm) electrodes. Start the root pass at the bottom of the pipe and weld uphill. Usually, when welding at the top and bottom of the coupon (flat and overhead positions), point the electrode toward the pipe axis. When welding up the coupon sides (vertical welding), use a slight electrode push angle (a maximum of 10 degrees) with a very short arc length to control the weld puddle.

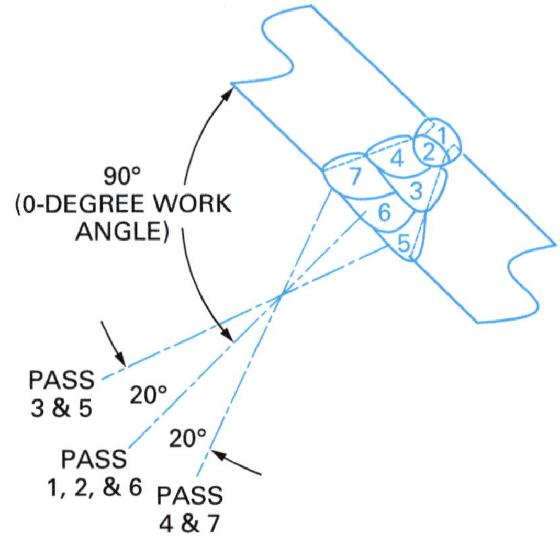

Figure 28 Multiple-pass 6G bead sequence and work angles.

Step 4 Chip, grind, file, or brush the root pass to remove trapped slag.

Step 5 Run the remaining passes at the proper work angles using 3⁄32" (2.4 mm) stainless steel electrodes. Overlap the passes and clean the weld after each pass.

Additional Resources

AWS A5.4, Specification for Stainless-Steel Electrodes for Shielded Metal Arc Welding. Miami, FL: American Welding Society.

AWS F4.1: Recommended Safe Practices for Preparation for Welding and Cutting of Containers and Piping. Miami, FL: American Welding Society.

Modern Welding Technology. Current Edition. Howard B. Cary and Scott Helzer. Englewood Cliffs, NJ: Prentice Hall.

OSHA 1926.351, Arc Welding and Cutting. Current Edition. Washington, DC: Occupational Safety & Health Administration.

The Procedure Handbook of Arc Welding. Current Edition. Cleveland, OH: The James F. Lincoln Arc Welding Foundation.

Welding Handbook, Volumes 1 through 4. Current Edition. Miami, FL: American Welding Society.

Welding Pressure Pipelines and Piping Systems (PDF). The Lincoln Electric Company, Cleveland, OH, USA.**www.lincolnelectric.com**. Last accessed May 29, 2015.

3.0.0 Section Review

1. Pipe axis positions can vary ±15 degrees from the basic position for the pipe positions of _____.
   a. 1G-ROTATED, 2G, and 6G
   b. 1G-ROTATED, 2G, and 5G
   c. 5G and 6G
   d. 1G-ROTATED and 6G

2. Never use stainless steel electrodes with _____.
   a. weld puddles
   b. stringer beads
   c. chipped or missing flux
   d. rod holders

3. To remove excess buildup or undercut after making a root pass, use a _____.
   a. hand grinder
   b. reamer
   c. cutoff wheel
   d. fine-grit sander

4. Practicing open-root V-groove welds on stainless steel plate will prepare the welder to make the more difficult _____.
   a. flat welds
   b. pipe welds
   c. vertical welds
   d. horizontal welds

5. The most common groove weld made on pipe is the _____.
   a. closed-root V-groove weld
   b. multi-pass 1G bead sequence
   c. overhead 4G plate weld
   d. open-root V-groove joint

SUMMARY

The various categories of stainless steel are different from carbon steels in terms of chemical composition and physical characteristics. When performing SMAW on stainless steel, welders must make appropriate preparations for the job, follow applicable safety precautions, and use the correct welding techniques.

This module describes the categories of stainless steel, and explained how to select and care for the SMAW electrodes used on stainless steel. Safety precautions that welders must observe when welding on stainless steel were emphasized, and welding techniques for stainless steel were introduced.

Although stainless steel can be welded using other methods, this module covered only the SMAW process. The dimensions and specifications in this module are representative of codes in general and may not be specific to any particular code. Always refer to the applicable code or site-specific specifications.

Review Questions

1. The AISI 200 and 300 series of stainless steels are a part of which group of stainless steels?
   a. Martensitic
   b. Crystalline
   c. Austenitic
   d. Ferritic

2. When a stabilized metal is heated to between 800°F (427°C) and 1,400°F (760°C), the columbium or titanium combines with the _____.
   a. iron in the metal
   b. zinc in the metal
   c. carbon in the metal
   d. nickel in the metal

3. Electrodes that have a thin carbon steel cylinder filled with powdered metal containing chromium, nickel, and other alloys, along with an external flux covering, are called _____.
   a. fast-fill electrodes
   b. fast-freeze electrodes
   c. coated tubular electrodes
   d. low-hydrogen electrodes

4. Usability designations for stainless steel SMAW electrodes include _____.
   a. -11 and -12
   b. -12 and -13
   c. -13 and -14
   d. -15 and -16

5. When the basic class alloy inside a stainless steel electrode has been modified with 0.04 to 0.08 percent carbon, the markings on the side of the electrode will include which additional letter(s) in the alloy modification designation position?
   a. L
   b. H
   c. HC
   d. Cb

6. Of the different stainless steel types, the one that is generally easiest to weld is _____.
   a. ferritic
   b. austenitic
   c. crystalline
   d. martensitic

7. For long-term cutting or welding of materials that contain zinc, cadmium, or manganese, a welder should always wear an approved _____.
   a. full-face, supplied-air respirator
   b. canister-style respirator
   c. standard respirator with a HEPA-rated filter
   d. multipurpose gas mask

8. When setting up a welding machine for use, check to be sure that the electrode holder is _____.
   a. adequately preheated
   b. set to the correct amperage
   c. connected to the workpiece lead
   d. not grounded

9. Feathering the end of a tack weld with a grinder helps to _____.
   a. align the pipe sections being welded
   b. fuse the root pass into the tack weld
   c. correct slightly out-of-round piping
   d. prevent carbide precipitation

10. When the axis of the pipe being welded is horizontal and the pipe is slowly rolled while being welded on the top, the weld position is called _____.
    a. 1G-ROTATED
    b. 2G
    c. 5G
    d. 6G

11. Never store stainless steel electrodes _____.
    a. in electric drying ovens
    b. loose on a table
    c. in portable electrode holders
    d. in hermetically sealed containers

12. The greatest challenge of an open-root V-groove weld is making the _____.
    a. bead
    b. overlap
    c. keyhole
    d. root pass

13. When performing vertical open-root V-groove welds on plate, use a(n) _____.
    a. downward travel direction with a 10- to 15-degree drag angle
    b. downward travel direction with a 40- to 50-degree push angle
    c. upward travel direction with a 0- to 10-degree push angle
    d. upward travel direction with a 10- to 15-degree drag angle

14. When making an open-root V-groove pipe weld of stainless steel pipe in the flat (1G-ROTATED) position, position the pipe weld coupon _____.
    a. vertically with the axis of the pipe
    b. horizontally with the axis of the pipe
    c. inclined 45-degrees from the vertical axis of the pipe
    d. horizontally on two sets of pipe rollers

15. When making an open-root V-groove pipe weld of stainless steel pipe in the multiple inclined (6G) position, a welder should start the root pass at the bottom of the pipe and weld _____.
    a. left to right
    b. right to left
    c. downhill
    d. uphill

Trade Terms Introduced in This Module

Annealing: A heat treatment process that reduces the brittleness of the work.

Carbide precipitation: The process of carbides combining with chromium to create chromium carbide. The result is a reduction of chromium and the corrosion resistance it provides. Intergranular corrosion results from the loss of chromium.

Heat-affected zone: The portion of the workpiece/base metal whose mechanical properties or microstructure have been or will be altered by the heat from welding, brazing, soldering, or thermal cutting processes.

Interpass temperature: The temperature of the base metal in the general area of the weld before the next pass is started. The definition of the area of the weld varies by the applicable welding code.

Sensitization: The process by which carbide precipitation takes place.

Appendix A

Stainless Steel SMAW Electrode Classes and Uses

AWS Classification	Intended Uses
E209-XX	These electrodes are most often used to weld AISI Type 209 (UNS S20910) base metals. The electrode is a nitrogen-strengthened austenitic stainless steel alloy with high strength and good toughness over a wide range of temperatures. Nitrogen alloying reduces the tendency for intergranular carbide precipitation in the weld area by inhibiting carbon diffusion. Nitrogen alloying coupled with the molybdenum content provides superior resistance to pitting and crevice corrosion in chloride solutions. These electrodes are also used to join dissimilar metals such as carbon steel to stainless steel and for direct overlay on carbon steel for corrosion applications.
E219-XX	These electrodes are most often used to weld AISI Type 219 (UNS S21900) base metals and have similar composition, characteristics, and uses as E209 electrodes.
E240-XX	These electrodes are most often used to weld AISI Type 240 and 241 base metals. They have similar composition, characteristics, and uses as E209 electrodes. Additionally, they exhibit an improvement in resistance to wear in particle-to-metal and metal-to-metal (galling) applications compared to the more conventional austenitic stainless steels like Type 304. Weldments in alloys AISI 240 and AISI 241, when compared to Type 304, exhibit improved resistance to transgranular stress corrosion cracking in hot chloride solutions.
E307-XX	These electrodes are used primarily for moderate strength welds with good crack resistance between dissimilar steels such as austenitic manganese steel and carbon steel forgings or castings.
E308-XX	These electrodes are most often used to weld base metal of similar composition such as AISI Types 301, 302, 304, and 305.
E308H-XX	These electrodes are the same as E308 except that the allowable carbon content has been increased to the highest portion of the range for E308. The high carbon content provides higher tensile and creep strengths at elevated temperatures. These electrodes are used for welding Type 304H base metal.
E308L-XX	These electrodes are the same as E308, except for restricted carbon content. The low carbon content of welds made by these electrodes reduces the possibility of intergranular carbide precipitation, which increases the resistance to intergranular corrosion without the use of stabilizers such as columbium (niobium) or titanium. The low carbon welds, however, are not as strong at elevated temperature as the columbium-stabilized alloys or 304H.
E308Mo-XX	These electrodes are the same as E308 except for the addition of molybdenum. The electrodes are recommended for welding ASTM CF8M stainless steel castings because they match the base metal with regard to chromium, nickel, and molybdenum. They can be used for welding wrought materials such as Type 316 stainless when increased ferrite is desired that cannot be attained with E316 electrodes.
E308MoL-XX	These electrodes are recommended for welding ASTM CF3M stainless steel castings, as they match the base metal with regard to chromium, nickel, and molybdenum. The electrodes may also be used for welding wrought materials such as Type 316L stainless when increased ferrite is desired that cannot be attained with E316L electrodes.
E309-XX	These electrodes are commonly used for welding similar alloys in wrought or cast form. They are used for welding dissimilar metals, such as joining Type 304 to carbon steel, welding the clad side of Type 304 clad steels, and applying stainless steel sheet linings to carbon steel shells. Occasionally, they are used to weld Type 304 and similar base metals where severe corrosion conditions exist requiring higher alloy weld metal.
E309L-XX	The weld metal is the same as E309 electrodes except for a restricted carbon content. The low carbon content of these weld deposits reduces the possibility of intergranular carbide precipitation, which increases the resistance to intergranular corrosion without the use of stabilizers such as columbium (niobium) and titanium. The low carbon welds are not as strong at elevated temperature as the columbium-stabilized alloys or higher carbon content Type 309 welds.

AWS Classification	Intended Uses
E309Cb-XX	The weld metal is the same as Type 309, except for the addition of columbium (niobium) and a reduction in the carbon limit. The columbium (niobium) provides resistance to carbide precipitation, which increases intergranular corrosion resistance and provides higher strength in elevated temperature service. These electrodes are also used for welding Type 347 clad steels or for the overlay of carbon steel.
E309Mo-XX	The weld metal is the same as E309 electrodes except for the addition of molybdenum and a small reduction in the carbon content. These electrodes are used for welding Type 316 clad steels or for the overlay of carbon steels.
E309MoL-XX	The weld metal is the same as E309Mo electrodes except for the restricted carbon content. The lower carbon content of the weld metal reduces the possibility of intergranular corrosion.
E310-XX	These electrodes are most often used to weld base metals of similar composition.
E310H-XX	The weld metal is the same as E310 electrodes, except that the allowable carbon content has been greatly increased. These electrodes are used primarily for welding or repairing high alloy heat and corrosion resistant castings designated as Type HK by the Alloy Castings Institute. The welds have high strength at temperatures over 1,700°F (927°C). The welds are not recommended for high sulfur atmospheres or where severe thermal shock is present. Long time exposure to temperatures in the approximate range of 1,400°F to 1,600°F (760°C to 871°C) may induce formation of sigma and secondary carbides that can result in reduced corrosion resistance, reduced ductility, or both.
E310Cb-XX	The weld metal is the same as E310 electrodes except for the addition of columbium (niobium) and a reduction in carbon content. These electrodes are used for the welding of heat resisting castings, Type 347 clad steels, or the overlay of carbon steels.
E310Mo-XX	The weld metal is the same as E310 electrodes except for the addition of molybdenum and a reduction in carbon content. These electrodes are used for the welding of heat resisting castings, Type 316 clad steels, or for the overlay of carbon steels.
E312-XX	These electrodes were originally designed to weld cast alloys of similar composition. They have been found to be valuable in welding dissimilar metals, especially if one of them is a stainless steel high in nickel. This electrode gives a two-phase weld deposit with substantial amounts of ferrite in an austenitic matrix. Even with considerable dilution by austenite-forming elements, such as nickel, the microstructure remains two-phase and highly resistant to weld metal cracks and fissures. Applications should be limited to service temperature below 800°F (420°C) to avoid formation of secondary brittle phases.
E316-XX	These electrodes are used for welding Type 316 and similar alloys. They have been used successfully in certain applications involving special base metals for high-temperature service. The presence of molybdenum in the weld metal provides creep resistance at elevated temperatures. Rapid corrosion of Type 316 weld metal may occur when the following three factors co-exist:
	1. The presence of a continuous or semi-continuous network of ferrite in the weld metal microstructure 2. A composition balance of the weld metal giving a chromium-to-molybdenum ratio of less than 8.2 to 1 3. The immersion of the weld metal in a corrosive medium Attempts to classify the media in which accelerated corrosion will take place by attack on the ferrite phase have not been entirely successful. Strong oxidizing and mildly reducing environments have been present where a number of corrosion failures were investigated and documented. The literature should be consulted for the latest recommendations.

AWS Classification	Intended Uses
E316H-XX	These electrodes are the same as E316 except that the allowable carbon content has been increased to the higher portion of the E316 range. Higher carbon content provides higher tensile and creep strengths at elevated temperatures. These electrodes are used for welding 316H base metal.
E316L-XX	These electrodes are the same as E316 except for restricted carbon content. The lower carbon content of the weld metal reduces the possibility of intergranular carbide precipitation, which increases the resistance to intergranular corrosion without the use of stabilizers such as columbium (niobium) or titanium. These electrodes are used principally for welding low carbon, molybdenum-bearing austenitic alloys. Tests have shown that the low carbon content in the weld metal gives adequate protection against intergranular corrosion in most cases. This low carbon electrode is not as strong at elevated temperatures as Type E316H.
E317-XX	The alloy content of weld metal deposited by these electrodes is somewhat higher than that of E316 electrodes, particularly in molybdenum. These electrodes are usually used for welding alloys of similar composition. They are also used in severely corrosive environments (such as those containing halogens) where crevice and pitting corrosion are of concern.
E317L-XX	The weld metal is the same as for E317 electrodes except for restricted carbon content. The lower carbon content of weld metal reduces the possibility of intergranular carbide precipitation, which increases the resistance to intergranular corrosion without the use of stabilizers such as columbium (niobium) or titanium. The low carbon welds are not as strong at elevated temperatures as the columbium (niobium)-stabilized alloys or the standard Type 317 weld metal with higher carbon content.
E318-XX	The weld metal is the same as for E316 electrodes except for the addition of columbium (niobium). Columbium (niobium) provides resistance to intergranular carbide precipitation and increases resistance to intergranular corrosion. These electrodes are used primarily for welding base metals of similar composition.
E320-XX	These electrodes have columbium (niobium) added to improve resistance to intergranular corrosion. The electrodes are primarily used to weld base metals of similar composition for applications where resistance to severe corrosion is required for a wide range of chemicals including sulfuric and sulfurous acids and their salts. These electrodes can be used to weld both castings and wrought alloys of similar composition without postweld beat treatment.
	A modification of this grade without columbium (niobium) is available for repairing castings that do not contain columbium. With this modified composition, solution annealing is required after welding.
E320LR-XX	The LR (low residuals) welds of E320LR electrodes are the same as E320 welds except that elements C, Si, P, and S are specified at lower maximum levels, and Cb (Nb) and Mn are controlled within narrower ranges. These changes reduce the weld metal fissuring (while maintaining the corrosion resistance) frequently encountered in fully austenitic stainless steel weld metals. Consequently, welding practices typically used to deposit ferrite-containing austenitic stainless steel weld metals can be used. Type E320LR weld metal has a lower minimum tensile strength than E320 weld metal.
E330-XX	These electrodes are commonly used where heat- and scale-resisting properties above 1,800°F (982°C) are required. However, high-sulfur environments may adversely affect performance at elevated temperature. Repairs of defects in alloy castings and the welding of castings and wrought alloys of similar composition are the most common applications.
E330H-XX	The weld metal is the same as E330 electrodes except for higher carbon content. These electrodes are used primarily for the welding and repairing of high alloy heat and corrosion resistant castings that are designated HT by the Alloy Castings Institute. The weld metal can be used in temperatures up to 2,100°F (1,149°C) in oxidizing atmospheres and at 2,000°F (1,093°C) in reducing atmospheres. However, high-sulfur environments may adversely affect performance at elevated temperatures.

AWS Classification	Intended Uses
E347-XX	These electrodes have columbium (niobium) or columbium (niobium) and tantalum added as a stabilizer. Either of these additions reduces the possibility of intergranular chromium carbide precipitation, which increases resistance to intergranular corrosion.
	These electrodes are usually used for welding chromium-nickel alloys that are stabilized with either columbium (niobium) or titanium. Electrodes depositing titanium as a stabilizing element are not commercially available because titanium is not readily transferred across the arc in shielded metal arc welding. Although columbium (niobium) is the stabilizing element usually specified in Type 347 alloys, it should be recognized that tantalum also is present. Tantalum and columbium (niobium) are almost equally effective in stabilizing carbon and in providing high-temperature strength. This specification recognizes the usual commercial practice of reporting columbium (niobium) as the sum of columbium (niobium) plus tantalum. If dilution by the base metal produces a low ferrite or fully austenitic weld metal deposit, crack sensitivity of the weld may increase substantially.
	Some applications, especially those involving high-temperature service, are adversely affected if the ferrite content is too high. Consequently, high ferrite content should not be specified for custom electrodes unless tests prove it necessary.
E349-XX	These electrodes are used for welding steels of similar composition such as AISI Type 651 or 652. The combination of columbium (niobium), molybdenum, and tungsten with chromium and nickel in the weld metal gives good high-temperature rupture strength. The weld metal also has an appreciable content of ferrite that increases the crack resistance of the weld metal.
E383-XX	These electrodes are used to weld base metal of a similar composition to itself and to other grades of stainless steel. Type E383 weld metal is recommended for sulfuric and phosphoric acid environments. The elements C, Si, P, and S are specified at low levels to minimize weld metal hot cracking and fissuring (while maintaining the corrosion resistance) frequently encountered in fully austenitic stainless steel weld metals.
E385-XX	These electrodes are used primarily for welding Type 904L materials for the handling of sulfuric acid and many chloride solutions. E385 electrodes are used to join Type 317L material where improved corrosion resistance in specific media is needed. E385 electrodes are also used for joining Type 904L base metal to other grades of stainless. The elements C, Si, P, and S are specified at low levels to minimize weld metal hot cracking and fissuring (while maintaining corrosion resistance) frequently encountered in fully austenitic weld metals.
E410-XX	This 12 Cr alloy electrode weld metal is an air-hardening steel. Preheat and postheat treatments are required to achieve welds of adequate ductility for many purposes. The most common application of these electrodes is for welding alloys of similar compositions. They are also used for surfacing of carbon steels to resist corrosion, erosion, or abrasion.
E410NiMo-XX	These electrodes are used for welding ASTM CA6NM castings or similar materials, as well as light gauge Type 410, 410S, and 405 base metals. The weld metal for these electrodes is modified to contain less chromium and more nickel than the weld metal of E410 electrodes. This is to eliminate ferrite in the microstructure because ferrite has a bad effect on the mechanical properties of the alloy. Final postweld heat treatment should not exceed 1,150°F. Higher temperatures may result in re-hardening due to untempered martensite in the microstructure after cooling to room temperature.
E430-XX	The weld metal of these electrodes contains between 15 and 18 Cr (wt.%). The composition is balanced by providing sufficient chromium to give adequate corrosion resistance for the usual applications and yet retain sufficient ductility in the heat-treated condition to meet the mechanical requirements of the specification. (Excessive chromium will result in lowered ductility.) Welding with an E430 electrode usually requires preheat and postheat. Optimum mechanical properties and corrosion resistance are obtained only when the weldment is heat-treated following the welding operation.

AWS Classification	Intended Uses
E630-XX	These electrodes are primarily designed for welding ASTM A564, Type 630, and some other precipitation-hardening stainless steels. The weld metal is modified to prevent the formation of ferrite networks in the martensite microstructure that could have a bad effect on mechanical properties. Dependent on the application and weld size, the weld may be used in an as-welded, precipitation-hardened, or solution-treated and precipitation-hardened condition.
E16-8-2-XX	These electrodes are used primarily for welding stainless steel, such as Types 16-8-2, 316, and 347, for high-pressure, high-temperature piping systems. The weld metal usually has a Ferrite Number no higher than 5 FN. The weld metal also has good hot ductility properties, which offer relative freedom from weld or crater cracking even under high-restraint conditions. The weld is usable in either the as-welded or solution-treated condition. These electrodes depend on a very carefully balanced chemical composition to develop their fullest properties. Corrosion tests indicate that Type 16-8-2 weld metal may have less corrosion resistance than Type 316 base metal depending on the corrosive media. Where the weldment is exposed to severe corrodents, surface layers should be deposited with a more corrosion-resistant weld metal.
E2209-XX	These electrodes are used primarily to weld duplex stainless steels that contain approximately 22% chromium. The weld metal of these electrodes has a duplex microstructure consisting of an austenite-ferrite matrix. E2209 electrodes combine increased tensile strength with improved resistance to pitting, corrosive attack, and stress corrosion cracking.
E2553-XX	These electrodes are used primarily to weld duplex stainless steels that contain approximately 25% chromium. The weld metal of these electrodes has a duplex microstructure consisting of an austenite-ferrite matrix. E2553 electrodes combine increased tensile strength with improved resistance to pitting, corrosive attack, and stress corrosion cracking.

Appendix B

Performance Accreditation Tasks

Performance Accreditation Tasks (PATs) provide specific acceptable criteria for performance and help to ensure a true competency-based welding program for students.

The following tasks are designed to evaluate your ability to run open-root V-groove welds on both plate and pipe in all positions using stainless steel electrodes and SMAW equipment. Perform each task when you are instructed to do so by your instructor. As you complete each task, take it to your instructor for evaluation. Do not proceed to the next task until told to do so.

For AWS 6G certifications, refer to *AWS EG4.0-96 Guide for the Training and Qualification of Welding Personnel: Level III – Expert Welders* for bend test requirements.

PATs 1 through 4 correspond to *AWS EG3.0-96, Section 3.3.6.1, Unit #1 SMAW, Learning Objective #7*. PAT 8 corresponds to *AWS EG4.0-96, Section 3.3.6.2, Unit #2 SMAW, Learning Objective #8* (with backing).

OPEN-ROOT V-GROOVE WELDS ON STAINLESS STEEL PLATE JOINTS IN THE FLAT (1G) POSITION

Note: It is preferable to use stainless steel base metal. However, substituting carbon steel is permitted by the codes.

Using stainless steel electrodes, make an open-root V-groove weld on plate in the flat (1G) position, as shown.

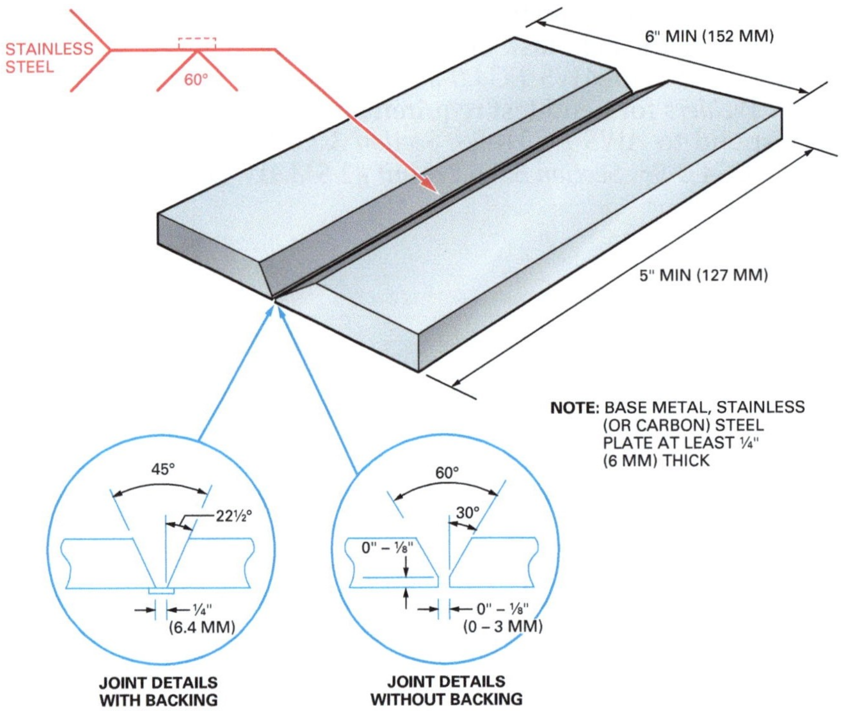

Criteria for Acceptance:

- Uniform appearance on the bead face ____________
- Craters and restarts filled to the full cross section of the weld ____________
- Acceptable weld profile in accordance with the applicable codes and standards ____________
- Smooth transition with complete fusion at the toes of the weld ____________
- Complete joint penetration and face reinforcement with the root reinforcement at least flush with the inside of the pipe, to a maximum buildup of ⅛" (3 mm) ____________
- No porosity ____________
- No overlap ____________
- No undercut ____________
- No inclusions ____________
- No cracks ____________
- No pinholes ____________

OPEN-ROOT V-GROOVE WELDS ON STAINLESS STEEL PLATE JOINTS IN THE HORIZONTAL (2G) POSITION

Using stainless steel electrodes, make an open-root V-groove weld on plate in the horizontal (2G) position, as shown.

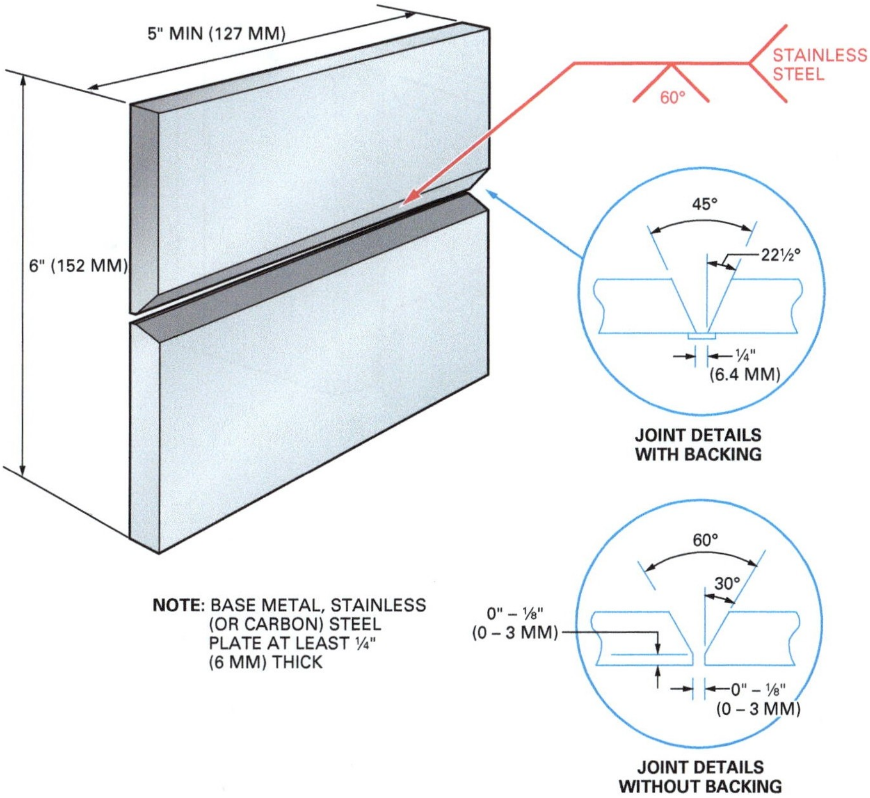

Criteria for Acceptance:

- Uniform appearance on the bead face ____________
- Craters and restarts filled to the full cross section of the weld ____________
- Acceptable weld profile in accordance with the applicable codes and standards ____________
- Smooth transition with complete fusion at the toes of the weld ____________
- Complete joint penetration and face reinforcement with the root reinforcement at least flush with the inside of the pipe, to a maximum buildup of ⅛" (3 mm) ____________
- No porosity ____________
- No overlap ____________
- No undercut ____________
- No inclusions ____________
- No cracks ____________
- No pinholes ____________

OPEN-ROOT V-GROOVE WELDS ON STAINLESS STEEL PLATE JOINTS IN THE VERTICAL (3G) POSITION

Using stainless steel electrodes, make an open-root V-groove weld on plate in the vertical (3G) position, as shown.

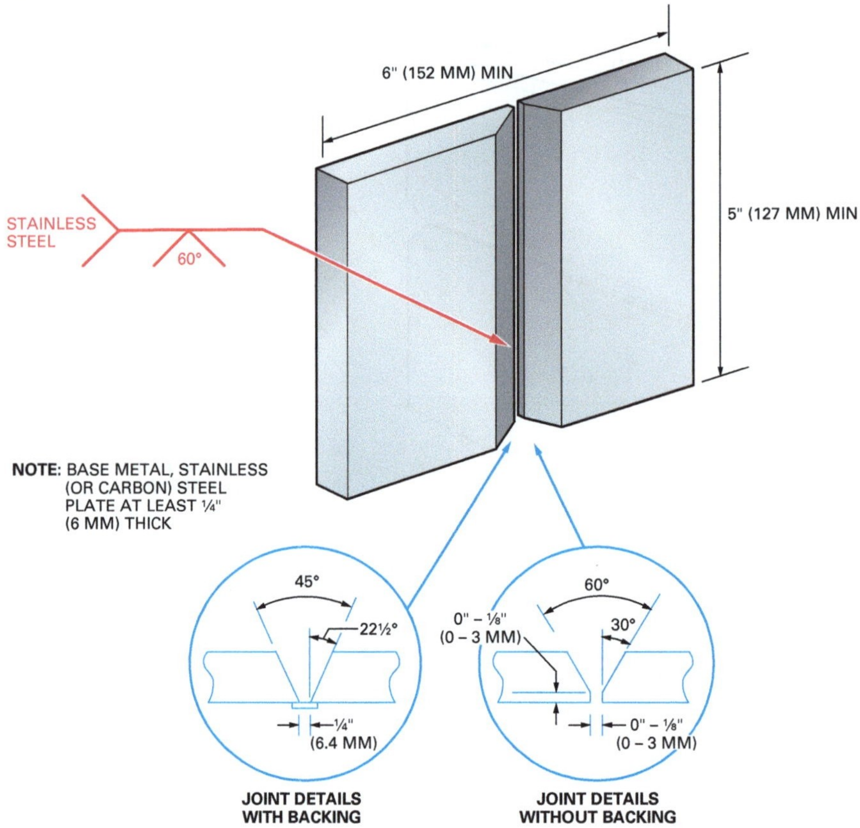

Criteria for Acceptance:

- Uniform appearance on the bead face __________
- Craters and restarts filled to the full cross section of the weld __________
- Acceptable weld profile in accordance with the applicable codes and standards __________
- Smooth transition with complete fusion at the toes of the weld __________
- Complete joint penetration and face reinforcement with the root reinforcement at least flush with the inside of the pipe, to a maximum buildup of ⅛" (3 mm) __________
- No porosity __________
- No overlap __________
- No undercut __________
- No inclusions __________
- No cracks __________
- No pinholes __________

OPEN-ROOT V-GROOVE WELDS ON STAINLESS STEEL PLATE JOINTS IN THE OVERHEAD (4G) POSITION

Using stainless steel electrodes, make an open-root V-groove weld on plate in the overhead (4G) position, as shown.

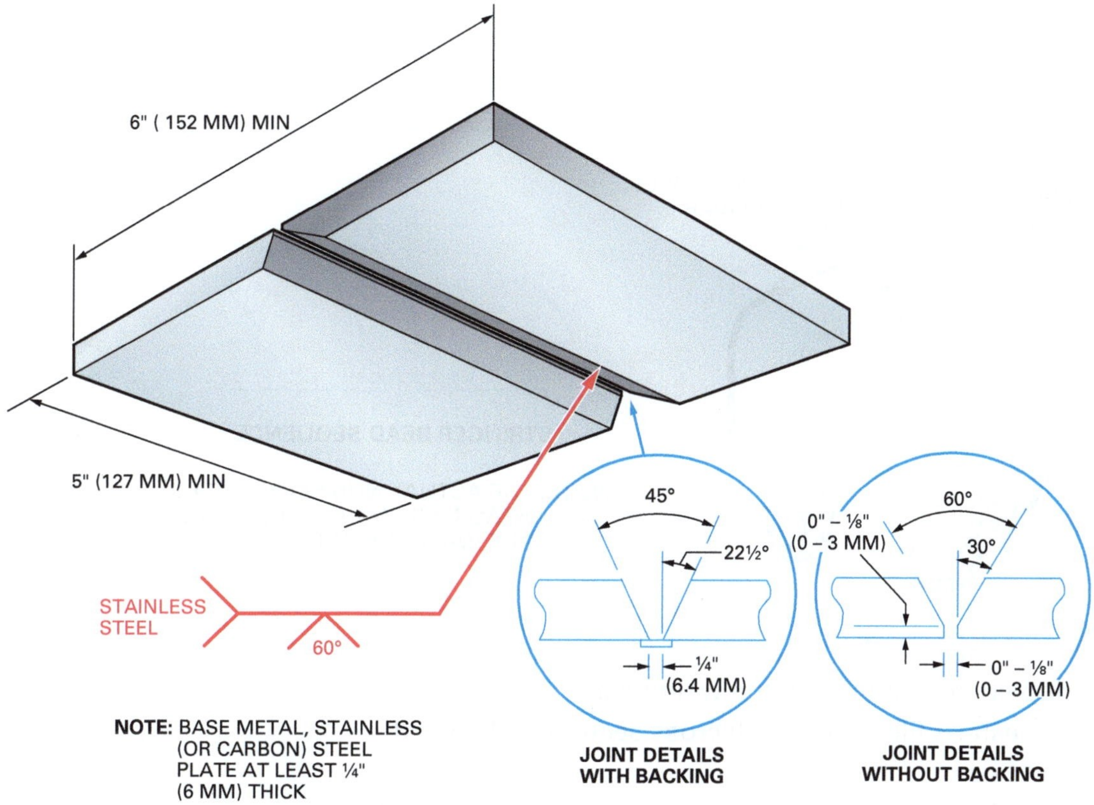

Criteria for Acceptance:

- Uniform appearance on the bead face ________
- Craters and restarts filled to the full cross section of the weld ________
- Acceptable weld profile in accordance with the applicable codes and standards ________
- Smooth transition with complete fusion at the toes of the weld ________
- Complete joint penetration and face reinforcement with the root reinforcement at least flush with the inside of the pipe, to a maximum buildup of ⅛" (3 mm) ________
- No porosity ________
- No overlap ________
- No undercut ________
- No inclusions ________
- No cracks ________
- No pinholes ________

OPEN-ROOT V-GROOVE STAINLESS STEEL PIPE WELD IN THE 1G-ROTATED POSITION

Using stainless steel electrodes, make an open-root V-groove weld on stainless steel pipe in the 1G-ROTATED position, as shown.

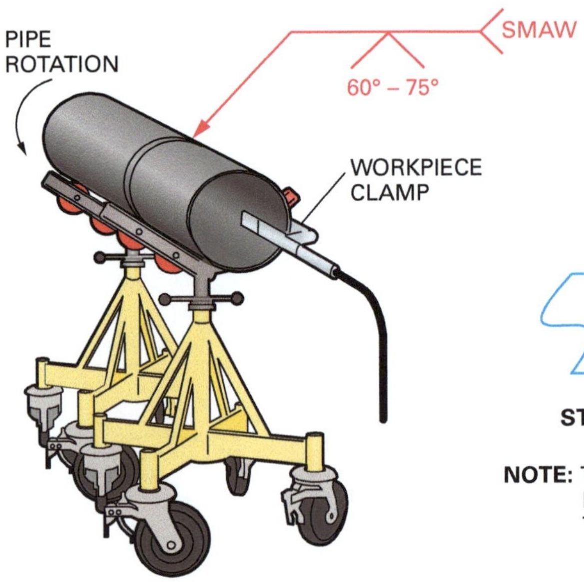

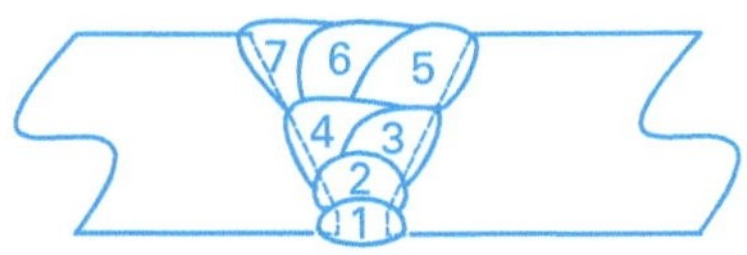

STRINGER BEAD SEQUENCE

NOTE: THE ACTUAL NUMBER OF WELD BEADS WILL VARY DEPENDING ON THE WALL THICKNESS.

Criteria for Acceptance:

- Uniform rippled appearance on the weld face __________
- Craters and restarts filled to the full cross section of the weld __________
- Uniform weld size __________
- Acceptable weld profile in accordance with the applicable codes and standards __________
- Smooth transition with complete fusion at the toes of the weld __________
- Complete joint penetration and face reinforcement with the root reinforcement at least flush with the inside of the pipe, to a maximum buildup of ⅛" (3 mm) __________
- No porosity __________
- No undercut __________
- No inclusions __________
- No cracks __________
- No pinholes __________

OPEN-ROOT V-GROOVE STAINLESS STEEL PIPE WELD IN THE 2G POSITION

Using stainless steel electrodes, make an open-root V-groove weld on stainless steel pipe in the 2G position, as shown.

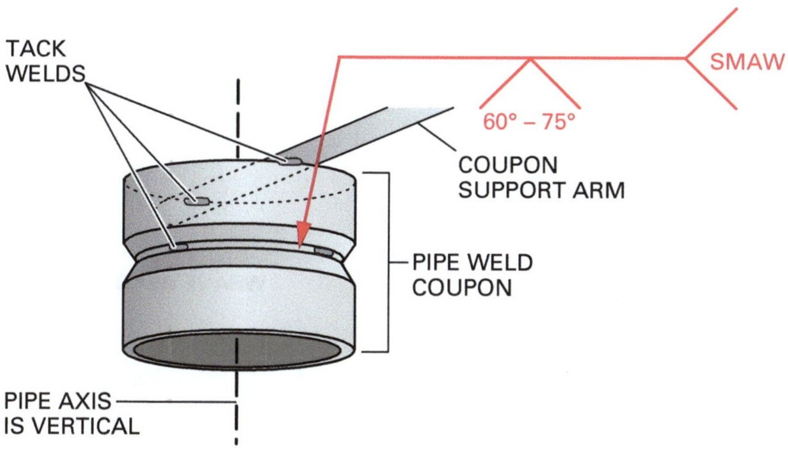

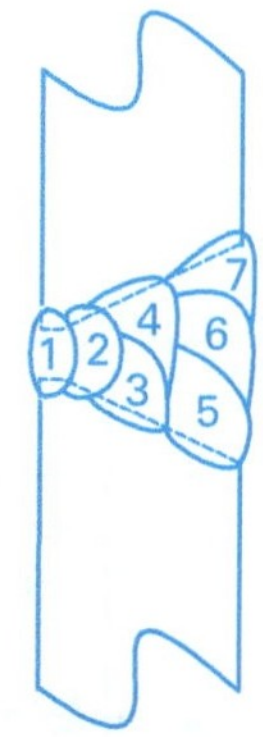

STRINGER BEAD SEQUENCE

Criteria for Acceptance:

- Uniform rippled appearance on the weld face ______
- Craters and restarts filled to the full cross section of the weld ______
- Uniform weld size ______
- Acceptable weld profile in accordance with the applicable codes and standards ______
- Smooth transition with complete fusion at the toes of the weld ______
- Complete joint penetration and face reinforcement with the root reinforcement at least flush with the inside of the pipe, to a maximum buildup of ⅛" (3 mm) ______
- No porosity ______
- No undercut ______
- No inclusions ______
- No cracks ______
- No pinholes ______

OPEN-ROOT V-GROOVE STAINLESS STEEL PIPE WELD IN THE 5G POSITION

Using stainless steel electrodes, make an open-root V-groove weld on stainless steel pipe in the 5G position, as shown.

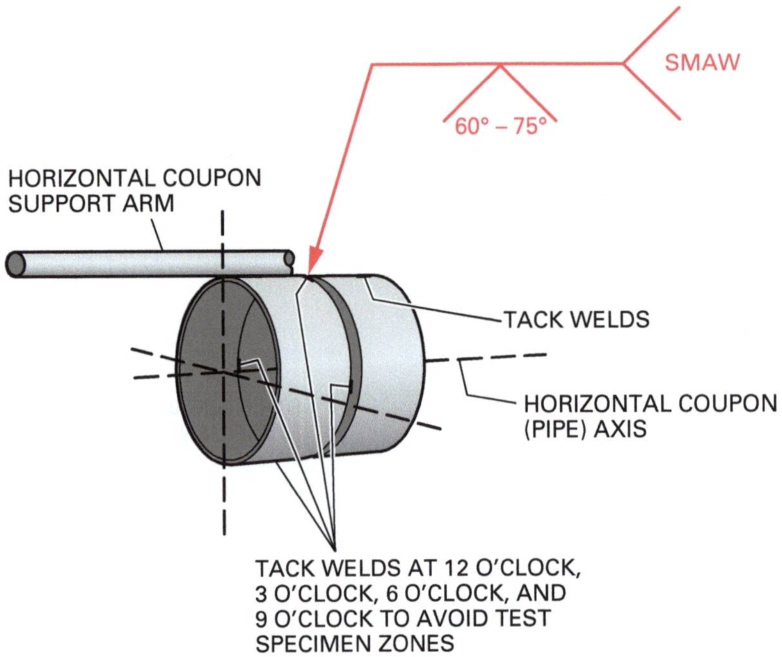

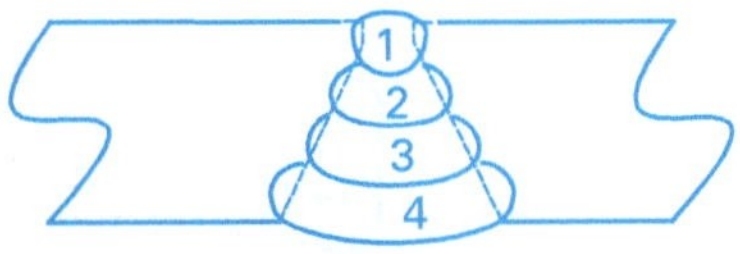

WEAVE BEAD SEQUENCE

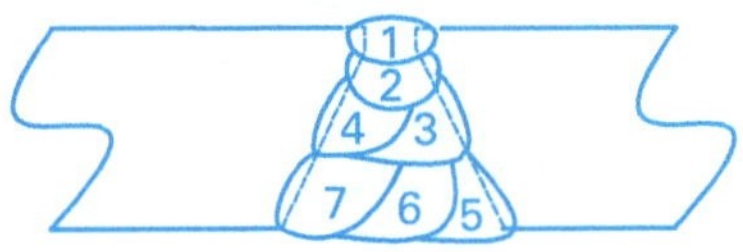

STRINGER BEAD SEQUENCE

Criteria for Acceptance:

- Uniform rippled appearance on the weld face ____________
- Craters and restarts filled to the full cross section of the weld ____________
- Uniform weld size ____________
- Acceptable weld profile in accordance with the applicable codes and standards ____________
- Smooth transition with complete fusion at the toes of the weld ____________
- Complete joint penetration and face reinforcement with the root reinforcement at least flush with the inside of the pipe, to a maximum buildup of ⅛" (3 mm) ____________
- No porosity ____________
- No undercut ____________
- No inclusions ____________
- No cracks ____________
- No pinholes ____________

OPEN-ROOT V-GROOVE STAINLESS STEEL PIPE WELDS IN THE MULTIPLE INCLINED (6G) POSITION

Using stainless steel electrodes, make an open-root V-groove weld on pipe in the multiple inclined (6G) position, as shown. If required, a restricting ring may be added to the 6G position to form a 6GR position coupon.

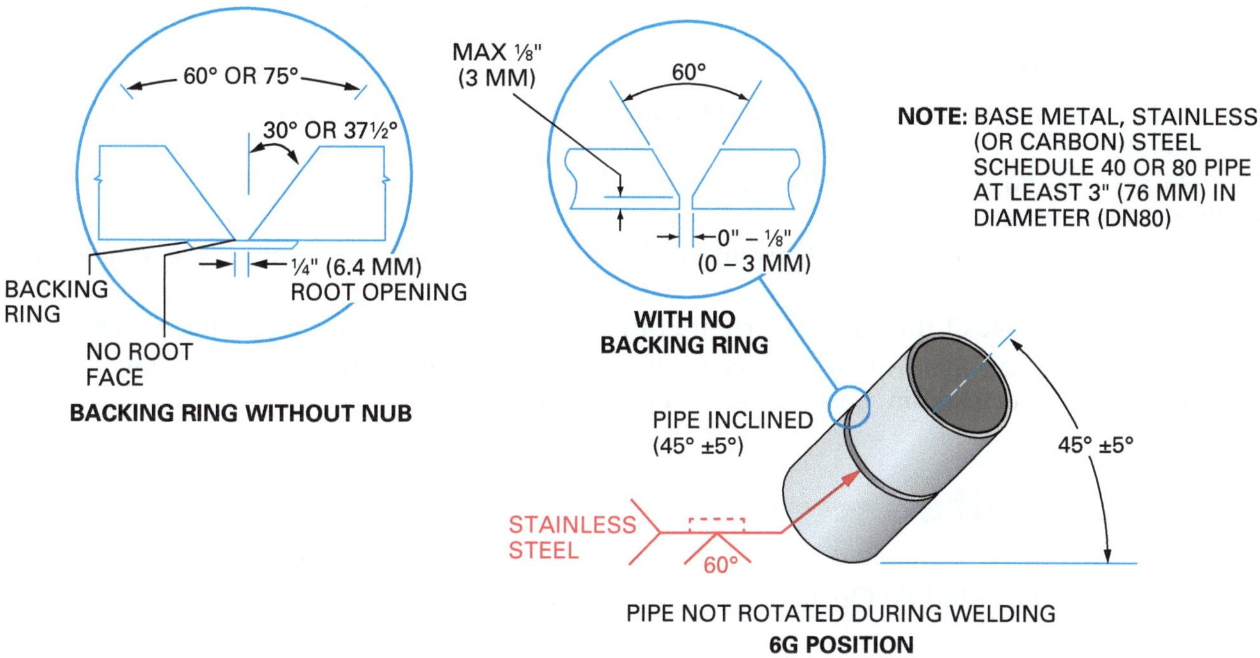

Criteria for Acceptance:

- Uniform rippled appearance on the weld face ____________
- Craters and restarts filled to the full cross section of the weld ____________
- Uniform weld size ____________
- Acceptable weld profile in accordance with the applicable codes and standards ____________
- Smooth transition with complete fusion at the toes of the weld ____________
- Complete joint penetration and face reinforcement with the root reinforcement at least flush with the inside of the pipe, to a maximum buildup of ⅛" (3 mm) ____________
- No porosity ____________
- No undercut ____________
- No inclusions ____________
- No cracks ____________
- No pinholes ____________

Additional Resources

This module presents thorough resource for task training. The following reference material is recommended for further study.

AWS A5.4, Specification for Stainless-Steel Electrodes for Shielded Metal Arc Welding. Miami, FL: American Welding Society.

AWS F4.1: Recommended Safe Practices for Preparation for Welding and Cutting of Containers and Piping. Miami, FL: American Welding Society.

Modern Welding Technology. Current Edition. Howard B. Cary and Scott Helzer. Englewood Cliffs, NJ: Prentice Hall.

OSHA 1926.351, Arc Welding and Cutting. Current Edition. Washington, DC: Occupational Safety & Health Administration.

The Procedure Handbook of Arc Welding. Current Edition. Cleveland, OH: The James F. Lincoln Arc Welding Foundation.

Welding Handbook, Volumes 1 through 4. Current Edition. Miami, FL: American Welding Society.

Welding Pressure Pipelines and Piping Systems (PDF). The Lincoln Electric Company, Cleveland, OH, USA.**www.lincolnelectric.com**. Last accessed May 29, 2015.

Figure Credits

Topaz Publications, Inc., Module Opener, Figure 3

Metallurgical Technologies, Inc. (**www.met-tech.com**), Figure 1

The Lincoln Electric Company, Cleveland, OH, USA, Figures 5–9

Terry Lowe, Figures 11 (photo), 12 (photo), 26 (photo), 27 (photo), 28 (photo)

Atlas Welding Accessories, Inc., Figure 24

Section Review Answer Key

Answer	Section Reference	Objective
Section One		
1. d	1.1.0	1a
2. d	1.2.0	1b
3. b	1.3.0	1c
4. a	1.4.0	1d
Section Two		
1. a	2.1.1	2a
2. d	2.2.3	2b
3. d	2.3.4	2c
Section Three		
1. b	3.1.2	3a
2. c	3.2.0	3b
3. a	3.3.0	3c
4. b	3.4.0	3d
5. d	3.5.0	3e

NCCER CURRICULA — USER UPDATE

NCCER makes every effort to keep its textbooks up-to-date and free of technical errors. We appreciate your help in this process. If you find an error, a typographical mistake, or an inaccuracy in NCCER's curricula, please fill out this form (or a photocopy), or complete the online form at **www.nccer.org/olf**. Be sure to include the exact module ID number, page number, a detailed description, and your recommended correction. Your input will be brought to the attention of the Authoring Team. Thank you for your assistance.

Instructors – If you have an idea for improving this textbook, or have found that additional materials were necessary to teach this module effectively, please let us know so that we may present your suggestions to the Authoring Team.

NCCER Product Development and Revision
13614 Progress Blvd., Alachua, FL 32615
Email: curriculum@nccer.org
Online: www.nccer.org/olf

❑ Trainee Guide ❑ Lesson Plans ❑ Exam ❑ PowerPoints Other ____________

Craft / Level: Copyright Date:

Module ID Number / Title:

Section Number(s):

Description:

Recommended Correction:

Your Name:

Address:

Email: Phone:

Glossary

Annealing: A heat treatment process that reduces the brittleness of the work.

Austenitic steel: These stainless steels are nonmagnetic in the annealed condition and non-hardenable by heat treatment; they can be hardened significantly by cold working. They combine excellent corrosion- and heat-resistance with good mechanical properties over a broad temperature range. Austenitic stainless steels make up the largest of the three common stainless steel groups and are the most likely to be encountered by welders.

Axis: The center line running the length of an object. In welding, the term axis is most often used in relation to an electrode or a section of pipe.

Carbide precipitation: The process of carbides combining with chromium to create chromium carbide. The result is a reduction of chromium and the corrosion resistance it provides. Intergranular corrosion results from the loss of chromium.

Carbide precipitation: The process of carbides combining with chromium to create chromium carbide. The result is a reduction of chromium and the corrosion-resistance it provides. Intergranular corrosion results from the loss of chromium.

Carburized: The process of iron or steel combining with carbon liberated from a metal when the metal is heated.

Composite wire electrode: A generic term for multicomponent filler metal electrodes. For GMAW, the tubular electrode contains a powdered metal inside a metal sheath. A flux-cored electrode contains flux inside a metal sheath.

Electrode extension: The length of the wire that extends beyond the tip of the welding gun contact tube.

Fast-freeze electrode: An electrode that solidifies quickly, produces very little slag, and gives good penetration into the root of the joint.

Feather: To grind the start and/or stop areas of a weld to a tapered edge.

Feather: To grind the start and/or stop areas of a weld to a tapered edge.

Ferritic steel: These stainless steels are always magnetic and can be hardened to some extent by cold working. They combine corrosion- and heat-resistance with fair mechanical properties over a narrower temperature range than austenitic steels. Ferritic stainless steels are straight chromium stainless steels containing 11.5 to 27 percent chromium, as well as other metals.

Flux-cored wire electrode: A wire electrode that is tube-like and filled with flux and/or degassing components with other elements to improve the chemical and mechanical characteristics of the weld.

Heat-affected zone: The portion of the workpiece/base metal whose mechanical properties or microstructure have been or will be altered by the heat from welding, brazing, soldering, or thermal cutting processes.

Hot pass: A nonstandard term for the weld pass that is made directly after the root pass.

Intergranular corrosion: A selective corrosion attack of the crystalline grain boundaries of a stainless steel. It results from chromium depletion which occurs during carbide precipitation.

Interpass temperature: The temperature of the base metal in the general area of the weld before the next pass is started. The definition of the area of the weld varies by the applicable welding code.

Keyhole: A hole that is formed when the root faces on two pipe edges are melted away by the arc.

Keyhole: A hole that is formed when the root faces on two pipe ends are melted away by the arc.

Low alloy steel: Steel that is alloyed with other elements, amounting to 1 to 8 percent of the steel by weight, to improve its mechanical properties.

Martensitic steel: These stainless steels comprise two groups: chromium-martensitic and martensitic. The chromium-martensitic stainless steels contain from 11.5 to 18 percent chromium, along with other metals. Martensitic stainless steels contain only 4 to 6 percent chromium; for this reason, they are not considered true stainless steels, although their corrosion-resistance is much greater than that of mild carbon steels. Martensitic stainless steels are magnetic and can be hardened by quenching and tempering.

Microwire: Generally considered to be wire electrode that is 0.045″ (1.1 mm) in diameter or smaller. However, due to the small size of most GMAW wire electrodes in comparison to SMAW electrodes and similar filler metals, some refer to all GMAW-related wire electrodes as microwire.

Pipe axis: An imaginary line that runs longitudinally (lengthwise) through the center of a pipe.

Pipe axis: An imaginary line that runs longitudinally (lengthwise) through the center of a pipe.

Remaining passes: A nonstandard term that is used to describe the fill and cover passes as a group.

Remaining passes: A nonstandard term that is used to describe the fill and cover passes as a group.

Root pass: A weld pass that is made to produce a root bead.

Root pass: The initial weld pass made to produce a root bead.

Sensitization: The process by which carbide precipitation takes place.

Standoff distance: The distance from the gas nozzle or insulating nozzle to the workpiece.

Stickout: The distance from the gas nozzle or insulating nozzle to the end of the electrode.